GRAVITATIONAL WAVES

Series in High Energy Physics, Cosmology and Gravitation

GRAVITATIONAL WAVES

Edited by

Ignazio Ciufolini

Department of Engineering,
University of Lecce

Vittorio Gorini, Ugo Moschella

Department of Chemical, Mathematical and Physical Sciences,
University of Insubria at Como

and

Pietro Fré

Department of Physics, University of Turin

CRC Press
Taylor & Francis Group
Boca Raton London New York

CRC Press is an imprint of the
Taylor & Francis Group, an **informa** business

CRC Press
Taylor & Francis Group
6000 Broken Sound Parkway NW, Suite 300
Boca Raton, FL 33487-2742

First issued in paperback 2019

© 2001 by Taylor & Francis Group, LLC
CRC Press is an imprint of Taylor & Francis Group, an Informa business

No claim to original U.S. Government works

ISBN-13: 978-0-367-39760-9

British Library Cataloguing-in-Publication Data

A catalogue record for this book is available from the British Library.

Library of Congress Cataloging-in-Publication Data are available

**Visit the Taylor & Francis Web site at
http://www.taylorandfrancis.com**

**and the CRC Press Web site at
http://www.crcpress.com**

Contents

Preface

Gravitational waves today represent a hot topic, which promises to play a central role in astrophysics, cosmology and theoretical physics.

Technological developments have led us to the brink of their direct observation, which could become a reality in the coming years.

The direct observation of gravitational waves will open an entirely new field; gravitational wave astronomy. This is expected to bring a revolution in our knowledge of the universe by allowing the observation of hitherto unseen phenomena such as coalescence of compact objects (neutron stars and black holes), fall of stars into supermassive black holes, stellar core collapses, big-bang relics and the new and unexpected.

During Spring 1999, the SIGRAV—Società Italiana di Relatività e Gravitazione (Italian Society of Relativity and Gravitation) sponsored the organization of a doctoral school on 'Gravitational Waves in Astrophysics, Cosmology and String Theory', which took place at the Center for Scientific Culture 'Alessandro Volta' located in the beautiful environment of Villa Olmo in Como, Italy.

This book brings together the courses given at the school and provides a comprehensive review of gravitational waves. It includes a wide range of contributions by leading scientists in the field. Topics covered are: the basics of GW with some recent advanced topics, GW detectors, the astrophysics of GW sources, numerical applications and several recent theoretical developments. The material is written at a level suitable for postgraduate students entering the field.

The main financial support for the School came from the University of Insubria at Como-Varese. Other contributors were the Department of Chemical, Physical and Mathematical Sciences of the same University, the Physics Departments of the Universities of Milan and Turin, and the Institute of Physics of Interplanetary Space—CNR, Frascati.

We are grateful to all the members of the scientific organizing committee and to the scientific coordinator of Centro Volta, Professor G Casati, for their invaluable help.

We also acknowledge the essential organizational support of the secretarial conference staff of Centro Volta, in particular of Chiara Stefanetti.

I Ciufolini, V Gorini, U Moschella and P Fré
Como
12 June 2000

Chapter 1

Gravitational waves, theory and experiment (an overview)

Ignazio Ciufolini[1] and Vittorio Gorini[2]
[1] *Dipartimento di Ingegneria dell'Innovazione, University of Lecce, Italy*
E-mail: ciufoli@nero.ing.uniroma1.it
[2] *Department of Chemical, Mathematical and Physical Sciences University of Insubria at Como, Italy*
E-mail: gorini@fis.unico.it

General relativity and electrodynamics display profound similarities and yet fundamental differences [1, 2]. In this connection, it may be interesting to point out some historical analogies between the two fields.

The enormous success of Maxwell's equations did not rest only in the fact that they incorporated, together with the Lorentz force equation, all the laws of electricity and magnetism, but also that on their basis James Clerk Maxwell (1831–1879) was able (in 1873) to predict the existence of a solution consisting of electric and magnetic fields changing in time, carrying energy and propagating with speed c in vacuum: the electromagnetic waves. Nevertheless, some distinguished physicists, such as Lord Kelvin, had serious doubts about the existence of such waves: 'The so-called "electromagnetic theory of light" has not helped us hitherto ... it seems to me that it is rather a backward step ... the one thing about it that seems intelligible to me, I do not think is admissible ... that there should be an electric displacement perpendicular to the line of propagation'. However, in 1887, eight years after Maxwell's death, electromagnetic waves were both generated and detected by Heinrich Hertz (1857–1894); then, in 1901, Guglielmo Marconi transmitted and received signals across the Atlantic Ocean.

In the twentieth century the detection and study of electromagnetic waves, other than visible light, opened a new era of dramatic changes in the knowledge of our universe: cosmic radio waves, discovered in the 1930s, revealed in subsequent decades colliding galaxies; quasars with dimensions of the order of the solar system but having luminosities orders of magnitude larger than our galaxy; enormous jets from galactic nuclei and quasars reaching lengths of hundreds of thousands of light years, rapidly rotating pulsars with rotational periods of a few milliseconds and, not least, the cosmic microwave background, a relic of the hot big bang. X-rays revealed accretion disks about black holes and neutron stars. Similarly, millimetre, infrared and ultraviolet radiation, and gamma rays opened other dramatic windows of knowledge on our universe.

In the same way, in general relativity [1, 2], Einstein's field equations (1915) not only described the gravitational interaction via the spacetime curvature generated by mass-energy, but also contained, through the Bianchi identities, the equations of motion of matter and fields, and on their basis Albert Einstein, in 1916, a few months after the formulation of the theory, predicted the existence of curvature perturbations propagating with speed c on a flat and empty spacetime; the gravitational waves [4]. Einstein's gravitational-wave theory was a linearized theory treating weak waves as weak perturbations of a flat background [1, 3, 5]. Similarly to what happened when electromagnetic waves were first predicted, some distinguished physicists had serious doubts about their existence. Arthur Eddington thought that these weak-field solutions of the wave equation obtained from Einstein's field equations were just coordinate changes which were 'propagating ... with the speed of thought' [6].

The linearized theory of gravitational waves had its limits because the linear approximation is not valid for sources where gravitational self-energy is not negligible. It was only in 1941 that Landau and Lifshitz [7] described the emission of gravitational waves by a self-gravitating system of slowly moving bodies. However, in the following years there were serious doubts about the reality of gravitational waves and not until 1957 did a gedanken experiment by Hermann Bondi show that gravitational waves do indeed carry energy [8].

This thought experiment was based on a system of two beads sliding on a stick with only a slight friction opposing their motion. If a plane gravitational wave impinges on this system, the beads move back and forth on the stick because of the change in the proper distance between them due to the change of the metric, i.e. to the gravitational-wave perturbation; this change is governed by the geodesic deviation equation and the proper dispacement between the two beads is a function of the gravitational-wave metric perturbation. Thus, the friction between beads and stick heats the system and thus increases the temperature of the stick. Therefore, since there is an energy transfer from gravitational waves to the system in the form of increased temperature of the system, this thought experiment showed that gravitational waves do indeed carry energy and are a real physical entity [1, 8].

It is interesting to note that in 1955 John Archibald Wheeler had devised the conceivable existence of a body with no 'mass' built up by gravitational or electromagnetic, radiation alone [9]. Indeed, an object can, in principle, be constructed out of gravitational radiation or electromagnetic radiation, or a mixture of the two, and may hold itself together by its own gravitational attraction. A collection of radiation held together in this way, is called a geon (gravitational electromagnetic entity) and studied from a distance, such an object would present the same kind of gravitational attraction as any other mass. Yet, nowhere inside the geon is there a place where there is 'mass' in the conventional sense of the term. In particular, for a geon made of pure gravitational radiation— a gravitational geon—there is no local measure of energy, yet there is global energy. The gravitational geon owes its existence to a dynamical localized—but everywhere regular—curvature of spacetime, and to nothing more. Thus, a geon is a collection of electromagnetic or gravitational-wave energy, or a mixture of the two, held together by its own gravitational attraction, that was described by Wheeler as 'mass without mass'.

In the 1960s, Joseph Weber began the experimental work to detect gravitational waves. He was essentially alone in this field of research [10]. Then, the theoretical work of Wheeler, Bondi, Landau and Lifshitz, Isaacson, Thorne and others and the experimental work of Weber, Braginski, Amaldi and others opened a new era of research in this field. In 1972 Steven Weinberg wrote '... gravitational radiation would be interesting even if there were no chance of ever detecting any, for the theory of gravitational radiation provides a crucial link between general relativity and the microscopic frontiers of physics' [11].

Today gravitational waves, both theory and experiment, are one of the main topics of research in general relativity and gravitation [3].

In the same way as electromagnetic waves other than visible light, that is radio, millimetre, infrared, ultraviolet, x-ray and gamma-ray astronomy opened new windows and brought radical changes in our knowledge of the universe, gravitational-wave astronomy is expected to bring a revolution in our knowledge of the universe by observing new exotic phenomena such as formation and collision of black holes, fall of stars into supermassive black holes, primordial gravitational waves emitted just after the big bang .... Nevertheless, today, about 85 years after the prediction of gravitational waves by Einstein, the only evidence for their actual existence is indirect and comes from the observation of the energy loss from the binary pulsar system PSR 1913+16, discovered in 1974 by Hulse and Taylor [12]. Quite remarkably, though of no surprise, the observed energy loss of the binary pulsar is in agreement with the theoretical prediction by general relativity for the energy loss by gravitational radiation emitted by a binary system, to within less than 0.3% error (in this respect, it might be interesting to note here that, in regard to the field that in general relativity is formally analogous to the magnetic field in electrodynamics, i.e. the so-called gravitomagnetic field, predicted by Lense and Thirring in 1916, the first evidence and measurement of the existence of such an effect on Earth's satellites, due to the Earth's rotation, was published only in 1996, that is 80 years after the derivation of the effect [13]). Thus, today, together with the enormous experimental efforts to detect gravitational waves, from bar detectors to laser interferometers on Earth, GEO-600, LIGO, VIRGO, ..., and from laser interferometers in space, LISA, to Doppler tracking of interplanetary spacecrafts, there is, aimed at increasing the chances of future detections, a strongly related theoretical and computational work to understand and predict the emission or gravitational waves from astrophysical systems in strong field conditions [3]. In this book contributions of leading experts in the field of gravitational waves, both theoretical and experimental, are presented.

The basic contribution by Bernard Schutz and Franco Ricci deals with the main features of gravitational waves, sources and detectors. The contribution is divided into six chapters and some chapters are followed by a few exercises. The first chapter describes the linearized theory and the fundamental properties of weak gravitational waves, perturbations of a flat background, analysed in the so-called transverse-traceless gauge. The second and third chapters deal with detectors and astrophysical sources; in particular an overview is presented of the most important detectors under construction (their physics, sensitivity and opportunity for the future) and the main expected sources of gravitational waves, such as binary systems, neutron stars, pulsars, γ-ray bursts, etc. The fourth chapter deals with the mathematical theory of waves in general, stress-energy tensor and energy carried by gravitational waves. The subsequent chapter describes radiation generation in linearized theory: mass- and current-quadrupole radiation, i.e. the quadrupole formulae for the outgoing flux of gravitational-wave energy emitted by a system characterized by slow motion. Finally, the last chapter describes some applications of radiation theory to some sources: binary systems and especially *r-modes* of neutron stars.

The contribution by Guido Pizzella deals with bar detectors of gravitational waves. A gravitational-wave resonant detector is usually a cylindrical bar of length L. The small change δL in the length of the whole bar at the fundamental resonance angular frequency, ω_0, can be described by the solution of the equation of a harmonic oscillator, with resonance angular frequency ω_0 (with a supplementary $4/\pi^2$ factor obtained by solving the problem of a continuous bar). In a gravitational-wave resonant detector the mechanical oscillations of the bar induced by a gravitational wave are converted by an electromechanical transducer into electric signals which are amplified with a low noise amplifier, such as a dc SQUID. Then the data analysis is performed. Using a resonant antenna one measures the Fourier component of the metric perturbation near the antenna resonance frequency ω_0. The typical damping time of the resonant detector is $2Q/\omega_0$, where Q is the so-called quality factor of the resonant detector. The ultimate sensitivity of bar antennae to a fractional change in dimension due to a short burst of gravitational radiation has been estimated to be of the order of 10^{-20} or 10^{-21}. Bar detectors, usually 3 m long aluminum bars, work at a typical frequency of about 10^3 Hz. Resonant antennae were first built by J Weber, around 1960, at the University of Maryland. Subsequently, gravitational-wave resonant detectors have been operated by the following universities: Beijing, Guangzhou, Louisiana, Maryland, Moscow, Rome, Padua, Stanford, Tokyo and Western Australia at Perth. The contribution of Pizzella deals with the bandwidth and the sensitivities of resonant detectors. It is shown that it might be possible to reach a frequency bandwidth up to 50 Hz. The sensitivity of five cryogenic bar detectors in operation, ALLEGRO, AURIGA, EXPLORER, NAUTILUS AND NIOBE is then discussed.

The paper by Angela Di Virgilio treats laser interferometers on Earth and in particular the Italian–French antenna VIRGO. Gravitational-wave laser-interferometers on Earth will operate in the frequency range between 10^4 Hz and a few tens of hertz. Various types of gravitational-wave laser interferometers have been proposed, among which are the standard Michelson and Fabry–Perot types. A Michelson-type gravitational-wave laser interferometer is essentially made of three masses suspended with wires at the ends of two orthogonal arms of length l. When a gravitational-wave with reduced wavelength $\lambda_{G\overline{W}} \gg l$ is impinging, for example, perpendicularly to this system, variations in the metric perturbation h due to the gravitational wave will, in turn, produce oscillations in the difference between the proper lengths of the two arms $\delta l(t)$ and therefore oscillations in the relative phase of the laser light at the beamsplitter; thus, they will finally produce oscillations in the intensity of the laser light measured by the photodetector. If the laser light will travel back and forth between the test masses $2N$ times ($N =$ number of round trips), then the variation of the difference between the proper lengths of the two arms will be (assuming $Nl \ll \lambda_{G\overline{W}}$): $\Delta l = 2Nlh(t)$, and therefore, the relative phase delay due to the variations in δl will be:

$$\Delta \phi = \frac{\Delta l}{\lambda_{\overline{L}}} = \frac{2Nl}{\lambda_{\overline{L}}} h(t),$$

where $\lambda_{\overline{L}}$ is the reduced wavelength of the laser light.

For most of the fundamental limiting factors of these Earth-based detectors, such as seismic noise, photon shot noise, etc ..., the displacement noise is essentially independent from the arms length l. Therefore, by increasing l one increases the sensitivity of the detectors.

Two antennae with 4 km arm lengths in the USA, the MIT and Caltech LIGOs, should reach sensitivities to bursts of gravitational radiation of the order of $h \sim 10^{-20}$–10^{-21} between 1000 and 100 Hz. GEO-600 is an underground 600 m laser interferometer built by the University of Glasgow and the Max-Planck-Institutes for Quantum Optics and for Astrophysics at Garching. TAMA is a 300 m antenna in Japan and ACIGA is a 3 km antenna planned in Australia. The paper of Di Virgilio describes the 3 km laser interferometer VIRGO, built by INFN of Pisa together with the University of Paris-Sud at Orsay, that should reach frequencies of operation as low as a few tens of hertz, using special filters to eliminate the seismic noise at these lower frequencies. The ultimate burst sensitivity for all of the above large interferometers is currently estimated to be of the order of 10^{-22} or 10^{-23} at frequencies near 100 Hz.

The paper of Peter Bender describes the space gravitational-wave detector LISA (Laser Interferometer Space Antenna). Below about 10 Hz the sensitivity of Earth-based gravitational-wave detectors is limited by gravity gradients variations. Even for perfect isolation of a detector from seismic and ground noise, an Earth-detector would still be affected by the time changes in the gravity field due to density variations in the Earth and its atmosphere. Due to this source of noise the sensitivity has been calculated to worsen as roughly the inverse fourth power of the frequency.

Therefore, to avoid this type of noise and to reduce noise from other sources, one should use an interferometer far from Earth and with very long arms. Indeed, to detect gravitational waves in the range of frequencies between about 10^{-4} and 1 Hz, Bender proposes to orbit in the solar system a space interferometer made of three spacecraft at a typical distance from each other of 5000 000 km.

Although the phase measurement system and the thermal stability are essential requirements, it is the main technological challenge of this experiment to keep very small the spurious accelerations of the test masses. A drag compensating system will be able to largely reduce these spurious accelerations.

Considering all the error sources, it has been calculated that, for periodic gravitational waves, with an integration time of about one year, LISA should reach a sensitivity able to detect amplitudes of $h \sim 10^{-23}$, in the range of frequencies between 10^{-3} and 10^{-2} Hz, amplitudes from $h \sim 10^{-20}$ to about 10^{-23} between 10^{-4} and 10^{-3} Hz, and amplitudes from $h \sim 10^{-22}$ to about 10^{-23} between 1 and 10^{-2} Hz.

Therefore, comparing the LISA sensitivity to the predicted theoretical amplitudes of gravitational radiation at these frequencies, LISA should be able to detect gravitational waves from galactic binaries, including ordinary main-sequence binaries, contact binaries, cataclysmic variables, close white dwarf binaries, neutron star and black hole binaries. The LISA sensitivity should also allow detection of possible gravitational pulses from distant galaxies from the inspiral of compact objects into supermassive black holes in Active Galactic Nuclei and from collapse of very massive objects to form black holes. LISA should also allow us to detect the stochastic background due to unresolved binary systems.

The contribution by Francesco Fucito treats spherical shape antennae and the detection of scalar gravitational waves. General relativity predicts only two independent states of polarization of a weak gravitational wave, the so-called '×' and '+' ones. Nevertheless, metric theories of gravity alternative to general relativity and non-metric theories of gravity predict different polarization states (up to six components in metric theories [14]). For example, the Jordan–Brans–Dicke theory predicts also an additional scalar component of a gravitational wave and, as the author explains, string theory could also imply the existence of other components. In this paper the possibility is described of placing limits on, or detecting, these additional polarizations of a gravitational wave, thus testing theories of gravity alternative to general relativity, by using spherical shape detectors. Spheroidal detectors of gravitational waves of two types are discussed, standard and hollow spherical ones.

The paper by Babusci, Foffa, Losurdo, Maggiore, Matone and Sturani treats stochastic gravitational waves. As the authors explain, the stochastic gravitational-wave background (SGWB) is a random background of gravitational waves without any specific sharp frequency component that might give information about the very early stages of our universe. It is important to note that relic cosmological gravitational waves emitted near the big bang might provide unique information on our universe at a very early stage. Indeed, as regards the cosmic microwave background radiation, electromagnetic waves decoupled a few 10^5 years after the big bang, whereas relic cosmological gravitational waves, the authors explain, might come from times as early as a few 10^{-44} s. The authors discuss that, in order to increase the chances of detecting a stochastic background of gravitational waves, the correlation of the outputs between two, or more, detectors would be convenient. Thus, after discussing three different detectors: laser interferometers, cylindrical bars and spherical antennae, the authors present various possibilities of correlation, between two laser interferometers (VIRGO, LIGOs, GEO-600 and TAMA-300), and between a laser interferometer and a cylindrical bar (AURIGA, NAUTILUS, EXPLORER) or a spherical antenna; they also discuss correlation between more than two detectors.

In the second part of this paper they discuss sources of the background of stochastic gravitational waves: topological defects in the form of points, lines or surfaces, called monopoles, cosmic strings and domain walls. In particular, they discuss cosmic strings and hybrid defects; inflationary cosmological models; string cosmology; and first-order phase transitions which occurred in the early stage of the expansion of the universe, for example in GUT-symmetry breaking and electroweak-symmetry breaking. Finally, they discuss astrophysical sources of stochastic gravitational waves. The conclusion is that the frequency domain of cosmological and astrophysical sources of stochastic gravitational waves might be very different and thus, the authors conclude, the astrophysical backgrounds might not mask the detection of a relic cosmological gravitational-wave background at the frequencies of the laser interferometers on Earth.

The contribution by Nicolai and Nagar deals with the symmetry properties of Einstein's vacuum field equations when the theory is reduced from four to two dimensions, namely in the presence of two independent spacelike commuting Killing vectors. Under these conditions, and using the vierbein formalism, the authors show that one can use a Kaluza–Klein ansatz to rewrite the Einstein–Hilbert Lagrangian in the form of two different two-dimensionally reduced Lagrangians named the Ehlers and Matzner–Misner ones, respectively, after the people who first introduced them. Each of these two Lagrangians represents two-dimensional reduced gravity in the conformal gauge as given by a part of pure two-dimensional gravity, characterized by a conformal factor and a dilaton field plus a 'matter part' given by two suitable bosonic fields. In either case, the matter part has a structure of a nonlinear sigma model with an SL(2,R)/SO(2) symmetry. These two different nonlinear symmetries can be combined into a unified infinite-dimensional symmetry group of the theory, called the Geroch group, whose Lie algebra is an affine Kac–Moody algebra, and whose action on the matter fields is both nonlinear and non-local. The existence of such an infinite-dimensional symmetry guarantees that the two-dimensionally reduced nonlinear field equations are integrable. This can be shown in a standard way by exploiting the symmetry to prove the equivalence of the theory to a system of linear differential equations whose compatibility conditions yield just the nonlinear equations that one wants to solve. As an example of the application of the method to the construction of exact solutions of the two-dimensionally reduced Einstein's equations, the results are employed to derive the exact expression of the metric which describe colliding plane gravitational waves with collinear and non-collinear polarization.

Gasperini's contribution deals with string cosmology and with the basic ideas of the so-called pre-big bang scenario of string cosmology. Then it treats the interesting problem of observable effects in different cosmological models, and in particular the so-called background of relic gravitational waves, comparing it with the expected sensitivities of the gravitational-wave detectors. The conclusion is that the sensitivity of the future advanced detectors of gravitational waves may be capable of detecting the background of gravitational waves predicted in the pre-big bang scenario of string cosmology and thus these detectors might test different cosmological models and also string theory models.

The paper by Bini and De Felice studies the problem of the behaviour of a test gyroscope on which a plane gravitational wave is impinging. The authors analyse whether there might be observable effects, i.e. a precession of the gyroscope with respect to a suitably defined frame of reference that is not Fermi–Walker transported.

The contribution by Luc Blanchet deals with the post-Newtonian computation of binary inspiral waveforms. In general relativity, the orbital phase of compact binaries, when gravitational radiation emitted is considered, is not constant as it is in the Newtonian calculation, but is a complex, nonlinear function of time, depending on small post-Newtonian corrections. For the data analysis on detectors, a formula containing at least the 3PN (third-post-Newtonian) order beyond the quadrupole formalism (see the contribution by Schutz and Ricci) is needed, that is a formula including terms of the order of $(v/c)^6$ (where v is a typical velocity in the source and c is the speed of light). Blanchet's paper thus treats the derivation of the third-post-Newtonian formula for the emission of gravitational radiation from a self-gravitating binary system.

The paper by Ed Seidel deals with numerical relativity. Among the astrophysical sources of gravitational radiation that might be detected by laser interferometers on Earth there is the spiralling coalescence of two black holes or neutron stars. However, gravitational waves are so weak at the detectors on Earth that, as Seidel explains in his paper, one needs to know the waveform in order to reliably detect them, in other words gravitational-wave signals can be interpreted and detected only by comparing the observational data with a set of theoretically determined 'waveform templates'. Unfortunately, we can solve the Einstein's field equations (coupled, nonlinear partial differential equations) only in especially simple cases. Thus, to find solutions of the Einstein's equations, for example in a system with emission of gravitational radiation, we need to find numerical solutions of these field equations, i.e. we need *numerical relativity*. Nevertheless, even the numerical approach to the emission of gravitational waves in strong field is extremely difficult and computer-time consuming. For example, as Seidel explains, the computer simulation of the coalescence of a compact object binary will require several years of super-computer time. However, special codes to solve the complete set of Einstein's equations have been designed that run very efficiently on large-scale parallel computers, in particular, one of these codes, the Cactus Computational Toolkit is presented in this paper. Then, after a description of the numerical formulation of the theory of general relativity, constraint equations and evolution equations, the numerical techniques for solving the evolution equations are reported and finally some recent applications, including gravitational waves and the evolution and collisions of black holes, are presented. It is important to note that there have been and there are large collaborations in numerical relativity, including: the NSF Black Hole Grand Challenge Project, the NASA Neutron Star Grand Challenge Project, the NCSA/Potsdam/Washington University numerical relativity collaboration and a EU European collaboration of ten institutions.

References

[1] Misner C W, Thorne K S and Wheeler J A 1973 *Gravitation* (New York: Freeman)
[2] Ciufolini I and Wheeler J A 1995 *Gravitation and Inertia* (Princeton, NJ: Princeton University Press)
[3] Thorne K S 1987 *Three Hundred Years of Gravitation* ed S W Hawking and W Israel (Cambridge: Cambridge University Press)
[4] Einstein A 1916 *Preuss. Akad. Wiss. Berlin, Sitzber.* 688
[5] Grishchuk L P and Polnarev A G 1980 *General Relativity and Gravitation, One Hundred Years After the Birth of Albert Einstein* ed A Held (New York: Plenum)
[6] Douglass D H and Braginsky V B 1979 *General Relativity, An Einstein Centenary Survey* ed S W Hawking and W Israel (Cambridge: Cambridge University Press)
[7] Landau L D and Lifshitz E M 1962 *The Classical Theory of Fields* (Reading, MA: Addison-Wesley)
[8] Bondi H 1957 *Nature* **179** 1072
    See also Bondi H and McCrea W H 1960 *Proc. Camb. Phil. Soc.* **56** 410
[9] Wheeler J A 1955 *Phys. Rev.* **97** 511
[10] Weber J 1961 *General Relativity and Gravitational Waves* (New York: Wiley–Interscience)
[11] Weinberg S 1972 *Gravitation and Cosmology* (New York: Wiley)
[12] Hulse R A and Taylor J H 1975 *Astrophys. J.* **195** L51

[13] Ciufolini I *et al* 1996 *Nuovo Cimento* A **109** 575
     See also Ciufolini I *et al* 1997 *Class. Quantum Grav.* **14** 2701
     Ciufolini I *et al* 1998 *Science* **279** 2100
[14] Will C M 1993 *Theory and Experiment in Gravitational Physics* (Cambridge:
     Cambridge University Press)

PART 1

GRAVITATIONAL WAVES, SOURCES AND DETECTORS

Bernard F Schutz[1] and Franco Ricci[2]

[1] *Max Planck Institute for Gravitational Physics (Albert Einstein Institute), Golm bei Potsdam, Germany and Department of Physics and Astronomy, Cardiff University, Wales*
[2] *University 'La Sapienza', Rome, Italy*

Synopsis

Gravitational waves and their detection are becoming increasingly important both for the theoretical physicist and the astrophysicist. In fact, technological developments have enabled the construction of such sensitive detectors (bars and interferometers) that the detection of gravitational radiation could become a reality during the next few years. In these lectures we give a brief overview of this interesting and challenging field of modern physics.

The topics covered are divided into six lectures. We begin (chapter 2) by describing gravitational waves in linearized general relativity, where one can examine most of the basic properties of gravitational radiation itself; propagation, gauge invariance and interactions with matter (and in particular with detectors).

The second lecture (chapter 3) deals with gravitational-wave detectors: how they operate, what their most important sources of noise are, and what mechanisms are used to overcome noise. We report here on the most important detectors planned or under construction (both ground-based and space-based ones), their likely sensitivity and their prospects for making detections. Other speakers will go into much more detail on specific detectors, such as LISA.

The third lecture (chapter 4) deals with the astrophysics of likely sources of gravitational waves: binary systems, neutron stars, pulsars, x-ray sources, supernovae/hypernovae, γ-ray bursts and the big bang. We estimate the expected wave amplitude h and the suitability of specific detectors for seeing waves from each source.

The fourth lecture (chapter 5) is much more theoretical. Here we develop the mathematical theory of gravitational waves in general, their effective stress-energy tensor, the energy carried by gravitational waves, and the energy in a random wave field (gravitational background generated by the big bang).

The fifth lecture (chapter 6) takes the theory further and examines the generation of gravitational radiation in linearized theory. We show in some detail how both mass-quadrupole and current-quadrupole radiation is generated, including how characteristics of the radiation such as its polarization are related to the motion of the source. Current-quadrupole radiation has become important very recently and may indeed be one of the first forms of gravitational radiation to be detected. We attempt to give a physical description of the way it is generated.

The final lecture (chapter 7) explores applications of the theory we have developed to various sources. We calculate the quadrupole moment of a binary system, the energy radiated in the Newtonian approximation and the back-reaction on the orbit. We conclude with a brief introduction to the current-quadrupole-driven instability in the r-modes of neutron stars.

Chapters 2 and 5 are followed by a few exercises to assist students. We presume the reader has some background in general relativity and its mathematical tools in differential geometry, at the level of the introductory chapters of Schutz (1985). A list of references is presented at the end of these lectures of sources suitable for further and background reading.

Chapter 2

Elements of gravitational waves

General relativity is a theory of gravity that is consistent with special relativity in many respects, and in particular with the principle that nothing travels faster than light. This means that changes in the gravitational field cannot be felt everywhere instantaneously: they must propagate. In general relativity they propagate at exactly the same speed as vacuum electromagnetic waves: the speed of light. These propagating changes are called gravitational waves.

However, general relativity is a nonlinear theory and there is, in general, no sharp distinction between the part of the metric that represents the waves and the rest of the metric. Only in certain approximations can we clearly define gravitational radiation. Three interesting approximations in which it is possible to make this distinction are:

- linearized theory;
- small perturbations of a smooth, time-independent background metric;
- post-Newtonian theory.

The simplest starting point for our discussion is certainly linearized theory, which is a weak-field approximation to general relativity, where the equations are written and solved in a nearly flat spacetime. The static and wave parts of the field cleanly separate. We idealize gravitational waves as a 'ripple' propagating through a flat and empty universe.

This picture is a simple case of the more general 'short-wave approximation', in which waves appear as small perturbations of a smooth background that is time dependent and whose radius of curvature is much larger than the wavelength of the waves. We will describe this in detail in chapter 5. This approximation describes wave propagation well, but it is inadequate for wave generation. The most useful approximation for sources is the post-Newtonian approximation, where waves arise at a high order in corrections that carry general relativity away from its Newtonian limit; we treat these in chapters 6 and 7.

For now we concentrate our attention on linearized theory. We follow the notation and conventions of Misner *et al* (1973) and Schutz (1985). In particular we choose units in which $c = G = 1$; Greek indices run from 0 to 3; Latin indices run from 1 to 3; repeated indices are summed; commas in subscripts or superscripts denote partial derivatives; and semicolons denote covariant derivatives. The metric has positive signature. These above two textbooks and others referred to at the end of these chapters give more details on the theory that we outline here. For an even simpler introduction, based on a scalar analogy to general relativity, see [1].

2.1   Mathematics of linearized theory

Consider a perturbed flat spacetime. Its metric tensor can be written as

$$g_{\alpha\beta} = \eta_{\alpha\beta} + h_{\alpha\beta}, \quad |h_{\alpha\beta}| \ll 1, \quad \alpha, \beta = 0, \dots, 3 \qquad (2.1)$$

where $\eta_{\alpha\beta}$ is the Minkowski metric $(-1, 1, 1, 1)$ and $h_{\alpha\beta}$ is a very small perturbation of the flat spacetime metric. Linearized theory is an approximation to general relativity that is correct to first order in the size of this perturbation. Since the size of tensor components depends on coordinates, one must be careful with such a definition. What we require for linearized theory to be valid is that there should exist a coordinate system in which equation (2.1) holds in a suitably large region of spacetime. Even though $\eta_{\alpha\beta}$ is not the true metric tensor, we are free to *define* raising and lowering indices of the perturbation with $\eta_{\alpha\beta}$, as if it were a tensor on flat spacetime. We write

$$h^{\alpha\beta} := \eta^{\alpha\gamma}\eta^{\beta\delta}h_{\gamma\delta}.$$

This leads to the following equation for the inverse metric, correct to first order (all we want in linearized theory):

$$g^{\alpha\beta} = \eta^{\alpha\beta} - h^{\alpha\beta}. \qquad (2.2)$$

The mathematics is simpler if we define the *trace-reversed* metric perturbation:

$$\bar{h}_{\alpha\beta} := h_{\alpha\beta} - \tfrac{1}{2}\eta_{\alpha\beta}h, \qquad (2.3)$$

where $h := \eta_{\alpha\beta}h^{\alpha\beta}$. There is considerable coordinate freedom in the components $h_{\alpha\beta}$, since we can wiggle and stretch the coordinate system with a comparable amplitude and change the components. This coordinate freedom is called *gauge freedom*, by analogy with electromagnetism. We use this freedom to enforce the *Lorentz (or Hilbert) gauge*:

$$\bar{h}^{\alpha\beta}{}_{,\beta} = 0. \qquad (2.4)$$

In this gauge the Einstein field equations (neglecting the quadratic and higher terms in $h^{\alpha\beta}$) are just a set of decoupled linear wave equations:

$$\left(-\frac{\partial^2}{\partial t^2} + \nabla^2\right)\bar{h}^{\alpha\beta} = -16\pi T^{\alpha\beta}. \qquad (2.5)$$

To understand wave propagation we look for the easiest solution of the vacuum gravitational field equations:

$$\Box \bar{h}^{\alpha\beta} \equiv \left(-\frac{\partial^2}{\partial t^2} + \nabla^2 \right) \bar{h}^{\alpha\beta} = 0. \tag{2.6}$$

Plane waves have the form:

$$\bar{h}_{\alpha\beta} = \mathcal{A} e_{\alpha\beta} \exp(ik_\gamma x^\gamma) \tag{2.7}$$

where the amplitude $\mathcal{A}$, polarization tensor $e^{\alpha\beta}$ and wavevector k^γ are all constants. (As usual one has to take the real part of this expression.)

The Einstein equations imply that the wavevector is 'light-like', $k^\gamma k_\gamma = 0$, and the gauge condition implies that the amplitude and the wavevector are orthogonal: $e^{\alpha\beta} k_\beta = 0$.

Linearized theory describes a classical gravitational field whose quantum description would be a massless spin 2 field that propagates at the speed of light. We expect from this that such a field will have only two independent degrees of freedom (helicities in quantum language, polarizations in classical terms). To show this classically we remember that $h_{\alpha\beta}$ is symmetric, so it has ten independent components, and that the Lorentz gauge applies four independent conditions to these, reducing the freedom to six. However, the Lorentz gauge does not fully fix the coordinates. In fact if we perform another infinitesimal coordinate transformation ($x^\mu \to x^\mu + \xi^\mu$ with $\xi^\mu{}_{,\nu} = O(h)$) and impose $\Box \xi^\mu = 0$, we remain in Lorentz gauge. We can use this freedom to demand:

$$e^{0\alpha} = 0 \implies e^{ij} k_j = 0 \quad \text{(transverse wave)}, \tag{2.8}$$

$$e^i{}_i = 0 \quad \text{(traceless wave)}. \tag{2.9}$$

These conditions can only be applied outside a sphere surrounding the source. Together they put the metric into the *transverse-traceless* (TT) gauge. We will explicitly construct this gauge in chapter 5.

2.2 Using the TT gauge to understand gravitational waves

The TT gauge leaves only *two independent polarizations* out of the original ten, and it ensures that $\bar{h}_{\alpha\beta} = h_{\alpha\beta}$. In order to understand the polarization degrees of freedom, let us take the wave to move in the z-direction, so that $k_z = \omega$, $k^0 = \omega$, $k_x = 0$, $k_y = 0$; the TT gauge conditions in equations (2.8) and (2.9) lead to $e^{0\alpha} = e^{z\alpha} = 0$ and $e^{xx} = -e^{yy}$. This leaves only two independent components of the polarization tensor, say e^{xx} and e^{xy} (which we denote by the symbols $\oplus$, $\otimes$).

A wave for which $e^{xy} = 0$ (pure $\oplus$ polarization) produces a metric of the form:

$$ds^2 = -dt^2 + (1 + h_+) dx^2 + (1 - h_+) dy^2 + dz^2, \tag{2.10}$$

Figure 2.1. Illustration of two linear polarizations and the associated wave amplitude.

where $h_+ = \mathcal{A}e^{xx}\exp[-i\omega(t-z)]$. Such a metric produces opposite effects on proper distance at the two transverse axes, contracting one while expanding the other.

If $e^{xx} = 0$ we have pure $\otimes$ polarization $h_\times$ which can be obtained from the previous case by a simple $45°$ rotation, as in figure 2.1. Since the wave equation and TT conditions are linear, a general wave will be a linear combination of these two polarization tensors. A circular polarization basis would be:

$$e_R = \frac{1}{\sqrt{2}}(e_+ + ie_\times), \quad e_L = \frac{1}{\sqrt{2}}(e_+ - ie_\times), \tag{2.11}$$

where e_+, $e_\times$ are the two linear polarization tensors and e_R and e_L are polarizations that rotate in the right-handed and left-handed directions, respectively. It is important to understand that, for circular polarization, the polarization pattern rotates around the central position, but test particles themselves rotate only in small circles relative to the central position.

Now we compute the effects of a wave in the TT gauge on a particle at rest in the flat background metric $\eta_{\alpha\beta}$ before the passage of the gravitational wave. The geodesic equation

$$\frac{d^2x^\mu}{d\tau^2} + \Gamma^\mu{}_{\alpha\beta}\frac{dx^\alpha}{d\tau}\frac{dx^\beta}{d\tau} = 0$$

implies in this case:

$$\frac{d^2x^i}{d\tau^2} = -\Gamma^i{}_{00} = -\frac{1}{2}(2h_{i0,0} - h_{00,i}) = 0, \tag{2.12}$$

so that the particle *does not move*. The TT gauge, to first order in $h_{\alpha\beta}$, represents a coordinate system that is comoving with freely-falling particles. Because $h_{0\alpha} = 0$, TT time is proper time on the clock of freely-falling particles at rest.

Tidal forces show the action of the wave independently of the coordinates. Let us consider the equation of geodesic deviation, which governs the separation of two neighbouring freely-falling test particles A and B. If the particles are initially at rest, then as the wave passes it produces an oscillating curvature tensor, and the separation ξ of the two particles is:

$$\frac{d^2\xi^i}{dt^2} = R^i{}_{0j0}\xi^j. \tag{2.13}$$

To calculate the component $R^i{}_{0j0}$ of the Riemann tensor in equation (2.13), we can use the metric in the TT gauge, because the Riemann tensor is gauge-invariant at linear order (see exercise (d) at the end of this chapter). Therefore, we can replace $R^i{}_{0j0}$ by $R^i{}_{0j0} = \frac{1}{2}h^{TTi}{}_{j,00}$ and write:

$$\frac{d^2\xi^i}{dt^2} = \frac{1}{2}h^{TTi}{}_{j,00}\xi^j. \tag{2.14}$$

This equation, with an initial condition $\xi^j_{(0)} = $ constant, describes the oscillations of Bs location as measured in the proper reference frame of A. The validity of equation (2.14) is the same as that of the geodesic deviation equation: geodesics have to be close to one another, in a neighbourhood where the change in curvature is small. In this approximation a gravitational wave is like an extra force, called a *tidal force*, perturbing the proper distance between two test particles. If there are other forces on the particles, so that they are not free, then as long as the gravitational field is weak, one can just add the tidal forces to the other forces and work as if the particle were in special relativity.

2.3 Interaction of gravitational waves with detectors

We have shown above that the TT gauge is a particular coordinate system in which the polarization tensor of a plane gravitational wave assumes a very simple form. This gauge is comoving for freely-falling particles and so it is not the locally Minkowskian coordinate system that would be used by an experimenter to analyse an experiment. In general relativity one must always be aware of how one's coordinate system is defined.

We shall analyse two typical situations:

- the detector is small compared to the wavelength of the gravitational waves it is measuring; and
- the detector is comparable to or larger than that wavelength.

In the first case we can use the geodesic deviation equation above to represent the wave as a simple extra force on the equipment. Bars detectors can always be analysed in this way. Laser interferometers on the Earth can be treated this way too. In these cases a gravitational wave simply produces a force to be measured. There is no more to say from the relativity point of view. The rest of the detection story is the physics of the detectors. Sadly, this is not as simple as gravitational wave physics!

In the second case, the geodesic deviation equation is not useful because we have to abandon the 'local mathematics' of geodesic deviation and return to the 'global mathematics' of the TT gauge and metric components $h^{TT}{}_{\alpha\beta}$. Space-based interferometers like LISA, accurate ranging to solar-system spacecraft and pulsar timing are all in this class. Together with ground interferometers, these are *beam detectors*: they use light (or radio waves) to register the waves.

To study these detectors, it is easiest to remain in the TT gauge and to calculate the effect of the waves on the (coordinate) speed of light. Let us consider, for example, the $\oplus$ metric from equation (2.10) and examine a null geodesic moving in the x-direction. The speed along this curve is:

$$\left(\frac{dx}{dt}\right)^2 = \frac{1}{1+h_+}. \tag{2.15}$$

This is only a *coordinate speed*, not a contradiction to special relativity.

To analyse the way in which detectors work, suppose one arm of an interferometer lies along the x-direction and the wave, for simplicity, is moving in the z-direction with a $\oplus$ polarization of *any* waveform $h_+(t)$ along this axis (it is a plane wave, so its waveform does not depend on x). Then a photon emitted at time t from the origin reaches the other end, at a fixed coordinate position $x = L$, at the coordinate time

$$t_{\text{far}} = t + \int_0^L \sqrt{1+h_+(t(x))}\,dx, \tag{2.16}$$

where the argument $t(x)$ denotes the fact that one must know the time to reach position x in order to calculate the wave field. This implicit equation can be solved in linearized theory by using the fact that h_+ is small, so we can use the first-order solution of equation (2.15) to calculate $h_+(t)$ to sufficient accuracy.

To do this we expand the square root in powers of h_+, and consider as a zero-order solution a photon travelling at the speed of light in the x-direction of a flat spacetime. We can set $t(x) = t + x$. The result is:

$$t_{\text{out}} = t + L + \tfrac{1}{2}\int_0^L h_+(t+x)\,dx. \tag{2.17}$$

In an interferometer, the light is reflected back, so the return trip takes

$$t_{\text{return}} = t + L + \tfrac{1}{2}\left[\int_0^L h_+(t+x)\,dx + \int_0^L h_+(t+x+L)\,dx\right]. \tag{2.18}$$

What one monitors is changes in the time taken by a return trip as a function of time at the origin. If there were no gravitational waves t_{return} would be constant because L is fixed, so changes indicate a gravitational wave.

The rate of variation of the return time as a function of the start time t is

$$\frac{dt_{\text{return}}}{dt} = 1 + \frac{1}{2}[h_+(t+2L) - h_+(t)]. \tag{2.19}$$

This depends only on the wave amplitude when the beam leaves and when it returns.

Let us consider now a more realistic geometry than the previous one, and in particular suppose that the wave travels at an angle θ to the z-axis in the x–z plane. If we redo this calculation, allowing the phase of the wave to depend on x in an appropriate way, and taking into account the fact that $h_+^{\mathrm{TT}xx}$ is reduced if the wave is not moving in a direction perpendicular to x, we find (see exercise (a) at the end of this chapter for the details of the calculation)

$$\frac{\mathrm{d}t_{\text{return}}}{\mathrm{d}t} = \frac{1}{2}\{(1 - \sin\theta)h_+^{xx}(t + 2L) - (1 + \sin\theta)h_+^{xx}(t)$$
$$+ 2\sin\theta\, h_+^{xx}[t + L(1 - \sin\theta)]\}. \tag{2.20}$$

This three-term relation is the starting point for analysing the response of all beam detectors. This is directly what happens in radar ranging or in transponding to spacecraft, where a beam in only one direction is used. In long-baseline interferometry, one must analyse the second beam as well. We shall discuss these cases in turn.

2.4 Analysis of beam detectors

2.4.1 Ranging to spacecraft

Both NASA and ESA perform experiments in which they monitor the return time of communication signals with interplanetary spacecraft for the characteristic effect of gravitational waves. For missions to Jupiter and Saturn, the return times are of the order 2–4×10^3 s. Any gravitational wave event shorter than this will leave an imprint on the delay time three times: once when the wave passes the Earth-based transmitter, once when it passes the spacecraft, and once when it passes the Earth-based receiver. Searches use a form of pattern matching to look for this characteristic imprint. There are two dominant sources of noise: propagation-time irregularities caused by fluctuations in the solar wind plasma, and timing noise in the clocks used to measure the signals. The plasma delays depend on the radio-wave frequency, so by using two transmission frequencies one can model and subtract the plasma noise. Then if one uses the most stable atomic clocks, it is possible to achieve sensitivities for h of the order 10^{-13}. In the future, using higher radio frequencies, such experiments may reach 10^{-15}. No positive detections have yet been made, but the chances are not zero. For example, if a small black hole fell into a massive black hole in the centre of the Galaxy, it would produce a signal with a frequency of about 10 mHz and an amplitude significantly bigger than 10^{-15}. Rare as this might be, it would be a dramatic event to observe.

2.4.2　Pulsar timing

Many pulsars, in particular old millisecond pulsars, are extraordinarily regular clocks, whose random timing irregularities are too small for even the best atomic clocks to measure. Other pulsars have weak but observable irregularities. Measurements of or even upper limits on any of these timing irregularities for single pulsars can be used to set *upper limits* on any background gravitational wave field with periods comparable to or shorter than the observing time. Here the three-term formula is replaced by a simpler two-term expression (see exercise (b) at the end of this chapter), because we only have a one-way transmission from the pulsar to Earth. Moreover, the transit time of a signal to Earth from the pulsar may be thousands of years, so we cannot look for correlations between the two terms in a given signal. Instead, the delay time is a combination of the effects of uncorrelated waves at the pulsar when the signal was emitted and at the Earth when it is received.

　　If one simultaneously observes two or more pulsars, the Earth-based part of the delay is correlated between them, and this offers a means of actually detecting long-period gravitational waves. Observations require a timescale of several years in order to achieve the long-period stability of pulse arrival times, so this method is suited to looking for strong gravitational waves with periods of several years.

2.4.3　Interferometry

An interferometer essentially measures changes in the difference in the return times along two different arms. It does this by looking for changes in the interference pattern formed when the returning light beams are superimposed on one another. The response of each arm will follow the three-term formula in equation (2.20), but with a different value of θ for each arm, depending in a complicated way on the orientation of the arms relative to the direction of travel and the polarization of the wave. Ground-based interferometers are small enough to use the small-L formulae we derived earlier. However, LISA, the space-based interferometer that is described by Bender in this book, is larger than a wavelength of gravitational waves for frequencies above 10 mHz, so a detailed analysis of its sensitivity requires the full three-term formula.

2.5　Exercises for chapter 2

Suggested solutions for these exercises are at the end of chapter 7.

(a)　1.　*Derive the full three-term return equation, reproduced here:*

$$\frac{\mathrm{d}t_{\text{return}}}{\mathrm{d}t} = \frac{1}{2}\{(1 - \sin\theta)h_+^{xx}(t + 2L) - (1 + \sin\theta)h_+^{xx}(t) + 2\sin\theta h_+^{xx}[t + L(1 - \sin\theta)]\}. \tag{2.21}$$

2. *Show that, in the limit where L is small compared to the wavelength of the gravitational wave, the derivative of the return time is the derivative of the excess proper distance $\delta L = Lh_+^{xx}(t)\cos^2\theta$ for small L. Make sure you know how to interpret the factor of $\cos^2\theta$.*

3. *Examine the limit of the three-term formula when the gravitational wave is travelling along the x-axis too ($\theta = \pm\frac{\pi}{2}$): what happens to light going parallel to a gravitational wave?*

(b) *Derive the two-term formula governing the delays induced by gravitational waves on a signal transmitted only one-way, for example from a pulsar to Earth.*

(c) *A frequently asked question is: if gravitational waves alter the speed of light, as we seem to have used here, and if they move the ends of an interferometer closer and further apart, might these effects not cancel, so that there would be no measurable effects on light? Answer this question. You may want to examine the calculation above: did we make use of the changing distance between the ends, and why or why not?*

(d) *Show that the Riemann tensor is gauge-invariant in linearized theory.*

Chapter 3

Gravitational-wave detectors

Gravitational radiation is a central prediction of general relativity and its detection is a key test of the integrity of the theoretical structure of Einstein's work. However, in the long run, its importance as a tool for observational astronomy is likely to be even more important. We have excellent observational evidence from the Hulse–Taylor binary pulsar system (described in chapter 4) that the predictions of general relativity concerning gravitational radiation are quantitatively correct. However, we have incomplete information from astronomy today about the likely sources of detectable radiation.

The gravitational wave spectrum is completely unexplored, and whenever a new electromagnetic waveband has been opened to astronomy, astronomers have discovered completely unexpected phenomena. This seems to me just as likely to happen again with gravitational waves, especially because gravitational waves carry some kinds of information that electromagnetic radiation cannot convey. Gravitational waves are generated by bulk motions of masses, and they encode the mass distributions and speeds. They are coherent and their low frequencies reflect the dynamical timescales of their sources.

In contrast, electromagnetic waves come from individual electrons executing complex and partly random motions inside their sources. They are incoherent, and individual photons must be interpreted as samples of the large statistical ensemble of photons being emitted. Their frequencies are determined by microphysics on length scales much smaller than the structure of the astronomical system emitting them. From electromagnetic observations we can make inferences about this structure only through careful modelling of the source. Gravitational waves, by contrast, carry information whose connection to the source structure and motion is fairly direct.

A good example is that of massive black holes in galactic nuclei. From observations that span the electromagnetic spectrum from radio waves to x-rays, astrophysicists have inferred that black holes of masses up to $10^9 M_\odot$ are responsible for quasar emissions and control the jets that power the giant radio emission regions. The evidence for the black hole is very strong but indirect: no other known object can contain so much mass in such a small volume. Gravitational wave observations will tell us about the dynamics of the holes themselves, providing unique signatures from which they can be identified, measuring their masses and spins directly from their vibrational frequencies. The interplay of electromagnetic and gravitational observations will enrich many branches of astronomy.

The history of gravitational-wave detection started in the 1960s with J Weber at the University of Maryland. He built the first *bar detector*: it was a massive cylinder of aluminium ($\sim 2 \times 10^3$ kg) operating at room temperature (300 K) with a resonant frequency of about 1600 Hz. This early prototype had a modest sensitivity, around 10^{-13} or 10^{-14}.

Despite this poor sensitivity, in the late 1960s Weber announced the detection of a population of coincident events between two similar bars at a rate far higher than expected from instrumental noise. This news stimulated a number of other groups (at Glasgow, Munich, Paris, Rome, Bell Laboratories, Stanford, Rochester, LSU, MIT, Beijing, Tokyo) to build and develop bar detectors to check Weber's results. Unfortunately for Weber and for the idea that gravitational waves were easy to detect, none of these other detectors found anything, even at times when Weber continued to find coincidences. Weber's observations remain unexplained even today. However, the failure to confirm Weber was in a real sense a confirmation of general relativity, because theoretical calculations had never predicted that reasonable signals would be strong enough to be seen by Weber's bars.

Weber's announcements have had a mixed effect on gravitational-wave research. On the one hand, they have created a cloud under which the field has laboured hard to re-establish its respectability in the eyes of many physicists. Even today the legacy of this is an extreme cautiousness among the major projects, a conservatism that will ensure that the next claim of a detection will be ironclad. On the other hand, the stimulus that Weber gave to other groups to build detectors has directly led to the present advanced state of detector development.

From 1980 to 1994 groups developed detectors in two different directions:

- *Cryogenic bar detectors*, developed primarily at Rome/Frascati, Stanford, LSU and Perth (Australia). The best of these detectors reach below 10^{-19}. They are the only detectors operating continuously today and they have performed a number of joint coincidence searches, leading to upper limits but no detections.
- *Interferometers*, developed at MIT, Garching (where the Munich group moved), Glasgow, Caltech and Tokyo. The typical sensitivity of these prototypes was 10^{-18}. The first long coincidence observation with interferometers was the Glasgow/Garching 100 hr experiment in 1989 [2].

In fact, interferometers had apparently been considered by Weber, but at that time the technology was not good enough for this kind of detector. Only 10–15 years later, technology had progressed. Lasers, mirror coating and polishing techniques and materials science had advanced far enough to allow the first practical interferometers, and it was clear that further progress would continue unabated. Soon afterwards several major collaborations were formed to build large-scale interferometric detectors:

- LIGO: Caltech and MIT (NSF) LIGO;
- VIRGO: France (CNRS) and Italy (INFN)
- GEO600: Germany (Max Planck) and UK (PPARC).

Later, other collaborations were formed in Australia (AIGO) and Japan (TAMA and JGWO). At present there is still considerable effort in building successors to Weber's original resonant-mass detector: ultra-cryogenic bars are in operation in Frascati and Padua, and they are expected to reach below 10^{-20}. Further, there are proposals for a new generation of spherical or icosahedral solid-mass detectors from the USA (LSU), Brazil, the Netherlands and Italy. Arrays of smaller bars have been proposed for observing the highest frequencies, where neutron star normal modes lie.

However, the real goal for the near future is to break through the 10^{-21} level, which is where theory predicts that it is not unreasonable to expect gravitational waves of the order of once per year (see the discussion in chapter 4 later). The first detectors to reach this level will be the large-scale interferometers that are now under construction. They have very long arms: LIGO, Hanford (WA) and Livingstone (LA), 4 km; VIRGO: Pisa, 3 km; GEO600: Hannover, 600 m; TAMA300: Tokyo, 300 m.

The most spectacular detector in the near future is the space-based detector LISA, which has been adopted by ESA (European Space Agency) as a Cornerstone mission for the twenty-first century. The project is now gaining a considerable amount of momentum in the USA, and a collaboration between ESA and NASA seems likely. This mission could be launched around 2010.

3.1   Gravitational-wave observables

We have described earlier how different gravitational-wave observables are from electromagnetic observables. Here are the things that we want to measure when we detect gravitational waves:

- $h_+(t)$, $h_\times(t)$, *phase(t)*: the amplitude and polarization of the wave, and the phase of polarization, as functions of time. These contain most of the information about gravitational waves.
- θ, ϕ: the direction on the sky of the source (except for observations of a stochastic background).

From this it is clear that gravitational-wave detection is not the same as electromagnetic-radiation detection. In electromagnetic astronomy one almost always rectifies the electromagnetic wave, while we can follow the oscillations of the gravitational wave. Essentially in electromagnetism one detects the power in the radiation, while for gravitational radiation, as we have said before, one detects the wave coherently.

Let us consider now what we can infer from a detection. If the gravitational wave has a short duration, of the order of the sampling time of the signal stream, then each detector will usually give just a single number, which is the amplitude of the wave projected on the detector (a projection of the two polarizations h_+ and $h_\times$). If the wave lasts more than one sampling time, then this information is a function of time.

If the signal lasts for a sufficiently long time, then both the amplitude and the phase of the wave can be affected by the motion of the detector, which moves and turns with the motion of the Earth. This produces an amplitude and phase modulation which is not intrinsic to the signal. If the signal's intrinsic form is understood, then this modulation can be used to determine the location of the source. We distinguish three distinct kinds of signals, from the point of view of observations.

Bursts have a duration so short that modulation due to detector motion is not observable. During the detection, the detector is effectively stationary. In this case we need at least three, and preferably four, interferometers to triangulate the positions of bursts on the sky and to find the two polarizations h_+ and $h_\times$. (See discussions in Schutz 1989.) A network of detectors is essential to extract all the information in this case.

Continuous waves by definition last long enough for the motion of the detector to induce amplitude and phase modulation. In this case, assuming a simple model for the intrinsic signal, we can use the information imprinted on the signal (the amplitude modulation and phase modulation) to infer the position and polarization amplitude of the source on the sky. A single detector, effectively, performs aperture synthesis, finding the position of the source and the amplitude of the wave entirely by itself. However, in order to be sure that the signal is not an artefact, it will be important that the signal is seen by a second or third detector.

Stochastic backgrounds can be detected just like noise in a single detector. If the detector noise is well understood, this excess noise may be detected as a stochastic background. This is closely analogous to the way the original microwave background detection was discovered.

A more reliable method for detecting stochastic radiation is the cross-correlation between two detectors, which experience the same cosmological noise but have a different intrinsic noise. Coherent cross-correlation between two detectors eliminates much detector noise and works best when detectors are closer than a wavelength.

In general, detection of gravitational waves requires joint observing by a network of detectors, both to increase the confidence of the detection and to provide accurate information on other physical observables (direction, amplitude and so on). Networks can be assembled from interferometers, bars, or both.

3.2   The physics of interferometers

Interferometric gravitational-wave detectors are the most sensitive instruments, and among the most complex, that have ever been constructed. They are remarkable for the range of physics that is important for their construction. Interferometer groups work at the forefront of the development in lasers, mirror polishing and coating, quantum measurement, materials science, mechanical isolation, optical system design and thermal science. In this section we shall only be able to take a fairly superficial look at one of the most fascinating instrumentation stories of our age. A good introduction to interferometer design is Saulson (1994).

Interferometers use laser light to compare the lengths of two perpendicular arms. The simplest design, originated by Michelson for his famous experiment on the velocity of light, uses light that passes up and down each arm once, as in the first panel in figure 3.1. Imagine such an instrument with identical arms defined by mirrors that hang from supports, so they are free to move horizontally in response to a gravitational wave. If there is no wave, the arms have the same length, and the light from one arm returns exactly in phase with that from the other. When the wave arrives, the two arms typically respond differently. The arms are no longer the same length, and so the light that arrives back at the centre from one arm will no longer be in phase with that arriving back from the other arm. This will produce a shift in the interference fringes between the two beams. This is the principle of detection.

Real detectors are designed to store the light in each arm for longer than just one reflection (see figure 3.1(*b*)). It is optimum to store the light for half of the period of the gravitational wave, so that on each reflection the light gains an added phase shift. Michelson-type *delay-line* interferometers store the light by arranging multiple reflections. *Fabry–Perot* interferometers store the light in cavities in each arm, allowing only a small fraction to escape for the interference measurement (figure 3.1(*e*)).

An advantage of interferometers as detectors is that the gravitational-wave-induced phase shift of the light can be made larger simply by making the arm length larger, since gravitational waves act by tidal forces. A detector with an arm length $l = 4$ km responds to a gravitational wave with an amplitude of 10^{-21} with

$$\delta l_{\text{gw}} \sim \tfrac{1}{2} h l \sim 2 \times 10^{-18} \text{ m} \tag{3.1}$$

where δl_{gw} is the change in the length of one arm. If the orientation of the interferometer is optimum, then the other arm will change by the same amount in the opposite direction, so that the interference fringe will shift by twice this length.

If the light path is folded or resonated, as in figure 3.1(*b*) and (*d*), then the effective number of bounces can be traded off against overall length to achieve a given desired total path length, or storage time. Shorter interferometers with many bounces have a disadvantage, however: even though they can achieve the

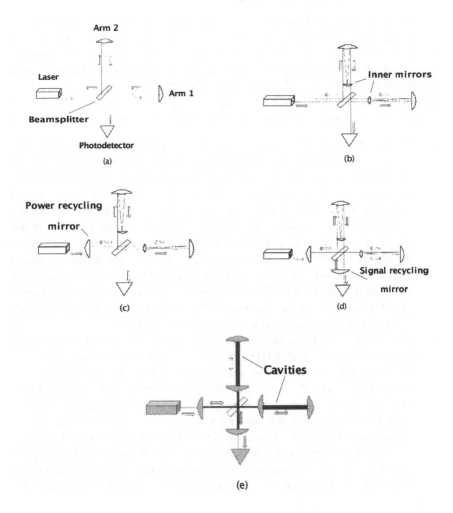

Figure 3.1. Five steps to a gravitational-wave interferometer. (*a*) The simple Michelson. Notice that there are two return beams: one goes toward the photodetector and the other toward the laser. (*b*) Delay line: a Michelson with multiple bounces in each arm to enhance the signal. (*c*) Power recycling. The extra mirror recycles the light that goes towards the laser, which would otherwise be wasted. (*d*) Signal recycling. The mirror in front of the photodetector recycles only the signal sidebands, provided that in the absence of a signal no light goes to the photodetector. (*e*) Fabry–Perot interferometer. The delay lines are converted to cavities with partially silvered interior mirrors.

same response as a longer interferometer, the extra bounces introduce noise from the mirrors, as discussed below. There is, therefore, a big advantage to long-arm interferometers.

There are three main sources of noise in interferometers: thermal, shot and vibrational. To understand the way they are controlled, it is important to think in frequency space. Observations with ground-based detectors will be made in a range from perhaps 10 Hz up to 10 kHz, and initial detectors will have a much smaller observing bandwidth within this. Disturbances by noise that occur at frequencies outside the observation band can simply be filtered out. The goal of noise control is to reduce disturbances in the observation band.

- *Thermal noise.* Interferometers work at room temperature, and vibrations of the mirrors and of the suspending pendulum can mask gravitational waves. To control this noise, scientists take advantage of the fact that thermal noise has its maximum amplitude at the frequency of the vibrational mode, and if the resonance of the mode is narrow (a high quality factor Q) then the amplitude at other frequencies is small. Therefore, pendulum suspensions are designed with the pendulum frequency at about 1 Hz, well below the observing window, and mirror masses are designed to have principal vibration modes above 1 kHz, well above the optimum observing frequency for initial interferometers. These systems are constructed with high values of Q (10^6 or more) to reduce the noise in the observing band. Even so, thermal noise is typically a dominant noise below 100 or 200 Hz.
- *Shot noise.* This is the principal limitation to sensitivity at higher frequencies, above 200–300 Hz. It arises from the quantization of photons. When photons form interference fringes, they arrive at random times and make random fluctuations in the light intensity that can look like a gravitational wave signal; the more photons one uses, the smoother will be the interference fringe. We can easily calculate this intrinsic noise. If N is the number of photons emitted by the laser during our measurement, then as a random process the fluctuation number δN is proportional to the square root of N. If we are using light with a wavelength λ (for example infrared light with $\lambda \sim 1\ \mu$m) one can expect to measure lengths to an accuracy of

$$\delta l_{\text{shot}} \sim \frac{\lambda}{2\pi \sqrt{N}}.$$

To measure a gravitational wave at a frequency f, one has to make at least $2f$ measurements per second, so one can accumulate photons for a time $1/2f$. If P is the light power, one has

$$N = \frac{P}{\frac{hc}{\lambda} \cdot \frac{1}{2f}}.$$

It is easy to work out from this that, for δl_{shot} to be equal to δl_{gw} in equation (3.1), one needs light power of about 600 kW. No continuous laser could provide this much light to an interferometer.

The key to reaching such power levels inside the arms of a detector is a technique called power recycling (see Saulson 1994) first proposed by

Drever and independently by Schilling. Normally, interferometers work on a 'dark fringe', that is they are arranged so that the light reaching the photodetector is zero if there is no gravitational wave. Then, as shown in figure 3.1(a), the whole of the input light must emerge from the interferometer travelling towards the laser. If one places another mirror, correctly positioned, between the laser and the beam splitter (figure 3.1(c)), it will reflect this wasted light back into the interferometer in such a way that it adds coherently in phase with light emerging from the laser. In this way, light can be recycled and the required power levels in the arms achieved.

Of course, there will be a maximum recycling gain, which is set by mirror losses. Light power builds up until the laser merely re-supplies the losses at the mirrors, due to scattering and absorption. The maximum power gain is

$$P = \frac{1}{1 - R^2}$$

where $1 - R^2$ is the total loss summed over all the optical surfaces. For the very high-quality mirrors used in these projects, $1 - R^2 \sim 10^{-5}$. This reduces the power requirement for the laser by the same factor, down to about 6 W. This is attainable with modern laser technology.

- *Ground vibration* and *mechanical vibrations* are another source of noise that must be screened out. Typical seismic vibration spectra fall sharply with frequency, so this is a problem primarily below 100 Hz. Pendulum suspensions are excellent mechanical filters above the pendulum frequency: it is a familiar elementary-physics demonstration that one can wiggle the suspension point of a pendulum vigorously at a high frequency and the pendulum itself remains undisturbed. Suspension designs typically involve multiple pendula, each with a frequency around 1 Hz. These provide very fat roll-off of the noise above 1 Hz. Interferometer spectra normally show a steep low-frequency noise 'wall': this is the expected vibrational noise amplitude.

In addition, there are noise sources that are not dominant in the present interferometers but will become important as sensitivity increases.

- *Quantum effects: uncertainty principle noise.* Shot noise is a quantum noise, but in addition there are other effects similar to those that bar detectors face, as described below: zero-point vibrations of suspensions and mirror surfaces, and back-action of light pressure fluctuations on the mirrors. These are small compared to present operating limits of detectors, but they may become important in five years or so. Practical schemes to reduce this noise have already been demonstrated in principle, but they need to be improved considerably. This is the subject of considerable theoretical work at the moment.

- *Gravity gradient noise.* Gravitational-wave detectors respond to any changes in the gradients (tidal forces) of the local gravitational field, not just those carried by waves. The environment always contains changes in the Newtonian fields of nearby objects. Besides obvious ones, like people, there are changes caused by density waves in ground vibrations, atmospheric pressure changes, and many other disturbances. Below about 1 Hz, these gravity gradient changes will be stronger than waves expected from astronomical objects, and they make it impossible to do observing at low frequencies from Earth. This is the reason that scientists have proposed the LISA mission, discussed later. Above 1 Hz, this noise does not affect the sensitivity of present detectors, but in ten years this could become a limiting factor.

Besides these noise sources, which are predictable and therefore can be controlled by detector design, it is possible that there will be unexpected or unpredicted noise sources. Interferometers will be instrumented with many kinds of environmental monitors, but there may occasionally be noise that is impossible to identify. For this reason, short bursts of gravitational radiation must be identified at two or more separated facilities. Even if detector noise is not at all understood, it is relatively easy to estimate from the observed noise profile of the individual detectors what the chances are of a coincident noise event between two detectors.

3.2.1 New interferometers and their capabilities

Interferometers work over a broad bandwidth and they do not have any natural resonance in their observing band. They are ideal for detecting *bursts*, since one can perform pattern-matching over the whole bandwidth and detect such signals optimally. They are also ideal for searching for unknown *continuous signals*, such as surveying the sky for neutron stars. And in observations of *stochastic signals* by cross-correlating two detectors, they can give information about the spectrum of the signal.

If an interferometer wants to study a signal with a known frequency, such as known pulsars, then there is another optical technique available to enhance its sensitivity in a narrow bandwidth, at the expense of sensitivity outside that band. This is called *signal recycling* [3]. In this technique, a further mirror is placed in front of the photodetector, where the signal emerges from the interferometer (see figure 3.1(*d*)). If the mirror is chosen correctly, it will build up the signal, but only in a certain bandwidth. This modifies the shot noise in the detector, but not other noise sources. Therefore, it can improve sensitivity only at the higher frequencies where shot noise is the limiting factor.

Four major interferometer projects are now under construction, and they could begin acquiring good data in the period between 2000–2003. They will all operate initially with a sensitivity approaching 10^{-21} over a bandwidth between 50–1000 Hz. Early detections are by no means certain, but recent work has made prospects look better for an early detection than when these detectors were funded.

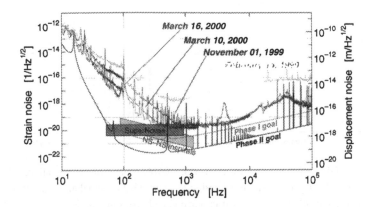

Figure 3.2. TAMA300 sensitivity as a function of frequency. The vertical axis is the 1σ noise level, measured in strain per root Hz. To get a limit on the gravitational wave amplitude h, one must multiply the height of the curve by the square root of the bandwidth of the signal. This takes into account the fact that the noise power at different frequencies is independent, so the power is proportional to bandwidth. The noise amplitude is therefore proportional to the square root of the bandwidth.

TAMA300 [4] (Japan) is located in Tokyo, and its arm length is 300 m. It began taking data without power recycling in 1999, but its sensitivity is not yet near 10^{-21}. Following improvements, especially power recycling, it should get to within a factor of ten of this goal. However, it is not planned as an observing instrument: it is a prototype for a kilometre-scale interferometer in Japan, currently called JGWO. By 2005 this may be operating, possibly with cryogenically cooled mirrors.

GEO600 [5] (Germany and Britain) is located near Hannover (Germany). Its arm length is 600 m and the target date for first good data is now the end of 2001. Unlike TAMA, GEO600 is designed as a leading-edge-technology detector, where high-performance suspensions and optical tricks like signal recycling can be developed and applied. Although it has a short baseline, it will have a similar sensitivity to the larger LIGO and VIRGO detectors at first. At a later stage, LIGO and VIRGO will incorporate the advanced methods developed in GEO, and at that point they will advance in sensitivity, leaving GEO behind.

As we can see from figure 3.3 the sensitivity of GEO600 depends on its bandwidth, which in its turn depends on the signal recycling factor. GEO600 can change its observing bandwidth in response to observing goals. By choosing low or high reflectivity for the signal recycling mirror, scientists can make GEO600 wide-band or narrow-band, respectively. The centre frequency of the observing band (in the right-hand panel of figure 3.3 it is ~600 Hz) can be tuned to any desired frequency by shifting the position of the signal recycling mirror, thus

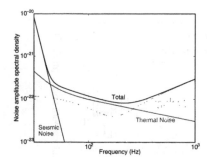 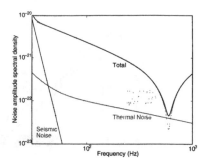

Figure 3.3. GEO600 noise curves. As for the TAMA curve, these are calibrated in strain per root Hz. The figure on the left-hand side shows GEO's wideband configuration; that on the right-hand side shows a possible narrowband operating mode.

changing the resonance frequency of the signal recycling cavity. This feature could be useful when interferometers work with bars or when performing wideband surveys.

LIGO [6] (USA) is building two detectors of arm length 4 km. One is located in Hanford WA and the other in Livingstone LA. The target date for observing is mid-2002. The two detectors are placed so that their antenna patterns overlap as much as possible and yet they are far enough apart that there will be a measurable time delay in most coincident bursts of gravitational radiation. This delay will give some directional information. The Hanford detector also contains a half-length interferometer to assist in coincidence searches. The two LIGO detectors are the best placed for doing cross-correlation for a random background of gravitational waves. LIGO's expected initial noise curve is shown in figure 3.4. These detectors have been constructed to have a long lifetime. With such long arms they can benefit from upgrades in laser power and mirror quality. LIGO has defined an upgrade goal called LIGO II, which it hopes to reach by 2007, which will observe at 10^{-22} or better over a bandwidth from 10 Hz up to 1 kHz.

VIRGO [7] (Italy and France) is building a 3 km detector near Pisa. Its target date for good data is 2003. Its expected initial noise curve is shown in figure 3.4. Like LIGO, it can eventually be pushed to much higher sensitivities with more powerful lasers and other optical enhancements. VIRGO specializes in sophisticated suspensions, and the control of vibrational noise. Its goal is to observe at the lowest possible frequencies from the ground, at least partly to be able to examine as many pulsars and other neutron stars as possible.

3.3   The physics of resonant mass detectors

The principle of operation of bar detectors is to use the gravitational tidal force of the wave to stretch a massive cylinder along its axis, and then to measure the elastic vibrations of the cylinder.

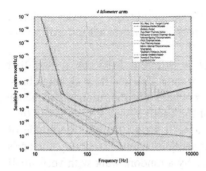
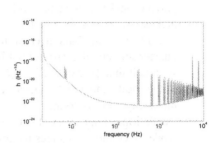

Figure 3.4. Noise curves of the initial LIGO (left) and VIRGO (right) detectors. The VIRGO curve is in strain per root Hz, as the GEO curves earlier. The LIGO curve is calibrated in metres per root Hz, so to convert to a limit on h one multiplies by the square root of the bandwidth and divides by the length of the detector arm, 4000 m.

Let us suppose we have a typical bar with length $L \sim 1$ m. (In the future, spheres may go up to 3 m.) Depending on the length of the bar and its material, the resonant frequency will be $f \sim 500$ Hz to 1.5 kHz and mass $M \sim 1000$ kg. A short burst gravitational wave h will make the bar vibrate with an amplitude

$$\delta l_{\text{gw}} \sim hl \sim 10^{-21} \text{ m.}$$

Unlike the interferometers, whose response is simply given by this equation, the bars respond in a complicated way depending on all their internal forces. However, if the duration of the wave is short, the amplitude will be of the same order as that given here. If the wave has long duration and is near the bars resonant frequency, then the signal can build up to much larger amplitudes. Normally, bar detector searches have been targeted at short-duration signals.

The main sources of noise that compete with this very small amplitude are:

- *Thermal noise.* This is the most serious source of noise. Interferometers can live with room-temperature thermal noise because their larger size makes their response to a gravitational wave larger, and because they observe at frequencies far from the resonant frequency, where the noise amplitude is largest. However, bars observe at the resonant frequency and have a very short length, so they must reduce thermal noise by going to low temperatures. The best ultra-cryogenic bars today operate at about $T = 100$ mK, where the rms amplitude of vibration is found by setting the kinetic energy of the normal mode, $M(\delta \dot{l})^2/2$, equal to $kT/2$, the equipartition thermal energy of a single degree of freedom. This gives then

$$\langle \delta l^2 \rangle_{\text{th}}^{\frac{1}{2}} = \left(\frac{kT}{4\pi^2 M f^2} \right)^{\frac{1}{2}} \sim 6 \times 10^{-18} \text{ m.}$$

This is far larger than the gravitational wave amplitude. In order to detect gravitational waves against this noise, bars are constructed to have a very high Q, of order 10^6 or better.

The reason that bars need a high Q is different from the reason that interferometers also strive for high-Q systems. To see how Q helps bars, we recall that Q is defined as $Q = f \cdot \tau$ where f is the resonant frequency of the mode and τ is the decay time of the oscillations. If Q is large, then the decay time is long. If the decay time is long, then the amplitude of oscillation changes very slowly in thermal equilibrium. Essentially, the bar's mode of vibration changes its amplitude by a random walk with very small steps, taking time $Q/f \sim 1000$ s to change by the full amount. On the other hand, a gravitational wave burst will cause an amplitude change in time of the order 1 ms, during which the thermal noise will have random walked to an expected amplitude change that is $Q^{\frac{1}{2}} = (\frac{1000 \text{ s}}{1 \text{ ms}})^{\frac{1}{2}}$ times smaller. In this case

$$\langle \delta l^2 \rangle^{\frac{1}{2}}_{\text{th: 1 ms}} = \left(\frac{kT}{4\pi^2 M f^2 Q} \right)^{\frac{1}{2}} \sim 6 \times 10^{-21} \text{ m}.$$

Thus, thermal noise only affects a measurement to the extent that it changes the amplitude of vibration during the time of the gravitational-wave burst. This change is similar to that produced by a gravitational wave of amplitude 6×10^{-21}. It follows that, if thermal noise were the only noise source, bars would be operating at around 10^{-20} today. Bar groups expect in fact to reach this level during the next few years, as they reduce the other competing sources of noise. Notice that the effect of thermal noise has nothing to do with the frequency of the disturbance, so it is not the reason that bars observe near their resonant frequency. In fact, both thermal impulses and gravitational-wave forces are mechanical forces on the bar, and the ratio of their induced vibrations is the same at all frequencies for a given applied impulsive force.

- *Sensor noise.* Because the oscillations of the bar are very small, bars require a *transducer* to convert the mechanical energy of vibration into electrical energy, and an *amplifier* that increases the electrical signal to a level where it can be recorded. If the amplifier were perfect, then the detector would in fact be broadband: it would amplify the smaller off-resonant responses just as well as the on-resonance ones. Conversely, real bars are narrow-band because of sensor noise, not because of their mechanical resonance.

Unfortunately sensing is not perfect: amplifiers introduce noise and this makes small amplitudes harder to measure. The amplitudes of vibration are largest in the resonance band near the resonant frequency f_0, so amplifier noise limits the detector sensitivity to frequencies near f_0. Now, the signal (a typical gravitational-wave burst) has a duration time $\tau_w \sim 1$ ms, so the amplifier's bandwidth should be at least $1/\tau_w$ in order for it to be able to record a signal every τ_w. In other words, bars require amplifiers with very small noise in a large bandwidth ($\sim$1000 Hz) near f_0 (note that this band is much larger than f/Q). Today typical bandwidths of realizable amplifiers are 1 Hz, but in the very near future it is hoped to extend these to 10 Hz, and eventually to 100 Hz.

• *Quantum limit*. According to the Heisenberg uncertainty principle, the zero-point vibrations of a bar with a frequency of 1 kHz have rms amplitude

$$\langle \delta l^2 \rangle^{\frac{1}{2}}_{\text{quant}} = \left(\frac{\hbar}{2\pi M f} \right)^{\frac{1}{2}} \sim 4 \times 10^{-21} \text{ m.}$$

This is bigger than the expected signal, and comparable to the thermal limit over 1 ms. It represents the accuracy with which one can measure the amplitude of vibration of the bar. So as soon as current detectors improve their thermal limits, they will run into the quantum limit, which must be overcome before a signal at 10^{-21} can be seen with such a detector. One way to overcome this limit is by increasing the size of the detector and even by making it spherical. This increases its mass dramatically, pushing the quantum limit down below 10^{-21}.

Another way around the quantum limit is to avoid measuring δl, but instead to measure other observables. After all, the goal is to infer the gravitational-wave amplitude, not to measure the state of vibration of the bar. It is possible to define a pair of conjugate observables that have the property that one of them can be measured arbitrarily accurately repeatedly, so that the resulting inaccuracy of knowing the conjugate variable's value does not disturb the first variable's value. Then, if the first variable responds to the gravitational wave, the gravitational wave may be measured accurately, even though the full state of the bar is poorly known. This method is called 'back reaction evasion'. The theory was developed in a classic paper by Caves et al [8]. However, no viable schemes to do this have been demonstrated for bar detectors so far.

3.3.1 New bar detectors and their capabilities

Resonant-mass detectors are limited by properties of materials and, as we have just explained, they have their best sensitivity in a narrow band around their resonant frequency. However, they can usefully explore higher frequencies (above 500 Hz), where the interferometer noise curves are rising (see earlier figures).

From the beginning, bars were designed to detect *bursts*. If the burst radiation carries significant energy in the bar's bandwidth, then the bar can do well. Standard assumptions about gravitational collapse suggest a signal with a broad spectrum to 1 kHz or more, so that most of the sensitive bars today would be suited to observe such a signal. Binary coalescence has a spectrum that peaks at low frequencies, so bars are not partiularly well suited for such signals. On the other hand, neutron-star and stellar-mass black-hole normal modes range in frequency from about 1 kHz up to 10 kHz, so suitably designed bars could, in principle, go after these interesting signals.

A bar gets all of its sensitivity in a relatively narrow bandwidth, so if a bar and an interferometer can both barely detect a burst of amplitude 10^{-20}, then the bar has much greater sensitivity than the interferometer in its narrow band, and much worse at other frequencies. This has led recently to interest in bars as detectors of *continuous signals*. If the signal frequency is in the observing band of the bar, it can do very well compared to interferometers. Signals from millisecond pulsars and possible signals from x-ray binaries are suitable if they have the right frequency. However, most known pulsars will radiate at frequencies rather low compared to the operating frequencies of present-day bars.

The excellent sensitivity of bars in their narrow bandwidth also suits them to detecting *stochastic signals*. Cross-correlations of two bars or of bars with interferometers can be better than searches with first-generation interferometers [9]. One gets no spectral information, of course, and in the long run expected improvements in interferometers will overtake bars in this regard.

Today's best bar detectors are orders of magnitude more sensitive than the original Weber bar. Two *ultra-cryogenic* bars have been built and are operating at thermodynamic temperatures below 100 mK: *NAUTILUS* [10] at Frascati, near Rome, and [11] in Legnaro. With a mass of several tons, these may be the coldest massive objects ever seen anywhere in the universe. These are expected soon to reach a sensitivity of 10^{-20} near 1 kHz. Already they are performing coincidence experiments with bars at around 4 K at Perth, Australia, and at LSU.

Proposals exist in the Netherlands, Brazil, Italy, and the USA for *spherical* or *icosahedral detectors* (see links from [10]). These detectors have more mass, so they could reach 10^{-21} near 1 kHz. Because of their shape, they have omnidirectional antenna patterns; if they are instrumented so that all five independent fundamental quadrupolar modes of vibration can be monitored, they can do all-sky observing and determine directions as well as verify detections using coincidences between modes of the same antenna.

3.4   A detector in space

As we have noted earlier, gravitational waves from astronomical objects at frequencies below 1 Hz are obscured by Earth-based gravity-gradient noise. Detectors must go into space to observe in this very interesting frequency range.

The *LISA* [12] mission is likely to be the first such mission to fly. LISA will be a triangular array of spacecraft, with arm lengths of 5×10^6 km, orbiting the Sun in the Earth's orbit, about 20° behind the Earth. The spacecraft will be in a plane inclined to the ecliptic by 60°. The three arms can be combined in different ways to form two independent interferometers. During the mission the configuration of spacecraft rotates in its plane, and the plane rotates as well, so that LISA's antenna pattern sweeps the sky.

LISA has been named a cornerstone mission of the European Space Agency (ESA), and NASA has recently formed its own team to study the same mission, with a view toward a collaboration with ESA. LISA will be sensitive in a range from 0.3 mHz to about 0.1 Hz, and it will be able to detect known binary star systems in the Galaxy and binary coalescences of supermassive black holes anywhere they occur in the universe. A joint ESA–NASA project looks very likely, aiming at a launch around 2010. A technology demonstration mission might be launched in 2005 or 2006.

LISA's technology is fascinating. We can only allude to the most interesting parts of the mission here. A full description can be found in the pre-Phase A study document [13]. The most innovative aspect of the mission is *drag-free control*. In order to guarantee that the interferometry is not disturbed by external forces, such as fluctuations in solar radiation pressure, the mirror that is the reference point for the interferometry is on a free mass inside the spacecraft. The spacecraft acts as an active shield, sensing the position of the free mass, firing jets to counteract external forces on itself and ensure that it does not disturb the free mass. The jets themselves are remarkable, in that they must be very weak compared to most spacecraft's control jets, and they must be capable of very precise control. They will work by expelling streams of ions, accelerated and controlled by a high-voltage electric field. Fuel for these jets is not a problem: 1 g will be enough for a mission lifetime of ten years!

LISA interferometry is not done with reflection from mirrors. When a laser beam reaches one spacecraft from the other, it is too weak to reflect: the sending spacecraft would only get the occasional photon! Instead, the incoming light is sensed, and an on-board laser is slaved to it, returning an amplified beam with the same phase and frequency as the incoming one. No space mission has yet implemented this kind of laser-transponding. The LISA team had to ensure that there was enough information in all the signals to compensate for inevitable frequency fluctuations among all six on-board lasers.

A further serious problem that the LISA team had to solve was how to compensate for the relative motions of the spacecraft. The laser signals converging on a single spacecraft from the other two corners will be Doppler shifted so that their fringes change at MHz frequencies. This has to be sensed on board and removed from the signal that is sent back to Earth, which can only be sampled a few tens of times per second.

When LISA flies it will, on a technical as well as a scientific level, be a worthy counterpart to its Earth-based interferometer cousins!

3.4.1 LISA's capabilities

In the low-frequency LISA window, most sources will be relatively long lived, at least a few months. During an observation, LISA will rotate and change its velocity by a significant amount. This will induce Doppler shifts into the signals, and modulate their amplitudes, so that LISA should be able to infer the position,

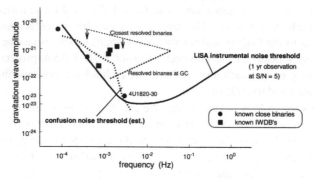

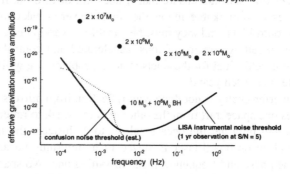

Figure 3.5. LISA sensitivity to binary systems in the Galaxy (top) and to massive black hole coalescences (bottom). The top figure is calibrated in the intrinsic amplitude of the signal, and the noise curve shows the detection threshold (5σ) for a one-year observation. It also shows the confusion limit due to unresolved binary systems. The bottom panel shows the effective amplitude of signals from coalescences of massive black holes. Since some such events last less than one year, what is shown is the expected signal-to-noise ratio of the observation.

polarization and amplitude of sources entirely from its own observations. Below about 1 mHz, this information weakens, because the wavelength of the radiation becomes comparable to or greater than the radius of LISA's orbit. The amplitude modulation is the only directional information in this frequency range.

3.5 Gravitational and electromagnetic waves compared and contrasted

To conclude this lecture it is useful to discuss the most important differences and similarities between gravitational waves and electromagnetic ones. We do this in the form of a table.

Table 3.1.

Electromagnetism	General relativity
Two signs of charges—large bodies usually neutral—waves usually emitted by single particles, often incoherently—waves carry 'local' information.	One sign of mass—gravity accumulates—waves emitted more strongly by larger body—waves carry 'global' information.
A genuine physical force, acting differently on different bodies. Detected by measuring accelerations.	Equivalence principle: gravity affects all bodies in the same way. Represented as a spacetime curvature rather than a force. Detected only by tidal forces—differential accelerations.
Maxwell's equations are *linear*. Physical field is $F_{\mu\nu}$ (E and B). Gauge field is vector potential A.	Einstein's equations are *nonlinear*. Physical field is Riemann curvature tensor $R_{\mu\nu\alpha\beta}$. Gauge fields are metric $g_{\mu\nu}$ and connection $\Gamma^\alpha_{\mu\nu}$. Gauge transformations are coordinate transformations.
Source is charge-current density J_μ. Charge creates electric field, current magnetic field.	Source is stress-energy tensor $T_{\mu\nu}$. Mass creates a Newtonian-like field, momentum as gravito-magnetic effects. Stress creates field too.
Moderately strong force on the atomic scale: $\frac{e^2/4\pi\epsilon_0}{Gm_p^2} = 10^{39}$.	Weaker than 'weak' interaction.
Wave generation for A_μ: $\partial^\beta\partial_\beta A_\mu = 4\pi\epsilon_0 J_\mu$ in a convenient gauge (Lorentz gauge).	Wave generation for $h_{\mu\nu} = g_{\mu\nu} - \eta_{\mu\nu}$: $\partial^\beta\partial_\beta(h_{\mu\nu} - \frac{1}{2}\eta_{\mu\nu}h^\alpha{}_\alpha) = 8\pi T_{\mu\nu}$ in a convenient gauge.
Propagate at the speed of light, amplitude falls as $1/r$.	Propagate at the speed of light, amplitude falls as $1/r$.
Conservation of charge $\Rightarrow$ radiation by low-velocity charges is dominated by dipole component.	Conservation of mass and momentum $\Rightarrow$ radiation by low-velocity masses is dominated by quadrupole component.

Table 3.1. (Continued)

Electromagnetism	General relativity
Simple detector: oscillating charge. Action is along a line, transverse to the directions of propagation. Spin $s = 1$ and two states of linear polarization that are inclined to each other at an angle of $90°$.	Simple detector: distorted ring of masses. Action is elliptic in a plane transverse to the direction of propagation. Spin $s = 2$ and two states of linear polarization that are inclined to each other at an angle of $45°$. Equivalence principle $\Rightarrow$ action depends only on $h_{\mu\nu}$, which is dimensionless.
Strength of force $\Rightarrow$ waves scatter and refract easily.	Weakness of gravity $\Rightarrow$ waves propagate almost undisturbed and transfer energy very weakly. Dimensionless amplitude h is small.
Local energy and flux well defined: Poynting vector etc.	Equivalence principle $\Rightarrow$ local energy density cannot be defined exactly. Only *global* energy balance is exact.
Multipole expansion in slow-motion limit is straightforward, radiation reaction well defined.	Multipole expansion different if fields are weak or strong. For quasi-Newtonian case fields are weak, and the resulting post-Newtonian expansion is delicate. Radiation reaction is still not fully understood.
Exact solutions, containing waves, are available and can guide the construction of approximation methods for more complicated situations.	Fully realistic exact solutions for dynamical situations of physical interest are not available. Extensive reliance on approximation methods.

Chapter 4

Astrophysics of gravitational-wave sources

There are a large number of possible gravitational-wave sources in the observable waveband, which spans eight orders of magnitude in frequency: from 10^{-4} Hz (lower bound of current space-based detector designs) to 10^4 Hz (frequency limit of likely ground-based detectors). Some of these sources are highly relativistic and not too massive, especially above 10 Hz: a black hole of mass $1000 M_\odot$ has a characteristic frequency of 10 Hz, and larger holes have lower frequencies in inverse proportion to the mass. Neutron stars have even higher characteristic frequencies. Other systems are well described by Newtonian dynamics, such as binary orbits.

For nearly-Newtonian sources the post-Newtonian approximation (see chapter 6) provides a good framework for calculating gravitational waves. More relativistic systems, and unusual sources like the early universe, require more sophisticated approaches (see chapter 7).

4.1 Sources detectable from ground and from space

4.1.1 Supernovae and gravitational collapse

The longest expected and still probably the least understood source, gravitational collapse is one of the most violent events known to astronomy. Yet, because we have little direct information about the deep interior, we cannot make reliable predictions about the gravitational radiation from it.

Supernovae are triggered by the gravitational collapse of the interior degenerate core of an evolved star. According to current theory the result should be a neutron star or black hole. The collapse releases an enormous amount of energy, about $0.15 M_\odot c^2$, most of which is carried away by neutrinos; an uncertain fraction is converted into gravitational waves. One mechanism for producing this radiation could be dynamical instabilities in the rapidly rotating core before it becomes a neutron star. Another likely source of radiation is the r-mode instability (see chapter 7). This could release $\sim 0.1 M_\odot c^2$ in radiation every time a neutron star is formed.

However, both kinds of mechanisms are difficult to model. The problem with gravitational collapse is that perfectly spherical motions do not emit gravitational waves, and it is still not possible to estimate in a reliable way the amount of asymmetry in gravitational collapse. Even modern computers are not able to perform realistic simulations of gravitational collapse in three dimensions, including all the important nuclear reactions and neutrino- and photon-transport. Similarly, it is hard to model the r-mode instability because its evolution depends on nonlinear hydrodynamics and on poorly known physics, such as the cooling and viscosity of neutron stars.

An alternative approach is to use general energy considerations. If, for example, we assume that 1% of the available energy is converted into gravitational radiation, then, from formulae we will derive in the next chapter, the amplitude h would be large enough to be detected by the first ground-based interferometers (LIGO/GEO600/VIRGO) at the distance of Virgo Cluster (18 Mpc) if the emission centres at 300 Hz. Moreover, bar and spherical-mass detectors with an effective sensitivity of 10^{-21} and the right resonant frequency could see these signals as well.

The uncertainties in our predictions have a positive aspect: it is clear that if we can detect radiation from supernovae, we will learn much that we do not know about the end stages of stellar evolution and about neutron-star physics.

4.1.2 Binary stars

Binary systems have given us our best proof of the reliability of general relativity for gravitational waves. The most famous example of such systems is the binary pulsar PSR1916+16, discovered by Hulse and Taylor in 1974; they were awarded the Nobel Prize for this discovery in 1993. From the observations of the modulation of the pulse period as the stars move in their orbits, one knows many important parameters of this system (orbital period, eccentricity, masses of the two stars, etc), and the data also show directly the decrease of the orbital period due to the emission of gravitational radiation. The observed value is 2.4×10^{-12} s/s. Post-Newtonian theory allows one to predict this from the other measured parameters of the system, without any free parameters (see chapter 7); the prediction is 2.38×10^{-12}, in agreement within the measurement errors.

Unfortunately the radiation from the Hulse–Taylor system will be too weak and of too low a frequency to be detectable by LISA.

4.1.3 Chirping binary systems

If a binary gives off enough energy for its orbit to shrink by an observable amount during an observation, it is said to *chirp*: as the orbit shrinks, the frequency and amplitude go up. LISA will see a few chirping binaries. If a binary system is compact enough to radiate above 10^{-3} Hz, it will always chirp within one year, provided its components have a mass above about $1 M_\odot$. If they are above

Table 4.1. The range for detecting a $2 \times 1.4M_\odot$ NS binary coalescence. The threshold for detection is taken to be 5σ. The binary and detector orientations are assumed optimum. The average S/N ratio for randomly oriented systems is reduced from the optimum by $1/\sqrt{5}$.

Detector:	TAMA300	GEO600	LIGO I	VIRGO	LIGO II
Range (S/N = 5)	3 Mpc	14 Mpc	30 Mpc	36 Mpc	500 Mpc

about $10^3 M_\odot$, the binary will go all the way to coalescence within the one-year observation.

Chirping binary systems are more easily detectable than gravitational collapse events because one can model with great accuracy the gravitational waveform during the inspiral phase. There will be radiation, possibly with considerable energy, during the poorly understood *plunge* phase (when the objects reach the last stable orbit and fall rapidly towards one another) and during the merger event, but the detectability of such systems rests on tracking their orbital emissions.

The major uncertainty about this kind of source is the event rate. Current pulsar observations suggest that there will be ~ 1 coalescence per year of a Hulse-Taylor binary out to about 200 Mpc. This is a *lower limit* on the event rate, since it comes from systems we actually observe. It is possible that there are other kinds of binaries that we have no direct knowledge of, which will boost the event rate.

Theoretical modelling of binary populations gives a wide spectrum of mutually inconsistent predictions. Some authors [14] suggest that there may be a large population that escapes pulsar surveys but brings the nearest neutron star coalescence in one year as far as 30 Mpc, only slightly farther than the Virgo cluster; but other models [15] put the rate near to the observational limit.

The most exciting motivation for detecting coalescing binaries is that they could be associated with gamma-ray bursts. The event rates are consistent, and neutron stars are able to provide the required energy. If gamma-bursts are associated with neutron-star coalescence, then observations of coalescence radiation should be followed within a second or so by a strong gamma-ray burst.

LISA will see a few chirping binaries in the Galaxy, but the sensitivity of the first generation of ground-based detectors is likely to be *too poor* to see many such events (see table 4.1).

A certain fraction of such systems could contain black holes instead of neutron stars. In fact black holes should be overrepresented in binary systems (relative to their birth rate) because their formation is much less likely to disrupt a binary system (there is much less mass lost) than the formation of a neutron star would be. Pulsar observations have not yet turned up a black-hole/neutron-star system, and of course one does not expect to see binary black holes

Table 4.2. The range for detecting a $10M_\odot$ black-hole binary. Conventions as in table 4.1.

Detector:	GEO600	LIGO I	VIRGO	LIGO II
Range (S/N = 5)	75 Mpc	160 Mpc	190 Mpc	2.6 Gpc

electromagnetically. So we can only make theoretical estimates, and there are big uncertainties.

Some evolution calculations [14] suggest that the coalescence rate of BH–BH systems may be of the same order as the NS–NS rate. Other models [15] suggest it could even be zero, because stellar-wind mass loss (significant in very massive stars) could drive the stars far apart before the second BH forms, leading to coalescence times longer than the age of the universe. A recent proposal identifies globular clusters as 'factories' for binary black holes, forming binaries by three-body collisions and then expelling them [16]. Gamma-ray bursts may also come from black-hole/neutron-star coalescences. If the more optimistic event rates are correct, then black-hole coalescences may be among the first sources detected by ground-based detectors (table 4.2).

4.1.4 Pulsars and other spinning neutron stars

There are a number of ways in which a spinning neutron star may give off a continuous stream of gravitational waves. They will be weak, so they will require long continuous observation times, up to many months. Here are some possible emission mechanisms for neutron stars.

The r-modes. Neutron stars are born hot and probably rapidly rotating. Before they cool (during their first year) they have a family of *unstable normal modes, the r-modes*. These modes are excited to instability by the emission of gravitational radiation, as predicted originally by Andersson [17]. They are particularly interesting theoretically because the radiation is gravitomagnetic, generated by mass currents rather than mass asymmetries. We will study the theory of this radiation in chapter 6. In chapter 7 we will discuss how the emission of this radiation excites the instability (the CFS instability mechanism).

Being unstable, young neutron stars will presumably radiate away enough angular momentum to reduce their spin and become stable. This could lower the spin of a neutron star to ~100 Hz within one year after its formation [18]. The energy emitted in this way should be a good fraction of the star's binding energy, so in principle this radiation could be detected from the Virgo Cluster by LIGO II, provided matched filtering can be used effectively.

We discuss a possible stochastic background of gravitational waves from the *r*-modes below.

Accreting neutron stars (figure 4.1) are the central objects of most of the

Figure 4.1. Accreting neutron star in a low-mass x-ray binary system.

binary x-ray sources in the Galaxy. Astronomers divide them into two distinct groups: the low-mass and high-mass binaries, according to the mass of the companion star. In these systems mass is pulled from the low- or high-mass giant by the tidal forces exerted by its neutron star companion. In low-mass x-ray binaries (LMXBs) the accretion lasts long enough to spin the neutron star up to the rotation rates of millisecond pulsars. Astronomers have therefore supposed for some time that the neutron stars in LMXBs would have a range of spins, from near zero (young systems) to near 500 or 600 Hz (at the end of the accretion phase). Until the launch of the Rossi X-ray Timing Explorer (RXTE), there was no observational evidence for the neutron star spins. However, in the last two years there has been an accumulation of evidence that most, if not all, of these stars have angular velocities in a narrow range around 300 Hz [19]. It is not known yet what mechanism regulates this spin, but a strong candidate is the emission of gravitational radiation.

A novel proposal by Bildsten [20] suggests that the temperature gradient across a neutron star that is accreting preferentially at its magnetic poles should lead to a composition and hence a density gradient in the deep crust. Spinning at 300 Hz, such a star could radiate as much as it accretes. It would then be a steady source for as long as accretion lasts, which could be millions of years.

In this model the gravitational-wave energy flux is proportional to the observed x-ray energy flux. The strongest source in this model is Sco X-1, which could be detected by GEO600 in a two-year-long narrow-band mode if the appropriate matched filtering can be done. LIGO II would have no difficulty in detecting this source.

Older stars may also be *lumpy*. For known pulsars, this is constrained by the rate of spindown: the energy radiated in gravitational waves cannot exceed the total energy loss. In most cases, this limit is rather weak, and stars would have to sustain strains in their crust of order 10^{-3} or more. It is unlikely that crusts could sustain this kind of strain, so the observational limits are probably significant overestimates for most pulsars. However, millisecond pulsars have much slower spindown rates, and it would be easier to account for the strain in their crusts, for example as a remnant Bildsten asymmetry. Such stars could, in principle, be radiating more energy in gravitational waves than electromagnetically.

Observations of individual neutron stars would be rich with information about astrophysics and fundamental nuclear physics. So little is known about the physics of these complex objects that the incentive to observe their radiation is great.

However, making such observations presents challenges for data analysis, since the motion of the Earth puts a strong phase modulation on the signal, which means that even if its rest-frame frequency is constant it cannot be found by simple Fourier analysis. More sophisticated pattern-matching (matched-filtering) techniques are needed, which track and match the signal's phase to within one cycle over the entire period of measurement. This is not difficult if the source's location and frequency are known, but the problem of doing a wide-area search for unknown objects is very challenging [21]. Moreover, if the physics of the source is poorly known, such as for LMXBs or r-mode spindown, the job of building an accurate family of templates is a difficult one. These questions are the subject of much research today, but they will need much more in the future.

4.1.5   Random backgrounds

The big bang was the most violent event of all, and it may have created a significant amount of gravitational radiation. Other events in the early universe may also have created radiation, and there may be backgrounds from more recent epochs. We have seen earlier, for example, that compact binary systems in the Galaxy will merge into a confusion-limited noise background in LISA observations below about 1 mHz.

Let us consider the r-modes as another important example. This process may have occurred in a good fraction of all neutron stars formed since the beginning of star formation. The sum of all of their r-mode radiation will be a stochastic background, with a spectrum that extends from a lower cut-off of about 200 Hz in the rest frame of the emitter to an upper limit that depends on the initial angular velocity of stars. If significant star formation started at, say, a redshift of five, then this background should extend down to about 25 Hz. If 10^{-3} of the mass of the Galaxy is in neutron stars, and each of them radiates 10% of its mass in this radiation, then the gravitational-wave background should have a density equal to 10^{-4} of the mean cosmological density of visible stars. Expressed as a fraction Ω_{gw} of the closure density of the universe, per logarithmic frequency interval, this converts to

$$\Omega_{gw}^{r\text{-modes}}(25\text{--}1000\,\text{Hz}) \approx 10^{-8}\text{--}10^{-7}.$$

This background would be easily detectable by LIGO II.

There may also be a cosmological background from either topological defects (e.g. cosmic strings) or from inflation (which amplifies initial quantum gravitational fluctuations as it does the scalar ones that lead to galaxy formation). Limits from COBE observations suggest that standard inflation could not produce a background stronger than $\Omega_{\text{gw}}^{\text{inflation}} \sim 10^{-14}$ today. This is too weak for any of the planned detectors to reach, but it remains an important long-range goal for the field. However, there could also be a component of background radiation that depends on what happened *before* inflation: string cosmological models, for example, predict spectra growing with frequency [22].

First-generation interferometers are not likely to detect these backgrounds: they may not be able to go below the upper limit set by the requirement that gravitational waves should not disturb cosmological nucleosynthesis, which is $\Omega_{\text{gw}} = 10^{-5}$. (This limit does not apply to backgrounds generated after nucleosynthesis, like the r-mode background.) Bar detectors may do as well or better than the first generation of interferometers for a broad-spectrum primordial background: as we have noted earlier, their noise levels within their resonance bands are very low. However, their frequencies are not right for the r-mode background.

Second-generation interferometers may be able to reach to 10^{-11} of closure or even lower, by cross-correlation of the output of the two detectors. However, they are unlikely to get to the inflation target of 10^{-14}. LISA may be able to go as low as 10^{-10} (if we have a confident understanding of the instrumental noise), but it is likely to detect only the confusion background of binaries, which is expected to be much stronger than a cosmological background in the LISA band.

4.1.6 The unexpected

At some level, we are bound to see things we did not expect. LISA, with its high signal-to-noise ratios for predicted sources, is particularly well placed to do this. Most of the universe is composed of dark matter whose existence we can infer only from its gravitational effects. It would not be particularly surprising if a component of this dark matter produced gravitational radiation in unexpected ways, such as from binaries of small exotic compact objects of stellar mass. We will have to wait to see!

Chapter 5

Waves and energy

Here we discuss wave-like perturbations $h_{\mu\nu}$ of a general background metric $g_{\mu\nu}$. The mathematics is similar to that of linearized theory: $h_{\mu\nu}$ is a tensor with respect to background coordinate transformations (as it was for Lorentz transformations in linearized theory) and it undergoes a gauge transformation when one makes an infinitesimal coordinate transformation. As in linearized theory, we will assume that the amplitude of the waves is small. Moreover, the waves must have a wavelength that is short compared to the radius of curvature of the background metric. These two assumptions allow us to visualize the waves as small ripples running through a curved and slowly changing spacetime.

5.1   Variational principle for general relativity

We start our analysis of the small perturbation $h_{\mu\nu}$ by introducing the standard Hilbert variational principle for Einstein's equations. The field equations of general relativity can be derived from an action principle using the Ricci scalar curvature as the Lagrangian density. The Ricci scalar (second contraction of the Riemann tensor) is an invariant quantity which contains in addition to $g_{\mu\nu}$ and its first derivatives also the second derivatives of $g_{\mu\nu}$, so our action can be written symbolically as:

$$I[g_{\mu\nu}] = \frac{1}{16\pi} \int R(g_{\mu\nu}, g_{\mu\nu,\alpha}, g_{\mu\nu,\alpha\beta})\sqrt{-g}\, \mathrm{d}^4 x \qquad (5.1)$$

where $\sqrt{-g}$ is the square root of the determinant of the metric tensor. As usual in variational principles, the metric tensor components are varied $g_{\mu\nu} \to g_{\mu\nu} + h_{\mu\nu}$, and one demands that the resulting change in the action should vanish to first order in any small variation $h_{\mu\nu}$ of compact support:

$$
\begin{aligned}
\delta I &= I[g_{\mu\mu} + h_{\mu\nu}] - I[g_{\mu\nu}] \\
&= \frac{1}{16\pi} \int \frac{\delta(R\sqrt{-g})}{\delta g_{\mu\nu}} h_{\mu\nu}\, \mathrm{d}^4 x + O(2)
\end{aligned}
\qquad (5.2)
$$

$$= -\frac{1}{16\pi} \int G^{\mu\nu} h_{\mu\nu} \sqrt{-g}\, \mathrm{d}^4 x + O(2) \tag{5.3}$$

where '$O(2)$' denotes terms quadratic and higher in $h_{\mu\nu}$. All the divergences obtained in the intermediate steps of this calculation integrate to zero since $h_{\mu\nu}$ is of compact support. This variational principle therefore yields the vacuum Einstein equations: $G^{\mu\nu} = 0$.

Let us consider how this changes if we include matter. This will help us to see how we can treat gravitational waves as a new kind of 'matter' field on spacetime.

Suppose we have a matter field, described by a variable Φ (which may represent a vector, a tensor or a set of tensors). It will have a Lagrangian density $L_m = L_m(\Phi, \Phi_{,\alpha}, \dots, g_{\mu\nu})$ that depends on the field and also on the metric. Normally derivatives of the metric tensor do not appear in L_m, since by the equivalence principle, matter fields should behave locally as if they were in flat spacetime, where of course there are no metric derivatives. Variations of L_m with respect to Φ will produce the field equation(s) for the matter system, but here we are more interested in variations with respect to $g_{\mu\nu}$, which is how we will find the matter field contribution to the gravitational field equations. The total action has the form:

$$I = \int (R + 16\pi L_m)\sqrt{-g}\, \mathrm{d}^4 x, \tag{5.4}$$

whose variation is

$$\delta I = \int \frac{\delta(R\sqrt{-g})}{\delta g_{\mu\nu}} h_{\mu\nu}\, \mathrm{d}^4 x + \int 16\pi \frac{\partial(L_m\sqrt{-g})}{\partial g_{\mu\nu}} h_{\mu\nu}\, \mathrm{d}^4 x. \tag{5.5}$$

This variation must yield full Einstein equations, so we must have the following result for the stress-energy tensor of matter:

$$T^{\mu\nu}\sqrt{-g} = 2 \frac{\partial L_m\sqrt{-g}}{\partial g_{\mu\nu}}, \tag{5.6}$$

leading to

$$\boxed{G^{\mu\nu} = 8\pi T^{\mu\nu}}. \tag{5.7}$$

This way of deriving the stress-energy tensor of the matter field has deep connections to the conservation laws of general relativity, to the way of constructing conserved quantities when the metric has symmetries and to the so-called pseudotensorial definitions of gravitational wave energy (see Landau and Lifshitz 1962) [23]. We shall use it in the latter sense.

5.2 Variational principles and the energy in gravitational waves

Before we introduce the mathematics of gravitational waves, it is important to understand which geometries we are going to examine. We have said that these geometries consist of a slowly and smoothly changing *background metric* which is altered by *perturbations* of small amplitude and high frequency. If L and λ are the characteristic lengths over which the background and 'ripple' metrics change significantly, we assume that the ratio λ/L will be very much smaller than unity and that $|h_{\mu\nu}|$ is of the same order of smallness as λ/L. In this way the total metric remains slowly changing on a macroscopic scale, while the high-frequency wave, when averaged over several wavelengths, will be the principal source of the curvature of the background metric. This is the 'short-wave' approximation [24]. Obviously this is a direct generalization of the treatment in chapter 2.

5.2.1 Gauge transformation and invariance

Consider an infinitesimal coordinate transformation generated by a vector field ξ^α,

$$x^\alpha \rightarrow x^\alpha + \xi^\alpha. \tag{5.8}$$

In the new coordinate system, neglecting quadratic and higher terms in $h^{\alpha\beta}$, it is not hard to show that the general gauge transformation of the metric is

$$h_{\mu\nu} \rightarrow h_{\mu\nu} - \xi_{\mu;\nu} - \xi_{\nu;\mu}, \tag{5.9}$$

where a semicolon denotes the covariant derivative. We assume that the derivatives of the coordinate displacement field are of the same order as the metric perturbation: $|\xi^{\alpha,\beta}| \sim |h^{\alpha\beta}|$.

Isaacson [24] showed that the gauge transformation of the Ricci and Riemann curvature tensors has the property

$$\bar{R}^{(1)}_{\mu\nu} - R^{(1)}_{\mu\nu} \approx \left(\frac{\lambda}{L}\right)^2 \tag{5.10}$$

$$\bar{R}^{(1)}_{\alpha\mu\beta\nu} - R^{(1)}_{\alpha\mu\beta\nu} \approx \left(\frac{\lambda}{L}\right)^2$$

where $R^{(1)}_{\mu\nu}$ and $R^{(1)}_{\alpha\mu\beta\nu}$ are the first order of Ricci and Riemann tensors (in powers of perturbation $h_{\mu\nu}$) and an overbar denotes their values after the gauge transformation. In our high-frequency limit, therefore, these tensors are gauge-invariant to linear order, just as in linearized theory.

5.2.2 Gravitational-wave action

Let us suppose that we are in vacuum so we have only the metric, no matter fields, but we work in the high-frequency approximation. The full metric is $g_{\mu\nu}$ (smooth *background* metric) $+h_{\mu\nu}$ (high-frequency perturbation). Our purpose is to show that the wave field can be treated as a 'matter' field, with a Lagrangian and its own stress-energy tensor. To do this we have to expand the action out to second order in the metric perturbation,

$$I[g_{\mu\mu} + h_{\mu\nu}] = \int R(g_{\mu\nu} + h_{\mu\nu}, g_{\mu\nu,\alpha} + h_{\mu\nu,\alpha}, \ldots)\sqrt{-g[g_{\mu\nu} + h_{\mu\nu}]}\,\mathrm{d}^4 x$$

$$= \int R(g_{\mu\nu}, \ldots)\sqrt{-g}\,\mathrm{d}^4 x + \int \frac{\delta(R\sqrt{-g})}{\delta g_{\mu\nu}} h_{\mu\nu}\,\mathrm{d}^4 x$$

$$+ \frac{1}{2}\int \left(\frac{\partial^2(R\sqrt{-g})}{\partial g_{\mu\nu}\partial g_{\alpha\beta}} h_{\mu\nu}h_{\alpha\beta} + 2\frac{\partial^2(R\sqrt{-g})}{\partial g_{\mu\nu}\partial g_{\alpha\beta,\gamma}} h_{\mu\nu}h_{\alpha\beta,\gamma} \right.$$

$$\left. + \frac{\partial^2(R\sqrt{-g})}{\partial g_{\mu\nu,\tau}\partial g_{\alpha\beta,\gamma}} h_{\mu\nu,\tau}h_{\alpha\beta,\gamma} + 2\frac{\partial^2(R\sqrt{-g})}{\partial g_{\mu\nu}\partial g_{\alpha\beta,\gamma\tau}} h_{\mu\nu}h_{\alpha\beta,\gamma\tau} \right)\mathrm{d}^4 x$$

$$+ O(3).$$

The first term is the action for the background metric $g_{\mu\nu}$. The second term vanishes (see equation (5.2)), since we assume that the background metric is a solution of the Einstein vacuum equation itself, at least to lowest order.

If we compare the above equation with equation (5.4), we can see that the third term, complicated as it seems, is an effective 'matter' Lagrangian for the gravitational field. Indeed, if one varies it with respect to $h_{\mu\nu}$ holding $g_{\mu\nu}$ fixed (as we would do for a physical matter field on the background), then the complicated coefficients are fixed and one can straightforwardly show that one gets exactly *the linear perturbation of the Einstein tensor itself.* Its vanishing is the equation for the gravitational-wave perturbation $h_{\mu\nu}$. In this way we have shown that, for a *small amplitude* perturbation, the gravitational wave can be treated as a 'matter' field with its own Lagrangian and field equations.

Given this Lagrangian, we should be able to calculate the effective stress-energy tensor of the wave field by taking the variations of the effective Lagrangian with respect to $g_{\mu\nu}$, holding the 'matter' field $h_{\mu\nu}$ fixed:

$$T^{(\mathrm{GW})\alpha\beta}\sqrt{-g} = 2\frac{\partial L^{(\mathrm{GW})}[g_{\mu\nu}, h_{\mu\nu}]\sqrt{-g}}{\partial g_{\alpha\beta}} \tag{5.11}$$

with

$$L^{(\mathrm{GW})}\sqrt{-g} = \frac{1}{32\pi}\left(\frac{\partial^2(R\sqrt{-g})}{\partial g_{\mu\nu}\partial g_{\alpha\beta}} h_{\mu\nu}h_{\alpha\beta} + 2\frac{\partial^2(R\sqrt{-g})}{\partial g_{\mu\nu}\partial g_{\alpha\beta,\gamma}} h_{\mu\nu}h_{\alpha\beta,\gamma} \right.$$

$$\left. + \frac{\partial^2(R\sqrt{-g})}{\partial g_{\mu\nu,\tau}\partial g_{\alpha\beta,\gamma}} h_{\mu\nu,\tau}h_{\alpha\beta,\gamma} + 2\frac{\partial^2(R\sqrt{-g})}{\partial g_{\mu\nu}\partial g_{\alpha\beta,\gamma\tau}} h_{\mu\nu}h_{\alpha\beta,\gamma\tau} \right). \tag{5.12}$$

This quantity is quadratic in the wave amplitude $h_{\mu\nu}$. It could be simplified further by integrations by parts, such as by taking a derivative off $h_{\alpha\beta,\gamma\tau}$. This would change the coefficients of the other terms. We will not need to worry about finding the 'best' form for the expression (4.12), as we now show.

As in linearized theory, so also in the general case, the quantity $h_{\mu\nu}$ behaves like a tensor with respect to background coordinate transformations, and so does $T_{\mu\nu}^{(GW)}$. However, it is not gauge-invariant and so it is not physically observable. Since the integral of the action *is* independent of coordinate transformations that have compact support, so too is the integral of the effective stress-energy tensor. In practical terms, this makes it possible to localize the energy of a wave to within a region of about one wavelength in size where the background curvature does not change significantly. In fact, if we restrict our gauge transformations to have a length scale of a wavelength, and if we average (integrate) the stress-energy tensor of the waves over such a region, then any gauge changes will be small surface terms.

By evaluating the effective stress-energy tensor on a smooth background metric in a Lorentz gauge, and performing the averaging (denoted by symbol $\langle \cdots \rangle$), one arrives at the *Isaacson tensor*:

$$T_{\alpha\beta}^{(GW)} = \frac{1}{32\pi} \langle h_{\mu\nu;\alpha} h^{\mu\nu}{}_{;\beta} \rangle. \tag{5.13}$$

This is a convenient and compact form for the gravitational stress-energy tensor. It localizes energy in short-wavelength gravitational waves to regions of the order of a wavelength. It is interesting to remind ourselves that our only experimental evidence of gravitational waves today is the observation of the effect on a binary orbit of the loss of energy to the gravitational waves emitted by the system. So this energy formula, or equivalent ones, is central to our understanding of gravitational waves.

5.3   Practical applications of the Isaacson energy

If we are far from a source of gravitational waves, we can treat the waves by linearized theory. Then if we adopt the TT gauge and specialize the stress-energy tensor of the radiation to a flat background, we get

$$T_{\alpha\beta}^{(GW)} = \frac{1}{32\pi} \langle h_{ij,\alpha}^{TT} h^{TTij}{}_{,\beta} \rangle. \tag{5.14}$$

Since there are only two components, a wave travelling with frequency f (wavenumber $k = 2\pi f$) and with a typical amplitude h in both polarizations carries an energy F_{gw} equal to (see exercise (f) at the end of this lecture)

$$F_{gw} = \frac{\pi}{4} f^2 h^2. \tag{5.15}$$

Putting in the factors of c and G and scaling to reasonable values gives

$$F_{gw} = 3 \text{ mW m}^{-2} \left[\frac{h}{1 \times 10^{-22}} \right]^2 \left[\frac{f}{1 \text{ kHz}} \right]^2, \tag{5.16}$$

which is a very large energy flux even for this weak a wave. It is twice the energy flux of a full moon! Integrating over a sphere of radius r, assuming a total duration of the event τ, and solving for h, again with appropriate normalizations, gives

$$h = 10^{-21} \left[\frac{E_{gw}}{0.01 M_\odot c^2} \right]^{\frac{1}{2}} \left[\frac{r}{20 \text{ Mpc}} \right]^{-1} \left[\frac{f}{1 \text{ kHz}} \right]^{-1} \left[\frac{\tau}{1 \text{ ms}} \right]^{-\frac{1}{2}}. \qquad (5.17)$$

This is the formula for the 'burst energy', normalized to numbers appropriate to a gravitational collapse occurring in the Virgo cluster. It explains why physicists and astronomers regard the 10^{-21} threshold as so important. However, this formula could also be applied to a binary system radiating away its orbital gravitational binding energy over a long period of time τ, for example.

5.3.1 Curvature produced by waves

We have assumed that the background metric satisfies the vacuum Einstein equations to linear order, but now it is possible to view the full action principle as a principle for the background with a wave field $h_{\mu\nu}$ on it, and to let the wave energy affect the background curvature [24]. This means that the background will actually solve, in a self-consistent way, the equation

$$G_{\alpha\beta}[g_{\mu\nu}] = 8\pi T_{\alpha\beta}^{GW}[g_{\mu\nu} + h_{\mu\nu}]. \qquad (5.18)$$

This does not contradict the vanishing of the first variation of the action, which we needed to use above, because now we have an Einstein tensor that is of quadratic order in $h_{\mu\nu}$, contributing a term of cubic order to the first-variation of the action, which is of the same order as other terms we have neglected.

5.3.2 Cosmological background of radiation

This self-consistent picture allows us to talk about, for example, a cosmological gravitational wave background that contributes to the curvature of the universe. Since the energy density is the same as the flux (when $c = 1$), we have

$$\varrho_{gw} = \frac{\pi}{4} f^2 h^2, \qquad (5.19)$$

but now we must interpret h in a statistical way. This will be treated in the contribution by Babusci *et al*, but basically it is done by replacing h^2 by a statistical mean square amplitude per unit frequency (Fourier transform power), so that the energy density per *unit frequency* is proportional to $f^2 |\tilde{h}|^2$. It is then conventional to talk about the energy density per unit logarithm of frequency, which means multiplying by f. The result, after being careful about averaging over all directions of the waves and all independent polarization components, is

$$\frac{d\varrho_{gw}}{d \ln f} = 4\pi^2 f^3 |\bar{h}(f)|^2. \qquad (5.20)$$

Finally, what is of the most interest is the energy density as a fraction of the closure or critical cosmological density, given by the Hubble constant H_0 as $\varrho_c = 3H_0^2/8\pi$. The resulting ratio is the symbol $\Omega_{\rm gw}(f)$ that we met in the previous lecture:

$$\Omega_{\rm gw}(f) = \frac{32\pi^3}{3H_0^2} f^3 |\bar{h}(f)|^2. \tag{5.21}$$

5.3.3 Other approaches

We finish this lecture by observing that there is no unique approach to defining energy for gravitational radiation or indeed for any solution of Einstein's equations. Historically this has been one of the most difficult areas for physicists to get to grips with. In the textbooks you will find discussions of pseudotensors, of energy measured at null infinity and at spacelike infinity, of Noether theorems and formulae for energy, and so on. None of these are worse than we have presented here, and in fact all of them are now known to be consistent with one another, if one does not ask them to do too much. In particular, if one wants only to localize the energy of a gravitational wave to a region of the size of a wavelength, and if the waves have short wavelength compared to the background curvature scale, then pseudotensors will give the same energy as the one we have defined here. Similarly, if one takes the energy flux defined here and evaluates it at null infinity, one gets the so-called Bondi flux, which was derived by H Bondi in one of the pioneering steps in the understanding of gravitational radiation. Many of these issues are discussed in the Schutz–Sorkin paper referred to earlier [23].

5.4 Exercises for chapter 5

(e) *In the notes above we give the general gauge transformation*

$$h_{\mu\nu} \to h_{\mu\nu} - \xi_{\mu;\nu} - \xi_{\nu;\mu}.$$

Use the formula for the derivation of Einstein's equations from an action principle,

$$\delta I = \frac{1}{16\pi} \int \frac{\delta(R\sqrt{-g})}{\delta g_{\mu\nu}} h_{\mu\nu}\, {\rm d}^4 x$$

with

$$\frac{\delta(R\sqrt{-g})}{\delta g_{\mu\nu}} = -G^{\mu\nu}\sqrt{-g},$$

but insert a pure gauge $h_{\mu\nu}$. Argue that since this is merely a coordinate transformation, the action should be invariant. Integrate the variation of the action to prove the contacted Bianchi identity

$$G^{\mu\nu}{}_{,\nu} = 0.$$

This shows that the divergence-free property of $G^{\mu\nu}$ is closely related to the coordinate invariance of Einstein's theory.

(f)   Suppose a plane wave, travelling in the z-direction in linearized theory, has both polarization components h_+ and $h_\times$. Show that its energy flux in the z-direction, $T^{(\mathrm{GW})0z}$, is

$$\langle T^{(\mathrm{GW})0z}\rangle = \frac{k^2}{32\pi}(A_+^2 + A_\times^2),$$

where the angle brackets denote an average over one period of the wave.

Chapter 6

Mass- and current-quadrupole radiation

In this lecture we focus on the wave amplitude itself, and how it and the polarization depend on the motions in the source. Consider an isolated source with a stress-energy tensor $T^{\alpha\beta}$. As in chapter 2, the Einstein equation is

$$\left(-\frac{\partial^2}{\partial t^2} + \nabla^2\right)\overline{h}^{\alpha\beta} = -16\pi T^{\alpha\beta} \tag{6.1}$$

$(\overline{h}^{\alpha\beta} = h^{\alpha\beta} - \frac{1}{2}\eta^{\alpha\beta}h$ and $\overline{h}^{\alpha\beta},_\beta = 0)$. Its general solution is the following retarded integral for the field at a position x^i and time t in terms of the source at a position y^i and the retarded time $t - R$:

$$\overline{h}^{\alpha\beta}(x^i, t) = 4\int \frac{1}{R}T^{\alpha\beta}(t - R, y^i)\,d^3y, \tag{6.2}$$

where we define

$$R^2 = (x^i - y^i)(x_i - y_i). \tag{6.3}$$

6.1 Expansion for the far field of a slow-motion source

Let us suppose that the origin of coordinates is in or near the source, and the field point x^i is far away. Then we define $r^2 = x^i x_i$ and we have $r^2 \gg y^i y_i$. We can, therefore, expand the term R in the dominator in terms of y^i. The lowest order is r, and all higher-order terms are smaller than this by powers of r^{-1}. Therefore, they contribute terms to the field that fall off faster than r^{-1}, and they are negligible in the far zone. Therefore, we can simply replace R by r in the dominator, and take it out of the integral.

The R inside the time-argument of the source term is not so simple. If we suppose that $T^{\alpha\beta}$ does not change very fast we can substitute $t - R$ by $t - r$ (the retarded time to the origin of coordinates) and expand

$$t - R = t - r + n^i y_i + O\left(\frac{1}{r}\right), \quad \text{with } n^i = \frac{x^i}{r}, \; n^i n_i = 1. \tag{6.4}$$

The two conditions $r \gg y^i y_i$ and the slow-motion source, can be expressed quantitatively as:

$$r \gg \bar{\lambda}$$
$$R \ll \bar{\lambda}$$

where $\bar{\lambda}$ is the reduced wavelength $\bar{\lambda} = \lambda/2\pi$ and R is the size of source.

The terms of order r^{-1} are negligible for the same reason as above, but the first term in this expansion must be taken into account. It depends on the direction to the field point, given by the unit vector n^i. We use this by making a Taylor expansion in time on the time-argument of the source. The combined effect of these approximations is

$$\bar{h}^{\alpha\beta} = \frac{4}{r} \int [T^{\alpha\beta}(t', y^i) + T^{\alpha\beta},_0(t', y^i)n^j y_j + \tfrac{1}{2}T^{\alpha\beta},_{00}(t', y^i)n^j n^k y_j y_k$$

$$+ \tfrac{1}{6}T^{\alpha\beta},_{000}(t', y^i)n^j n^k n^l y_j y_k y_l + \cdots] \, d^3 y. \tag{6.5}$$

We will need all the terms of this Taylor expansion out to this order.

The integrals in expression (5.5) contain moments of the components of the stress-energy. It is useful to give these names. Use M for moments of the density T^{00}, P for moments of the momentum T^{0i} and S for the moments of the stress T^{ij}. Here is our notation:

$$M(t') = \int T^{00}(t', y^i) \, d^3 y, \quad M^j(t') = \int T^{00}(t', y^i) y^j \, d^3 y,$$

$$M^{jk}(t') = \int T^{00}(t', y^i) y^j y^k \, d^3 y, \quad M^{jkl}(t') = \int T^{00}(t', y^i) y^j y^k y^l \, d^3 y,$$

$$P^l(t') = \int T^{0l}(t', y^i) \, d^3 y, \quad P^{lj}(t') = \int T^{0l}(t', y^i) y^j \, d^3 y,$$

$$P^{ljk}(t') = \int T^{0l}(t', y^i) y^j y^k \, d^3 y,$$

$$S^{lm}(t') = \int T^{lm}(t', y^i) \, d^3 y, \quad S^{lmj}(t') = \int T^{lm}(t', y^i) y^j \, d^3 y.$$

These are the moments we will need.

Among these moments there are some identities that follow from the conservation law in linearized theory, $T^{\alpha\beta},_\beta = 0$, which we use to replace time derivatives of components of T by divergences of other components and then integrate by parts. The identities we will need are

$$\dot{M} = 0, \quad \dot{M}^k = P^k, \quad \dot{M}^{jk} = P^{jk} + P^{kj}, \quad \dot{M}^{jkl} = P^{jkl} + P^{klj} + P^{ljk},$$
$$\tag{6.6}$$

$$\dot{P}^j = 0, \quad \dot{P}^{jk} = S^{jk}, \quad \dot{P}^{jkl} = S^{jkl} + S^{jlk}. \tag{6.7}$$

These can be applied recursively to show, for example, two further very useful relations

$$\frac{d^2 M^{jk}}{dt^2} = 2S^{jk}, \qquad \frac{d^3 M^{jkl}}{dt^3} = 6\dot{S}^{(jkl)} \tag{6.8}$$

where the round brackets on indices indicate full symmetrization.

Using these relations and notations it is not hard to show that

$$\bar{h}^{00}(t, x^i) = \frac{4}{r}M + \frac{4}{r}P^j n_j + \frac{4}{r}S^{jk}(t')n_j n_k + \frac{4}{r}\dot{S}^{jkl}(t')n_j n_k n_l + \cdots \tag{6.9}$$

$$\bar{h}^{0j}(t, x^i) = \frac{4}{r}P^j + \frac{4}{r}S^{jk}(t')n_k + \frac{4}{r}\dot{S}^{jkl}(t')n_k n_l + \cdots \tag{6.10}$$

$$\bar{h}^{jk}(t, x^i) = \frac{4}{r}S^{jk}(t') + \frac{4}{r}\dot{S}^{jkl}(t')n_l + \cdots . \tag{6.11}$$

In these three formulae there are different orders of time-derivatives, but in fact they are evaluated to the same final order in the slow-motion approximation. One can see that from the gauge condition $\bar{h}^{\alpha\beta}{}_{,\beta} = 0$, which relates time-derivatives of some components to space-derivatives of others.

In these expressions, one must remember that the moments are evaluated at the retarded time $t' = t - r$ (except for those moments that are constant in time), and they are multiplied by components of the unit vector to the field point $n^j = x^j/r$.

6.2   Application of the TT gauge to the mass quadrupole field

In the expression for the amplitude that we derived so far, the final terms are those that represent the current-quadrupole and mass-octupole radiation. The terms before them represent the static parts of the field and the mass-quadrupole radiation. In this section we treat just these terms, placing them into the TT gauge. This will be simpler than treating it all at once, and the procedure for the next terms will be a straightforward generalization.

6.2.1   The TT gauge transformations

We are already in Lorentz gauge, and this can be checked by taking derivatives of the expressions for the field that we have derived above. However, we are manifestly not in the TT gauge. Making a gauge transformation consists of choosing a vector field ξ^α and modifying the metric by

$$h_{\alpha\beta} \rightarrow h_{\alpha\beta} - \xi_{\alpha,\beta} - \xi_{\beta,\alpha}. \tag{6.12}$$

The corresponding expression for the potential $\bar{h}^{\alpha\beta}$ is

$$\bar{h}^{\alpha\beta} \rightarrow \bar{h}^{\alpha\beta} + \xi^{\alpha,\beta} + \xi^{\alpha,\beta} - \eta^{\alpha\beta}\xi^\mu{}_{,\mu}. \tag{6.13}$$

For the different components this implies changes

$$\delta\bar{h}^{00} = \xi^{0,0} + \xi^j{}_{,j} \tag{6.14}$$

$$\delta\bar{h}^{0j} = \xi^{0,j} + \xi^{j,0} \tag{6.15}$$

$$\delta\bar{h}^{jk} = \xi^{j,k} + \xi^{k,j} - \delta^{jk}\xi^\mu{}_{,\mu} \tag{6.16}$$

where δ^{jk} is the Kronecker delta (unit matrix). In practice, when taking derivatives, the algebra is vastly simplified by the fact that we are keeping only r^{-1} terms in the potentials. This means that spatial derivatives do not act on $1/r$ but only on $t' = t - r$. It follows that $\partial t'/\partial x^j = -n_j$, and $\partial h(t')/\partial x^j = -\dot{h}(t')n_j$.

It is not hard to show that the following vector field puts the metric into the TT gauge to the order we are working:

$$\xi^0 = \frac{1}{r}P^k{}_k + \frac{1}{r}P^{jk}n_jn_k + \frac{1}{r}S^l{}_{lk}n^k + \frac{1}{r}S^{ijk}n_in_jn_k, \tag{6.17}$$

$$\xi^i = \frac{4}{r}M^i + \frac{4}{r}P^{ij}n_j - \frac{1}{r}P^k{}_kn^i - \frac{1}{r}P^{jk}n_jn_kn^i + \frac{4}{r}S^{ijk}n_jn_k$$

$$\qquad - \frac{1}{r}S^l{}_{lk}n^kn^i - \frac{1}{r}S^{jlk}n_jn_ln_kn^i. \tag{6.18}$$

6.2.2   Quadrupole field in the TT gauge

The result of applying this gauge transformation to the original amplitudes is

$$\bar{h}^{TT00} = \frac{4M}{r}, \tag{6.19}$$

$$\bar{h}^{TT0i} = 0, \tag{6.20}$$

$$\bar{h}^{TTij} = \frac{4}{r}\left[\perp^{ik}\perp^{jl}S_{lk} + \frac{1}{2}\perp^{ij}(S_{kl}n^kn^l - S^k{}_k)\right]. \tag{6.21}$$

Remember that here we are not including $\dot{S}^{jkl}$, because it is a third-order effect.

The notation $\perp^{ik}$ represents the projection operator perpendicular to the direction n^i to the field point.

$$\perp^{jk} = \delta^{jk} - n^jn^k. \tag{6.22}$$

It can be verified that this tensor is transverse to the direction n^i and is a projection, in the sense that it projects to itself

$$\perp^{jk}n_k = 0, \quad \perp^{jk}\perp_k{}^l = \perp^{jl}. \tag{6.23}$$

The spherical component of the field is not totally eliminated in this gauge transformation: the time–time component of the metric must contain the constant Newtonian field of the source. (In fact we have succeeded in eliminating the dipole, or momentum part of the field, which is also part of the non-wave solution. Our gauge transformation has incorporated a Lorentz transformation that has put us into the rest frame of the source.) The time-dependent part of the field is now purely spatial, transverse (because everything is multiplied by ⊥), and traceless (as can be verified by explicit calculation).

The expression for the spatial part of the field actually does not depend on the trace of S_{jk}, as can be seen by constructing the trace-free part of the tensor, defined as:

$$\tilde{S}^{jk} = S^{jk} - \tfrac{1}{3}\delta^{jk} S^l{}_l. \tag{6.24}$$

In fact, it is more conventional to use the mass moment here instead of the stress, so we also define

$$\tilde{M}^{jk} = M^{jk} - \frac{1}{3}\delta^{jk} M^l_l, \quad \tilde{S}^{jk} = \frac{1}{2}\frac{d^2 \tilde{M}^{jk}}{dt^2}. \tag{6.25}$$

In terms of $\tilde{M}$ the far field is

$$\bar{h}^{\mathrm{TT}ij} = \frac{2}{r}\left(\perp^{ik}\perp^{jl}\ddot{\tilde{M}}_{kl} + \frac{1}{2}\perp^{ij}\ddot{\tilde{M}}_{kl}n^l n^k\right). \tag{6.26}$$

This is the usual formula for the mass-quadrupole field. In textbooks the notation is somewhat different than we have adopted here. In particular, our tensor $\tilde{M}$ is what is called $\mathcal{I}$ in Misner *et al* (1973) and Schutz (1985). It is the basis of most gravitational-wave source estimates. We have derived it only in the context of linearized theory, but remarkably its form is identical if we go to the post-Newtonian approximation, where the gravitational waves are a perturbation of the Newtonian spacetime rather than of flat spacetime.

Given this powerful formula, it is important to try to interpret it and understand it as fully as possible. One obvious conclusion is that the dominant source of radiation, at least in the slow-motion limit, is the second time-derivative of the second moment of the mass density T^{00} (the mass-quadrupole moment). This is a very important difference between gravitational waves and electromagnetism, in which the most important source is the electric-dipole. In our case the mass-dipole term is not able to radiate because it is constant, reflecting conservation of the linear momentum of the source. In electromagnetism, however, if the dipole term is absent for some reason (all charges positive, for example) then the quadrupole term dominates and it looks very similar to equation (6.26).

6.2.3 Radiation patterns related to the motion of sources

The projection operators in equation (6.26) show that the radiative field is transverse, as we expect. However, the form of equation (6.26) hides two equally important messages:

- the only motions that produce the radiation are the ones transverse to the line of sight; and
- the induced motions in a detector mirror the motions of the source projected onto the plane of the sky.

To see why these are true, we define the *transverse traceless quadrupole tensor*

$$M_{ij}^{\text{TT}} = \perp^k_{\ i} \perp^l_{\ j} M_{kl} - \tfrac{1}{2} \perp_{ij} \perp^{kl} M_{kl}. \tag{6.27}$$

(Notice that some of our definitions of tracelessness involve subtracting $\tfrac{1}{3}$ of the trace, as in equation (6.24), and sometimes $\tfrac{1}{2}$ of the trace, as in equation (6.27). The appropriate factor is determined by the effective dimensionality (rank) of the tensor. Although we have three spatial dimensions, the projection tensor $\perp$ projects the mass-quadrupole tensor onto a two-dimensional plane, where the trace involves only two components, not three.)

Now, if in equation (6.26) we replace $\tilde{M}_{ij}$ by its definition in terms of M_{ij}, and then collect terms appropriately, it is not hard to show that the equation simplifies to its most natural form:

$$\bar{h}^{\text{TT}ij} = \frac{2}{r}\ddot{M}^{\text{TT}ij}. \tag{6.28}$$

This could of course have been derived directly by applying the TT operation to equations (6.9)–(6.11). Since this equation involves only the TT part of M, our first assertion above is proved. According to this equation, in order to calculate the quadrupole radiation that a particular observer will receive, one need only compute the mass-quadrupole tensor's second time-derivative, project it onto the plane of the sky as seen by the observer looking toward the source, take away its trace, and rescale it by a factor $2/r$. In particular, the TT tensor that describes the action of the wave (as in the polarization diagram in figure 2.1) is a copy of the TT tensor of the mass distribution. This proves our second assertion above.

Looking again at figure 2.1 we imagine a detector consisting of two free masses whose separation is being monitored. If the wave causes them to oscillate relative to one another along the x-axis (the $\oplus$ polarization), this means that the source motion contained a component that did the same thing. If the source is a binary, then the binary orbit projected onto the sky must involve motion of the stars back and forth along either the x- or the y-axis.

It is possible from this to understand many aspects of quadrupole radiation in a simple way. Consider a binary star system with a circular orbit. Seen by a distant observer in the orbital plane, the projected source motion is linear, back and forth. The received polarization will be linear, the polarization ellipse aligned with the orbit. Seen by a distant observer along the axis of the orbit of the binary, the projected motion is circular, which is a superposition of two linear motions separated in phase by 90°. The received radiation will also have circular polarization. Because both linear polarizations are present, the amplitude of the wave emitted up the axis is twice that emitted in the plane. In this way we can completely determine the radiation pattern of a binary system.

Notice that, when viewed at an arbitrary angle to the axis, the radiation will be elliptically polarized, and the degree of ellipticity will directly measure the inclination of the orbital plane to the line of sight. This is a very special kind of information, which one cannot normally obtain from electromagnetic observations of binaries. It illustrates the complementarity of the two kinds of observing.

6.3 Application of the TT gauge to the current-quadrupole field

Now we turn to the problem of placing next-order terms of the wave field, the current quadrupole and mass octupole, into the TT gauge. Our interest here is to understand current-quadrupole radiation in the same physical way as we have just done for mass-quadrupole radiation. So we shall put the field into the TT gauge and then see how to separate the current-quadrupole part from the mass-octupole, which we will discard from the present discussion.

6.3.1 The field at third order in slow-motion

The next order terms in the non-TT metric bear a simple relationship to the mass-quadrupole terms (see equations (6.9)–(6.11)). In each of the metric components, just replace S^{jk} by $\dot{S}^{jkl}n_l$ to go from one order to the next.

This means that we can just skip to the end of the application of the gauge transformations in equations (6.17) and (6.18) and write the next order of the final field, only using S again, not M:

$$\bar{h}^{\text{TT}ij} = \frac{4}{r}\left[\perp^{ik}\perp^{jl}\dot{S}_{lkm}n^m + \frac{1}{2}\perp^{ij}(\dot{S}_{klm}n^k n^l n^m - \dot{S}^k{}_{kl}n^l)\right], \qquad (6.29)$$

or more compactly

$$\bar{h}^{\text{TT}ij} = \frac{4}{r}\left(\perp^{ik}\perp^{jl}\dot{\tilde{S}}_{klm}n^m + \frac{1}{2}\perp^{ij}\dot{\tilde{S}}_{klm}n^l n^k n^m\right). \qquad (6.30)$$

The tilde on S represents a trace-free operation *on the first two indices*.

$$\dot{\tilde{S}}_{klm} = \dot{S}_{lkm} - \tfrac{1}{3}\delta_{kl}\dot{S}^i{}_{im}.$$

These are the indices that come from the indices of T^{jk}, so the tensor is symmetric on these. By analogy with the quadrupole calculation, we can also define the TT part of S_{ijk} by doing the TT projection on the first two indices,

$$S^{\text{TT}}_{ijm} = \perp^k{}_i \perp^l{}_j S_{klm} - \tfrac{1}{2}\perp_{ij}\perp^{kl} S_{klm}. \qquad (6.31)$$

The TT projection of the equation for the metric is

$$h^{\text{TT}ij} = \frac{4}{r} \dot{S}^{\text{TT}ijk} n_k. \tag{6.32}$$

6.3.2 Separating the current quadrupole from the mass octupole

The last equation is compact, but it does not have the ready interpretation that we have at quadrupole order. This is because the moment of the stress, S_{ijk}, does not have such a clear physical interpretation. We see from equations (6.6)–(6.8) that S_{ijk} is a complicated mixture of moments of momentum and density. To gain more physical insight into radiation at this order, we need to separate these different contributions. It is straightforward algebra to see that the following identity follows from the earlier ones:

$$\dot{S}^{ijk} = \tfrac{1}{6} \dddot{M}^{ijk} + \tfrac{2}{3} \ddot{P}^{[jk]i} + \tfrac{2}{3} \ddot{P}^{[ik]j}, \tag{6.33}$$

where square brackets around indices mean antisymmetrization:

$$A^{[ik]} := \tfrac{1}{2}(A^{ik} - A^{ki}).$$

This is a complete separation of the mass terms (in M) from the momentum terms (in P) because the only identities relating the momentum moments to the mass moments involve the symmetric part of P^{ijk} on its first two indices, and this is absent from equation (6.33).

The first term in equation (6.33) is the third moment of the density, and this is the source of the *mass-octupole* field. It produces radiation through the third time-derivative. Since we are in a slow-motion approximation, this is smaller than the mass-quadrupole radiation by typically a factor v/c. Unless there were some very special symmetry conditions, one would not expect the mass octupole to be anything more than a small correction to the mass quadrupole. For this reason we will not treat it here.

The second and third terms in equation (6.33) involve the second moment of the momentum, and together they are the source of the *current-quadrupole* field. It involves two time-derivatives, just as the mass quadrupole does, but these are time-derivatives of the momentum moment, not the mass moment, so these terms produce a field that is also v/c smaller than the typical mass-quadrupole field. However, it requires less of an accident for the mass quadrupole to be absent and the current quadrupole present. It just requires motions that leave the density unchanged to lowest order. This happens in the r-modes. Therefore, the current quadrupole deserves more attention, and we will work exclusively with these terms from now on.

The terms in equation (6.33) that we need are the ones involving $\ddot{P}^{ijk}$. These are antisymmetrized on the first two indices, which involves effectively a vector product between the momentum density (first index) and one of the moment indices. This is essentially the angular momentum density. To make the angular momentum explicit and to simplify the expression, we introduce the angular momentum and the first moment of the angular momentum density

$$J^i := \epsilon^{ijk} P_{jk},$$

(6.34)

$$J^{il} := \epsilon^{ijk} P_{jk}{}^l,$$

(6.35)

where ϵ^{ijk} is the fully antisymmetric (Levi-Civita) symbol in three dimensions. It follows from this that

$$P^{[jk]l} = \tfrac{1}{2}\epsilon^{jki} J_i{}^l.$$

These terms enter the TT projection of the field (6.32) with the last index of S always contracted with the direction n^i to the observer from the source. According to equation (6.33), this contraction always occurs on one of the antisymmetrized indices, or if we use the form in the previous equation then we will always have a contraction of n^i with ϵ^{ijk}. This is a simple object, which we call

$$\perp\epsilon^{jk} := n_i\epsilon^{ijk}.$$

(6.36)

This is just the two-dimensional Levi-Civita object in the plane perpendicular to n^i, which is the plane of the sky as seen by the observer. These quantities will be used in the current-quadrupole field, which contains projections on all the indices. Therefore, the only components of J^{jk} that enter are those projected onto the sky, and so it will simplify formulae to define the sky-projected moment of the angular momentum $\perp J$

$$\perp J^{ij} := \perp^i{}_l \perp^j{}_m J^{lm}.$$

(6.37)

Using this assembled notation, the current-quadrupole field is

$$h^{\mathrm{TT}ij} = \frac{4}{3r}(\perp\epsilon^{ik} \perp\ddot{J}_k{}^j + \perp\epsilon^{jk} \perp\ddot{J}_k{}^i + \perp^{ij} \perp\epsilon^{km} \perp\ddot{J}_{km}).$$

(6.38)

This is similar in form and complexity to the mass-quadrupole field expression. The interpretation of the contributions is direct. Only components of the angular momentum in the plane of the sky contribute to the field. Similarly only moments of this angular momentum transverse to the line of sight contribute. If one wants, say, the xx component of the field, then the $\perp\epsilon$ factor tells us it is determined by the y-component of momentum, i.e. the component perpendicular to the x-direction in the sky. In fact, it is much simpler just to write out the actual components, assuming that the wave travels toward the observer along the z-direction. Then we have

$$h^{\mathrm{TT}xx} = \frac{4}{3r}(\ddot{J}^{xy} + \ddot{J}^{yx}),$$

(6.39)

$$h^{\mathrm{TT}xy} = \frac{4}{3r}(\ddot{J}^{yy} - \ddot{J}^{xx}),$$

(6.40)

and the remaining components can be found from the usual symmetries of the TT-metric. I have dropped the prefix $\perp$ on J because in this coordinate system the given components are already transverse.

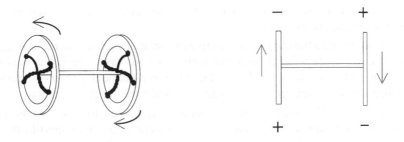

Figure 6.1. A simple current-quadrupole radiator. The left-hand panel shows how the two wheels are connected with blade springs to a central axis. The wheels turn in opposite directions, each oscillating back and forth about its rest position. The right-hand panel shows the side view of the system, and the arrows indicate the motion of the near side of the wheels at the time of viewing. The + signs indicate where the momentum of the mass of the wheel is toward the viewer and the − signs indicate where it is away from the viewer.

The simplicity of these expressions is striking. There are two basic cases where one gets current-quadrupole radiation.

- If there is an oscillating angular momentum distribution with a dipole moment along the angular momentum axis, as projected onto the sky, then in an appropriate coordinate system $\ddot{J}^{xx}$ will be nonzero and we will have $\otimes$ radiation. To have a non-vanishing dipole moment, the angular momentum density could, for example, be symmetrical under reflection through the origin along its axis, so that it points in opposite directions on opposite sides.
- If there is an oscillating angular momentum distribution with a dipole moment along an axis perpendicular to the angular momentum axis, as projected onto the sky, then in an appropriate coordinate system $\ddot{J}^{xy}$ will be nonzero and we will have $\oplus$ radiation.

6.3.3 A model system radiating current-quadrupole radiation

To see that the first of these two leads to physically sensible results, let us consider a simple model system that actually bears a close resemblance to the r-mode system. Imagine, as in the left panel of figure 6.1, two wheels connected by an axis, and the wheels are sprung on the axis in such a way that if a wheel is turned by some angle and then released, it will oscillate back and forth about the axis. Set the two wheels into oscillation with opposite phases, so that when one wheel rotates clockwise, the other rotates anticlockwise, as seen along the axis.

Then when viewed along the axis, the angular momentum has no component transverse to the line of sight, so there is no radiation along the axis. This is sensible, because when projected onto the plane of the sky the two wheels are performing exactly opposite motions, so the net effect is that there is zero projected momentum density.

When viewed from a direction perpendicular to the axis, with the axis along the x-direction, then the angular momentum is transverse, and it has opposite direction for the two wheels. There is therefore an x-moment of the x-component of angular momentum, and the radiation field will have the $\otimes$ orientation.

To see that this has a physically sensible interpretation, look back again at the polarization diagram, figure 2.1, and look at the bottom row of figures illustrating the $\otimes$ polarization. See what the particles on the x-axis are doing. They are moving up and down in the y-direction. What motions in the source could be producing this?

At first one might guess that it is the up-and-down motion of the mass in the wheels as they oscillate, because in fact the near side of each wheel does exactly what the test particles at the observer are doing. However, this cannot be the explanation, because the far side of each wheel is doing the opposite, and when they both project onto the sky they cancel. What in fact gives the effect is that at the *top* of the wheel the momentum density is first positive (towards the observer) and then negative, while at the *bottom* of the wheel it is first negative and then positive. On the other wheel, the signs are reversed.

Current-quadrupole radiation is produced, at least in simple situations like the one we illustrate here, by (the second time-derivative derivative of) the component of source momentum along the line of sight. If this is positive in the sense that it is towards the observer, then the momentum density acts as a positive gravitational 'charge'. If negative, then it is a negative 'charge'. The wheels have an array of positive and negative spots that oscillates with time, and the test particles in the polarization diagram are drawn toward the positive ones and pushed away from the negative ones. Interestingly, in electromagnetism, magnetic dipole and magnetic quadrupole radiation are also generated by the component of the electric current along the line of sight.

This is a rather simple physical interpretation of some rather more complex equations. It is possible to re-write equation (6.38) to show explicitly the contribution of the line-of-sight momentum, but the expressions become even more complicated. Instead of dwelling on this, I will turn to the question of calculating the total energy radiated by the source.

6.4   Energy radiated in gravitational waves

We have calculated the energy flux in equation (5.14), and we now have the TT wave amplitudes. We need only integrate the flux over a distant sphere to get the total luminosity. We do this for the mass and current quadrupoles in separate sections.

6.4.1 Mass-quadrupole radiation

The mass-quadrupole radiation field in equation (6.26) must be put into the energy flux formula, and the dependence on the direction n^i can then be integrated over a sphere. It is not a difficult calculation, but it does require some angular integrals over multiple products of the vector n^i, which depends on the angular direction on the sphere. By symmetry, integrals of odd numbers of factors of n^i vanish. For even numbers of factors, the result is essentially determined by the requirement that after integration the result must be fully symmetric under interchange of any two indices and it cannot have any special directions (so it must depend only on the Kronecker delta $\delta^i{}_j$). The identities we need are

$$\int n^i n^j \, d\Omega = \frac{4\pi}{3} \delta^{ij}, \tag{6.41}$$

$$\int n^i n^j n^k n^l \, d\Omega = \frac{4\pi}{15} (\delta^{ij} \delta^{kl} + \delta^{ik} \delta^{jl} + \delta^{il} \delta^{jk}). \tag{6.42}$$

Using these, one gets the following simple formula for the total luminosity of mass-quadrupole radiation

$$\boxed{L_{\text{gw}}^{\text{mass}} = \tfrac{1}{5} \langle \dddot{M}^{jk} \dddot{M}_{jk} \rangle}. \tag{6.43}$$

Here we still preserve the angle brackets of equation (5.14), because this formula only makes sense in general if we average in time over one cycle of the radiation.

6.4.2 Current-quadrupole radiation

The energy radiated in the current quadrupole is nearly as simple to obtain as the mass-quadrupole formula. The extra factor of n^i in the radiation field makes the angular integrals longer, and requires two further identities:

$$\int n^i n^j n^k n^l n^p n^q \, d\Omega = \frac{4\pi}{7} \delta^{(ij} \delta^{kl} \delta^{pq)}, \tag{6.44}$$

$$\epsilon^{ijk} \epsilon^{i'j'k'} = \delta^{ii'} \delta^{jj'} \delta^{kk'} + \delta^{ij'} \delta^{jk'} \delta^{ki'} + \delta^{ik'} \delta^{ji'} \delta^{kj'}$$
$$- \delta^{ii'} \delta^{jk'} \delta^{kj'} - \delta^{ij'} \delta^{ji'} \delta^{kk'} - \delta^{ik'} \delta^{jj'} \delta^{ki'}, \tag{6.45}$$

where the round brackets indicate full symmetrization on all indices. The expression is simplest if we define

$$\tilde{J}^{jk} := \epsilon^{jlm} \tilde{P}_{lm}{}^k + \epsilon^{klm} \tilde{P}_{lm}{}^j,$$

where

$$\tilde{P}^{kij} := P^{kij} - \tfrac{1}{3} \delta^{ij} P^{kl}{}_l.$$

The result of the integration of the flux formula over a distant sphere is [18, 25], in our notation,

$$L_{\text{gw}}^{\text{current}} = \tfrac{4}{5} \langle \dddot{\tilde{J}}^{jk} \dddot{\tilde{J}}_{jk} \rangle. \tag{6.46}$$

6.5   Radiation in the Newtonian limit

The calculation so far has been within the assumptions of linearized theory. Real sources are likely to have significant self-gravity. This means, in particular, that there will be a significant component of the source energy in gravitational potential energy, and this must be taken into account.

In fact a more realistic equation than equation (6.1) would be

$$\Box \bar{h}^{\alpha\beta} = -16\pi(T^{\alpha\beta} + t^{\alpha\beta,}) \tag{6.47}$$

where $t^{\alpha\beta}$ is the stress-energy *pseudotensor* of gravitational waves. This is hard to work with: equation (6.47) is an implicit equation because $t^{\alpha\beta}$ depends on $\bar{h}^{\alpha\beta}$.

Fortunately, the formulae that we have derived are more robust than they seem. It turns out that the *leading order* radiation field from a Newtonian source has the same formula as in linearized theory. By leading order we mean the dominant radiation. If there is mass-quadrupole radiation, then the mass-octupole radiation from a Newtonian source will not be given by the formulae of the linearized theory. On the other hand, current-quadrupole and mass-quadrupole radiation can coexist, because they have different symmetries, so the work we have done here is generally applicable.

More details on how one calculates radiation to higher order in the Newtonian limit will be given in Blanchet's contribution in this book. This is particularly important for computing the radiation to be expected from coalescing binary systems, whose orbits become highly relativistic just before coalescence and which are, therefore, not well described by linearized theory.

Chapter 7

Source calculations

Now that we have the formulae for the radiation from a system, we can use them for some simple examples.

7.1 Radiation from a binary system

The most numerous sources of gravitational waves are binary stars systems. In just half an orbital period, the non-spherical part of the mass distribution returns to its original configuration, so the angular frequency of the emitted gravitational waves is twice the orbital angular frequency.

We shall calculate here the mass-quadrupole moment for two stars of masses m_1 and m_2, orbiting in the x–y plane in a circular orbit with angular velocity Ω, governed by Newtonian dynamics. We take their total separation to be R, which means that the orbital radius of mass m_1 is $m_2 R/(m_1 + m_2)$ while that of mass m_2 is $m_1 R/(m_1 + m_2)$. We place the origin of coordinates at the centre of mass of the system. Then, for example, the xx-component of M^{ij} is

$$M_{xx} = m_1 \left(\frac{m_2 R \cos(\Omega t)}{m_1 + m_2} \right)^2 + m_2 \left(\frac{m_1 R \cos(\Omega t)}{m_1 + m_2} \right)^2$$
$$= \mu R^2 \cos^2(\Omega t), \tag{7.1}$$

where $\mu := m_1 m_2/(m_1 + m_2)$ is the reduced mass. By using a trigonometric identity and throwing away the part that does not depend on time (since we will use only time-derivatives of this expression) we have

$$M_{xx} = \tfrac{1}{2}\mu R^2 \cos(2\Omega t). \tag{7.2}$$

By the same methods, the other nonzero components are

$$M_{yy} = -\tfrac{1}{2}\mu R^2 \cos(2\Omega t), \quad M_{xy} = \tfrac{1}{2}\mu R^2 \sin(2\Omega t).$$

This shows that the radiation will come out at twice the orbital frequency.

In this case the trace-free moment $\tilde{M}^{ij}$ differs from M^{ij} only by a constant, so we can use these values for M^{ij} to calculate the field and luminosity.

As an example of calculating the field, let us compute $\bar{h}^{\text{TT}xx}$ as seen by an observer at a distance r from the system along the y-axis, i.e. lying in the plane of the orbit. We first need the TT part of the mass-quadrupole moment, from equation (6.27):

$$M^{\text{TT}xx} = M^{xx} - \tfrac{1}{2}(M^{xx} + M^{zz}).$$

However, since $M^{zz} = 0$, this is just $M^{xx}/2$. Then from equation (6.28) we find

$$\bar{h}^{\text{TT}xx} = -2\frac{\mu}{r}(R\Omega)^2 \cos[2\Omega(t - r)]. \tag{7.3}$$

Similarly, the result for the luminosity is

$$L_{\text{gw}} = \tfrac{32}{5}\mu^2 R^4 \Omega^6. \tag{7.4}$$

The various factors in these two equations are not independent, because the angular velocity is determined by the masses and separations of the stars. When observing such a system, we cannot usually measure R directly, but we can infer Ω from the observed gravitational-wave frequency, and we may often be able to make a guess at the masses (we will see below that we can actually measure an important quantity about the masses). So we eliminate R using the Newtonian orbit equation

$$R^3 = \frac{m_1 + m_2}{\Omega^2}. \tag{7.5}$$

If in addition we use the gravitational-wave frequency $\Omega_{\text{gw}} = 2\Omega$, we get

$$\bar{h}^{\text{TT}xx} = -2^{1/3}\frac{\mathcal{M}^{5/3}\Omega_{\text{gw}}^{2/3}}{r}\cos[\Omega_{\text{gw}}(t - r)], \tag{7.6}$$

$$L_{\text{gw}} = \frac{4}{5 \times 2^{1/3}}(\mathcal{M}\Omega_{\text{gw}})^{\frac{10}{3}}, \tag{7.7}$$

where we have introduced the symbol for the *chirp mass* of the binary system:

$$\mathcal{M} := \mu^{3/5}(m_1 + m_2)^{2/5}.$$

Notice that both the field and the luminosity depend only on $\mathcal{M}$, not on the individual masses in any other combination.

The power represented by L_{gw} must be supplied by the orbital energy, $E = -m_1 m_2 / 2R$. By eliminating R as before we find the equation

$$E = -\frac{1}{2^{5/3}}\mathcal{M}^{5/3}\Omega_{\text{gw}}^{2/3}.$$

This is remarkable because it too involves only the chirp mass $\mathcal{M}$. By setting the rate of change of E equal to the (negative of the) luminosity, we find an equation for the rate of change of the gravitational-wave frequency

$$\dot{\Omega}_{\text{gw}} = \frac{12 \times 2^{1/3}}{5}\mathcal{M}^{5/3}\Omega_{\text{gw}}^{11/3}. \tag{7.8}$$

As we mentioned in chapter 4, since the frequency increases, the signal is said to 'chirp'.

These results show that the chirp mass is the only mass associated with the binary that can be deduced from observations of its gravitational radiation, at least if only the Newtonian orbit is important. Moreover, if one can measure the field amplitude (e.g. $h^{\text{TT}xx}$) plus Ω_{gw} and $\dot{\Omega}_{\text{gw}}$, one can deduce from these the value of $\mathcal{M}$ *and* the distance r to the system! *A chirping binary with a circular orbit, observed in gravitational waves,* is a standard candle: *one can infer its distance purely from the gravitational-wave observations.* To do this one needs the full amplitude, not just its projection on a single detector, so one generally needs a network of detectors or a long-duration observation with a single detector to get enough information.

It is very unusual in astronomy to have standard candles, and they are highly prized. For example, one can, in principle, use this information to measure Hubble's constant [26].

7.1.1 Corrections

In the calculation above we made several simplifying assumptions. For example, how good is the assumption that the orbit is circular? The Hulse–Taylor binary is in a highly eccentric orbit, and this turns out to enhance its gravitational-wave luminosity by more than a factor of ten, since the elliptical orbit brings the two stars much nearer to one another for a period of time than a circular orbit with the same period would do. So there are big corrections for this system.

However, systems emitting at frequencies observable from ground-based interferometers are probably well approximated by circular orbits, because they have arrived at their very close separation by gravitational-wave-driven in-spiral. This process removes eccentricity from the orbit faster than it shrinks the orbital radius, so by the time they are observed they have insignificant eccentricity.

Another assumption is that the orbit is well described by Newtonian theory. This is not a good assumption in most cases. Post-Newtonian orbit corrections will be very important in observations. This is not because the stars eventually approach each other closely. It is because they spend a long time at wide separations where the small post-Newtonian corrections accumulate systematically, eventually changing the phase of the orbit by an observable amount. So it is very important for observations that we match signals with a template containing high-order post-Newtonian corrections, as described in Blanchet's contribution. But even so, the information contained in the Newtonian part of the radiation is still there, so all our conclusions above remain important.

7.2 The *r*-modes

We consider rotating stars in Newtonian gravity and look at the effect that the emission of gravitational radiation has on their oscillations. One might expect that the loss of energy to gravitational waves would damp out any perturbations, and indeed this is normally the case. However, it was a remarkable discovery of Chandrasekhar [27] that the opposite sometimes happens.

A rotating star is idealized as an axially symmetric perfect-fluid system. In the Newtonian theory the pulsations of a perturbed fluid can be described by normal modes which are the solutions of perturbed Euler and gravitational field equations. If the star is stable, the eigenfrequencies σ of the normal modes are real; if the star is unstable, there is at least one pair of complex-conjugate frequencies, one of which represents an exponentially growing mode and the other a decaying mode. (We take the convention that the time-dependence of a mode is $\exp(i\sigma t)$.)

In general relativity, the situation is, in principle, the same, except that there is a boundary condition on the perturbation equations that insists that gravitational waves far away be outgoing, i.e. that the star loses energy to gravitational waves. This condition forces all eigenfrequencies to be complex. The sign of the imaginary part of the frequency determines stability or instability.

The loss of energy to gravitional radiation can destabilize a star that would otherwise (i.e. in Newtonian theory) be stable. This is because it opens a pathway to lower-energy configurations that might not be accessible to the Newtonian star. This normally happens because gravitational radiation also carries away angular momentum, a quantity that is conserved in the Newtonian evolution of a perturbation.

The sign of the angular momentum lost by the star is a critical diagnostic for the instability. A wave that moves in the positive angular direction around a star will radiate positive angular momentum to infinity. A wave that moves in the opposite direction, as seen by an observer at rest far away, will radiate negative angular momentum. In a spherical star, both actions result in the damping of the perturbation because, for example, the positive-going wave has intrinsically positive angular momentum, so when it radiates its angular momentum decreases and so its amplitude decreases. Similarly, the negative-going wave has negative angular momentum, so when it radiates negative angular momentum its amplitude decreases.

The situation can be different in a rotating star, as first pointed out by Friedman and Schutz [28]. The angular momentum carried by a wave depends on its pattern angular velocity *relative to the star's angular velocity*, not relative to an observer far away. If a wave pattern travels backwards relative to the star, it represents a small effective slowing down of the star and therefore carries negative angular momentum. This can lead to an anomalous situation: if a wave travels backwards relative to the star, but forwards relative to an inertial observer (because its angular velocity relative to the star is smaller than the star's angular velocity), then it will have negative angular momentum but it will radiate positive angular momentum. The result will be that its intrinsic angular momentum will get more negative, and its amplitude will grow.

This is the mechanism of the Chandrasekar–Friedman–Schutz (CFS) instability. In an ideal star, it is always possible to find pressure-driven waves of short enough wavelength around the axis of symmetry (high enough angular eigenvalue m) that satisfy this condition. However, it turns out that even a small amount of viscosity can damp out the instability in such waves, so it is not clear that pressure-driven waves will ever be significantly unstable in realistic stars.

However, in 1997 Andersson [17] pointed out that there was a class of modes called r-modes (Rossby modes) that no-one had previously investigated, and that were formally unstable in all rotating stars. Rossby waves are well known in oceanography, where they play an important role in energy transport around the Earth's oceans. They are hard to detect, having long wavelengths and very low-density perturbations. They are mainly *velocity perturbations* of the oceans, whose restoring force is the Coriolis effect, and that is their character in neutron stars too. Because they have very small density perturbation, the gravitational radiation they emit is dominated by the current-quadrupole radiation.

For a slowly-rotating, nearly-spherical Newtonian star, the following velocity perturbation is characteristic of r-modes:

$$\delta v^a = \varsigma(r)\epsilon^{abc}\nabla_b r \nabla_c Y_{lm}, \qquad (7.9)$$

where $\varsigma(r)$ is some function of r determined by the mode equations. This velocity is a curl, so it is divergence-free; since it has no radial component, it does not change the density. If the star is perfectly spherical, these perturbations are simply a small rotation of some of the fluid, and it continues to rotate. They have no oscillation, and have zero frequency.

If we consider a star with a small rotational angular velocity Ω, then the frequency σ is no longer exactly zero and a Newtonian calculation to first order in Ω shows that there is a mode with *pattern speed* $\omega_p = -\sigma/m$ equal to

$$\omega_p = \Omega\left[1 - \frac{2}{l(l+1)}\right]. \qquad (7.10)$$

These modes are now oscillating currents that move (approximately) along the equipotential surfaces of the rotating star.

For $l \geqslant 2$, ω_p is positive but slower than the speed of the star, so by the CFS mechanism these modes are unstable to the emission of gravitational radiation for an arbitrarily slowly rotating star.

The velocity pattern given in equation (7.9) for ($l = 2$, $m = 2$) is closely related to the wheel model we described for current-quadrupole radiation in figure 6.1. Take two such wheels and orient their axes along the x- and y-axes, with the star rotating about the z-axis. Choose the sense of rotation so that the wheels at positive-x and positive-y are spinning in the opposite sense at any time, i.e. so that their adjacent edges are always moving in the *same* direction. Then this relationship will be reproduced for all other adjacent pairs of wheels: adjacent edges move together.

When seen from above the equatorial plane, the line-of-sight momenta of the wheels reinforce each other, and we get the same kind of pattern that we saw when looking at one wheel from the side. However, in this case the pattern rotates with the angular velocity $2\Omega/3$ of equation (7.10). Since the pattern of line-of-sight momenta repeats itself every half rotation period, the gravitational waves are circularly polarized with frequency $4\Omega/3$. Seen along the x-axis, the wheel along the x-axis contributes nothing, but the other wheel contributes fully, so the radiation amplitude in this direction is half that going out the rotation axis. Seen along a line at 45° to the x-axis, the line-of-sight momenta of the wheels on the front part of the star cancel those at the back, so there is no radiation. Thus, along the equator there is a characteristic series of maxima and zeros, leading to a standard $m = 2$ radiation pattern. This pattern also rotates around the star, but the radiation in the equator remains linearly polarized because there is only the $\otimes$ component, not the $\oplus$. Again, the radiation frequency is twice the pattern speed because the radiation goes through a complete cycle in half a wave rotation period.

This discussion cannot go into the depth required to understand the r-modes fully. There are many issues of principle: what happens beyond linear order in Ω; what happens if the star is described in relativity and not Newtonian gravity; what is the relation between r-modes and the so-called g-modes that can have similar frequencies; what happens when the amplitude grows large enough that the evolution is nonlinear; what is the effect of magnetic fields on the evolution of the instability? The literature on r-modes is developing rapidly. We have included references where some of the most basic issues are discussed [17, 18, 29–31], but the interested student should consult the current literature carefully.

7.2.1 Linear growth of the *r*-modes

We have seen how the r-mode becomes unstable when coupled to gravitational radiation, and now we turn to the practical question: is it important? This will depend on the balance between the growth rate of the mode due to relativistic effects and the damping due to viscosity.

When coupled to gravitational radiation and viscosity, the mode has a complex frequency. If we define $\Im(\sigma) := 1/\tau$, then τ is the characteristic damping time. When radiation and viscosity are treated as small effects, their contributions to the eigenfrequencies add, so we have that the total damping is given by

$$\frac{1}{\tau(\Omega)} = \frac{1}{\tau_{\text{GR}}} + \frac{1}{\tau_{\text{v}}}, \quad \frac{1}{\tau_{\text{v}}} = \frac{1}{\tau_{\text{s}}} + \frac{1}{\tau_{\text{b}}}, \quad (7.11)$$

where $1/\tau_{\text{GR}}$, $1/\tau_{\text{v}}$ are the contributions due to gravitational radiation emission and viscosity, and where the latter has been further divided between shear viscosity ($\frac{1}{\tau_{\text{s}}}$) and bulk viscosity ($\frac{1}{\tau_{\text{b}}}$).

If we consider a 'typical' neutron star with a polytropic equation of state $p = k\rho^2$ (for which k has been chosen so that a $1.5M_\odot$ model has a radius $R = 12.47$ km), and if we express the angular velocity in terms of the scale for

Table 7.1. Gravitational radiation and viscous timescales, in seconds. Negative values indicate instability, i.e. a growing rather than damping mode.

l	m	τ_{gw} (s)	p_{gw}	τ_{bv} (s)	p_{bv}	τ_{sv} (s)
2	2	-20.83	5.93	9.3×10^{10}	1.77	2.25×10^{8}
3	3	-316.1	7.98	1.89×10^{10}	1.83	3.53×10^{7}

the approximate maximum speed $\sqrt{\pi G \bar{\rho}}$ and the temperature in terms of 10^9 K, then it can be shown that [30]

$$\frac{1}{\tau} = \frac{1}{\tau_{gw} \left(\frac{1 \text{ ms}}{P} \right)^{p_{gw}} + \frac{1}{\tau_{bv}} \left(\frac{1 \text{ ms}}{P} \right)^{p_{bv}}} \left(\frac{10^9 \text{ K}}{T} \right)^6 + \frac{1}{\tau_{sv}} \left(\frac{T}{10^9 \text{ K}} \right)^2, \quad (7.12)$$

where the scaling parameters $\tilde{\tau}_{sv}$, $\tilde{\tau}_{bv}$, $\tilde{\tau}_{gw}$ and the exponents p_{gw} and p_{bv} have to be calculated numerically. Some representative values relevant to the r-modes with $2 \leqslant l \leqslant 6$ are in table 7.1 [30].

The physics of the viscosity is interesting. It is clear from equation (7.12) that gravitational radiation becomes a stronger and stronger destabilizing influence as the angular velocity of a star increases, but the viscosity is much more complicated. There are two contributions: shear and bulk. Shear viscosity comes mainly from electrons scattering off protons and other electrons. This effect falls with increasing temperature, just as does viscosity of everyday materials. So a cold, slowly rotating star will not have the instability, where a hotter star might. However, at high temperatures, bulk viscosity becomes dominant. This effect arises in neutron stars from nuclear physics. Neutron-star matter always contains some protons and electrons. When it is compressed, some of these react to form neutrons, emitting a neutrino. When it is expanded, some of the neutrons beta-decay to protons and electrons, again emitting a neutrino. The emitted neutrino is not trapped in the star; within a short time, of the order of a second or less, it escapes. This irreversible loss of energy each time the star is compressed creates a bulk viscosity. Now, bulk viscosity acts only due to the density perturbation, which is small in r-modes. So the effect of bulk viscosity only dominates at very high temperatures.

The balance of the viscous and gravitational effects is illustrated in figure 7.1 [30]. This is indicative, but not definitive: much more work is needed on the physics of viscosity and the structure of the modes at large values of Ω (small P).

7.2.2 Nonlinear evolution of the star

Our description so far is only a linear approximation. To understand the full evolution of the r-modes we have to treat the nonlinear hydrodynamical effects

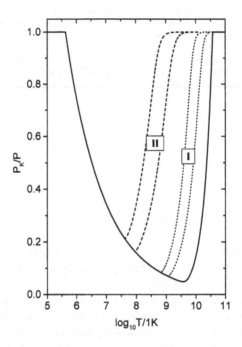

Figure 7.1. The balance of viscous and gravitational radiation effects in the *r*-modes is illustrated in a diagram of rotation speed, showing the ratio of the maximum period P_k to the rotation period *P* versus the temperature of the star. The solid curve indicates the boundary between viscosity-dominated and radiation-dominated behaviour: stars above the line are unstable. The dashed curves illustrate possible nonlinear evolution histories as a young neutron star cools.

that become important as the modes grow. This could only be done with a numerical simulation, which some groups are now working on. However, it is possible to make simple estimates analytically.

We characterize the initial configuration with just two paramters: the uniform angular velocity Ω, and the amplitude α of the *r*-modes perturbation. The star is assumed to cool at the accepted cooling rate for neutron stars, independently of whether it is affected by the *r*-mode instability or not. The star is assumed to lose angular momentum to gravitational radiation at a rate given by the linear radiation field, with its large amplitude α. This loss is taken to drive the star through a sequence of equilibrium states of lower and lower angular momentum. Details of this approximation are in [31], here we report only the results. The evolution turns out to have three phases.

(a) Initially the angular velocity Ω of the hot rapidly rotating neutron star is nearly constant, evolving on the viscous timescale $1/\tau_v$, while the amplitude α grows exponentially on the gravitational radiation timescale $1/\tau_{GR}$.

(b) After a short time nonlinear effects become important and stop the growth of the amplitude α. Most of the initial angular momentum of the star is radiated away by gravitational radiation. The star spins down and evolves to a point where the angular velocity Ω and the temperature is sufficiently low that the r-mode is stable.

(c) Finally gravitational radiation and viscosity damp out the r-mode and drive the star into its final equilibrium configuration.

This may take about a year, a timescale governed by the cooling time of the star. During this year, the star would radiate away most of its angular momentum and rotational kinetic energy. This could be a substantial fraction of a solar mass in energy.

7.2.3 Detection of r-mode radiation

The large amount of energy radiated into the r-modes makes them attractive for detection, but detection will not be trivial. The r-mode event occurs at the rate of supernovae: some fraction (hopefully large) of all supernovae leave behind a rapidly spinning neutron star that spins down over a one-year period. This means we should have sufficient sensitivity to reach the Virgo Cluster (20 Mpc distance). Estimates [31] suggest that a neutron star in the Virgo Cluster could be detected by second generation of LIGO and VIRGO gravitational-wave detectors with an amplitude signal-to-noise of about eight, provided one can use matched filtering (exact template matching).

It will not be easy to use matched filtering, since one must follow all cycles of the signal as the star spins down, and we will not know this well because of many uncertainties: initial temperature, initial spin distribution, detailed physics of viscosity, and so on. However, it would be helpful to have a parametrized model to take account of the uncertainties, so that we could look for a significant fit to one or more of the parameters.

In addition, it is likely that, if a significant proportion of all neutron stars went through the r-mode instability, then the universe has been filled by their radiation. There should be a background with an energy density Ω_{gw} that is a good fraction of the closure density. Its lower frequency limit should be around 200 Hz in the rest frame of the star. When we see radiation cosmologically, its lower frquency limit will indicate the epoch at which star formation began.

It is clear that the discovery of this new source of gravitational waves will open several prospects for astronomy. Observations could be used as supernovae detectors, revealing supernovae hidden in clouds of dust, identifying them about a year after they are formed. The existence of the radiation raises several prospects and questions about the physics of neutron stars, not least the interaction of magnetic fields with the instability.

7.3   Conclusion

These lectures (chapters 2 to 7) have taken us through the basic theory of gravitational radiation and its applications in astrophysics, so far as we can understand and predict them now. In a few years, perhaps as little as two, perhaps as many as eight, we will start to make observations of gravitational radiation from astrophysical sources. If gravitational-wave astronomy follows other branches of observational astronomy, it will not be long before completely unexpected signals are seen, or unexpected features in long-predicted signals. To interpret these will require joining a physical understanding of the relationship between gravitational radiation and its source to a wide knowledge of astronomical phenomena. We encourage the students who have attended these lectures, and others who may study them, to get themselves ready to contribute to this activity. It will be an exciting time!

References

We have divided the references into two sections. The first gives generally useful references—books, conference summaries, etc—that interested students should go to for background and a more complete discussion of the theory. The second section contains specific references to the research literature. General references are indicated in the text by the author name plus year, as Misner *et al* (1973). The specific references are indicated by numbers, for example [1,3].

General references

General relativity

There are a number of good text books on GR. The following cover, at different levels of difficulty and completeness, linearized theory, gauges and the definition of energy:

Ciufolini I and Wheeler J L 1995 *Gravitation and Inertia* (Princeton, NJ: Princeton University Press)
Landau L and Lifshitz E M 1962 *The Classical Theory of Fields* (New York: Pergamon)
Misner C W, Thorne K S and Wheeler J L 1973 *Gravitation* (San Francisco, CA: Freeman)
Ruffini R and Ohanian H C 1997 *Gravitazione e Spazio-Tempo* (Bologna: Zanichelli)
Schutz B F 1995 *A First Course in General Relativity* (Cambridge: Cambridge University Press)
Wald R M 1994 *General Relativity* (Chicago, IL: Chicago University Press)
Weinberg S 1972 *Gravitation and Cosmology* (New York: Wiley)

Gravitational-wave detectors

Conference volumes on detector progress appear more than once per year these days. You can find progress reports on detectors on the web sites of the different groups, which you will find in the list of literature references below. The two references below are more tutorial, aimed at introducing the subject.

Blair D G 1991 *The Detection of Gravitational Waves* (Cambridge: Cambridge University Press)
Saulson P R 1994 *Fundamentals of Interferometric Gravitational Wave Detectors* (Singapore: World Scientific)

Sources of gravitational waves

Again, there have been a number of conference publications on this subject. The third and fourth references are recent reviews of sources. The first two survey the problem of data analysis.

Schutz B F (ed) 1989 *Gravitational Wave Data Analysis* (Dordrecht: Kluwer)
——1997 The detection of gravitational waves *Relativistic Gravitation and Gravitational Radiation* ed J A Marck and J P Lasota (Cambridge: Cambridge University Press) pp 447–75
Thorne K S 1987 Gravitational radiation *300 Years of Gravitation* ed S W Hawking and W Israel (Cambridge: Cambridge University Press) pp 330–458
——1995 *Proc. 1994 Summer Study on Particle and Nuclear Astrophysics and Cosmology* ed E W Kolb and R Peccei (Singapore: World Scientific)

Text references

[1] Schutz B F 1984 Gravitational waves on the back of an envelope *Am. J. Phys.* **52** 412–19
[2] Nicholson D *et al* 1996 Results of the first coincident observations by 2 laser-interferometric gravitational-wave detectors *Phys. Lett.* A **218** 175–80
[3] Meers B J 1988 Recycling in laser-interferometric gravitational wave detectors *Phys. Rev.* D **38** 2317–26
[4] TAMA300 project website, http://tamago.mtk.nao.ac.jp/
[5] GEO600 project website, http://www.geo600.uni-hannover.de/
[6] LIGO project website, http://ligo.caltech.edu/
[7] VIRGO project website, http://virgo4p.pg.infn.it/virgo/
[8] Caves C M *et al* 1980 *Rev. Mod. Phys.* **52** 341–92
[9] Astone P, Lobo J A and Schutz B F 1994 Coincidence experiments between interferometric and resonant bar detectors of gravitational waves *Class. Quantum Grav.* **11** 2093–112
[10] Rome gravitational wave group website, http://www.roma1.infn.it/rog/rogmain.html
[11] Auriga detector website, http://axln01.lnl.infn.it/
[12] LISA project websites, http://www.lisa.uni-hannover.de/ and http://lisa.jpl.nasa.gov/
[13] Bender P *et al LISA: Pre-Phase A Report* MPQ 208 (Max-Planck-Institut für Quantenoptik, Garching, Germany) (also see the 2nd edn, July 1998)
[14] Lipunov V M, Postnov K A and Prokhorov M E 1997 Formation and coalescence of relativistic binary stars: the effect of kick velocity *Mon. Not. R. Astron. Soc.* **288** 245–59
[15] Yungelson L R and Swart S F P 1998 Formation and evolution of binary neutron stars *Astron. Astrophys.* **332** 173–88
[16] Zwart S F P and McMillain S L W 2000 Black hole mergers in the universe *Astrophys. J.* **528** L17–20 (gr-qc/9910061)
[17] Andersson N 1998 A new class of unstable modes of rotating relativistic stars *Astrophys. J.* **502** 708–13 (gr-qc/9706075)
[18] Lindblom L, Owen B J and Morsink S M 1998 Gravitational radiation instability in hot young neutron stars *Phys. Rev. Lett.* **80** 4843–6 (gr-qc 9803053)

[19] van der Klis M 1998 *The Many Faces of Neutron Stars* ed R Buccheri, J van Paradijs and M A Alpar (Dordrecht: Kluwer Academic)

[20] Bildsten L 1998 Gravitational radiation and rotation of accreting neutron stars *Astrophys. J.* **501** L89–93 (astro-ph/9804325)
See also the recent paper Ushomirsky G, Cutler C and Bildsten L 1999 Deformations of accreting neutron star crusts and gravitational wave emission *Preprint* (astro-ph/0001136)

[21] Brady P R, Creighton T, Cutler C and Schutz B F 1998 Searching for periodic sources with LIGO *Phys. Rev.* D **57** 2101–16

[22] Brustein R, Gasperini M, Giovannini M and Veneziano G 1996 Gravitational radiation from string cosmology *Proc. Int. Europhysics Conf. (HEP 95)* ed C Van der Velde, F Verbeure and J Lemonne (Singapore: World Scientific)

[23] Schutz B F and Sorkin R 1977 Variational aspects of relativistic field theories, with application to perfect fluids *Ann. Phys., NY* **107** 1–43

[24] Isaacson R *Phys. Rev.* **166** 1263–71
Isaacson R 1968 *Phys. Rev.* **166** 1272–80

[25] Thorne K S 1980 Multipole expansions of gravitational radiation *Rev. Mod. Phys.* **52** 299–340

[26] Schutz B F 1986 Determining the Hubble constant from gravitational wave observations *Nature* **323** 310–11

[27] Chandrasekar S 1970 *Phys. Rev. Lett.* **24** 611

[28] Friedman J L and Schutz B F 1978 *Ap. J.* **222** 281

[29] Friedman J L and Morsink S M 1998 Axial instability of rotating relativistic stars *Astrophys. J.* **502** 714–20 (gr-qc/9706073)

[30] Andersson N, Kokkotas K and Schutz B F 1999 Gravitational radiation limit on the spin of young neutron stars *Astrophys. J.* **510** 846–8539 (astro-ph/9805225)

[31] Owen B, Lindblom L, Cutler C, Schutz B F, Vecchio A and Andersson N 1998 Gravitational waves from hot young rapidly rotating neutron stars *Phys. Rev.* D **5808** 4020 (gr-qc/9804044)

Solutions to exercises

Chapter 2

Exercise (a)

1. Let us take the form of the wave to be

$$h^{\mathrm{TT}jk} = e_{\oplus}^{jk} h_+(t - \hat{n} \cdot \hat{x})$$

where $e_{\oplus}^{jk}$ is the polarization tensor for the $\oplus$ polarization, and where $\hat{n}$ is the unit vector in the direction of travel of the wave. We will let h_+ be an arbitrary function of its phase argument.

If the wave travels in the x–z plane at an angle θ to the z-direction, then the unit vector in our coordinates is

$$\hat{n}^i = (\sin\theta, 0, \cos\theta).$$

We need to calculate the polarization tensor's components in x, y, z coordinates. We do this by rotating the $\oplus$ polarization tensor from its TT form in coordinates parallel to the wavefront to its form in our coordinates. This requires a simple rotation around the y-axis. The transformation matrix is:

$$\Lambda^{j'}{}_k = \begin{pmatrix} \cos\theta & 0 & \sin\theta \\ 0 & 1 & 0 \\ -\sin\theta & 0 & \cos\theta \end{pmatrix}.$$

The polarization tensor in our coordinates (primed indices) becomes:

$$e^{j'k'} = \Lambda^{j'}{}_l \Lambda^{k'}{}_m e^{lm}$$
$$= \begin{pmatrix} \cos^2\theta & 0 & -\sin\theta\cos\theta \\ 0 & -1 & 0 \\ -\sin\theta\cos\theta & 0 & \sin^2\theta \end{pmatrix}.$$

Notice that the new polarization tensor is again traceless.

The gravitational wave will be, at an arbitrary time t and position (x, z) in our (x, z)-plane,

$$h^{\mathrm{TT}j'k'} = e^{j'k'} h_+(t - x\sin\theta - z\cos\theta).$$

For this problem we need the xx-component because the photon is propagating along this direction, and we will always stay at $z = 0$, so we have

$$h^{TTxx} = \cos^2\theta h_+(t - x\sin\theta).$$

We see that for this geometry the wave amplitude is reduced by a factor of $\cos^2\theta$.

Generalizing the argument in the text, the relation between time and position for the photon on its trip outwards along the x-direction is $t = t_0 + x$, where t_0 is the starting time. The analogous relation after the photon is reflected is $t = t_0 + L + (L - x)$, since in this case x decreases in time from L to 0. If we put these into the equation for the linearized corrections to the return time, we get

$$t_{\text{return}} = t_0 + 2L + \tfrac{1}{2}\cos^2\theta \left\{ \int_0^L h_+[t_0 + (1 - \sin\theta)x]\,dx \right.$$
$$\left. \times \int_0^L h_+[t_0 + 2L - (1 + \sin\theta)x]\,dx \right\}.$$

This expression must be differentiated with respect to t_0 to find the variation of the return time as a function of the start time. The key point is how to handle differentiation within the integrals. Consider, for example, the function $h_+[t_0 + (1 - \sin\theta)x]$. It is a function of a single argument,

$$\xi := t_0 + (1 - \sin\theta)x$$

so derivatives with respect to t_0 can be converted to derivatives with respect to x as follows

$$\frac{dh_+}{dt_0} = \frac{dh_+}{d\xi}\frac{d\xi}{dt_0} = \frac{dh_+}{d\xi};$$
$$\frac{dh_+}{dx} = \frac{dh_+}{d\xi}\frac{d\xi}{dx} = (1 - \sin\theta)\frac{dh_+}{d\xi}.$$

It follows that

$$\frac{dh_+}{dt_0} = \frac{dh_+}{dx}/(1 - \sin\theta).$$

On the return trip the factor will be $-(1 + \sin\theta)$. So when we differentiate we can convert the derivatives with respect to t_0 inside the integrals into derivatives with respect to x. Taking account of the factor $\cos^2\theta = (1 - \sin\theta)(1 + \sin\theta)$ in front of the integrals, the result is

$$\frac{dt_{\text{return}}}{dt_0} = 1 + \frac{1}{2}(1 + \sin\theta)\int_0^L \frac{dh_+}{dx}[t_0 + (1 - \sin\theta)x]\,dx$$
$$+ \frac{1}{2}(1 - \sin\theta)\int_0^L \frac{dh_+}{dx}[t_0 + 2L - (1 + \sin\theta)x]\,dx.$$

The integrals can now be done, since they simply invert the differentiation by x. Evaluating the integrands at the end points of the integrals gives equation (2.20):

$$\frac{dt_{\text{return}}}{dt_0} = 1 - \tfrac{1}{2}(1 + \sin\theta)h_+(t_0) + \sin\theta h_+[t_0 + (1 - \sin\theta)L]$$

$$+ \tfrac{1}{2}(1 - \sin\theta)h_+(t_0 + 2L).$$

2. If we Taylor-expand this equation in powers of L about $L = 0$, the leading term vanishes, and the first-order term is:

$$\frac{dt_{\text{return}}}{dt_0} = L\sin\theta(1 - \sin\theta)\dot{h}_+(t_0) + L(1 - \sin\theta)\dot{h}_+(t_0),$$

$$= L\cos^2\theta\dot{h}_+(t_0).$$

This is just what was required. The factor of $\cos^2\theta$ comes, as we saw above, from the projection of the TT field on the x-coordinate direction.

3. All the terms cancel and there is no effect on the return time.

Exercise (b)

This is part of the calculation in the previous example. All we need is the segment where the light travels from the distant end to the centre:

$$t_{\text{out}} = t_0 + \tfrac{1}{2}\cos^2\theta \int_0^L h_+[t_0 + L - (1 + \sin\theta)x]\,dx$$

and so dt_{out}/dt_0 is:

$$\frac{dt_{\text{out}}}{dt_0} = 1 + \tfrac{1}{2}(1 - \sin\theta)[h_+(t_0 - \sin\theta L) - h_+(t_0 + L)].$$

Exercise (c)

This question is frequently asked, but not by people who have done the calculation. The answer is that the two effects occur in different gauges, not in the same one. So they cannot cancel. The apparent speed of light changes in the TT gauge, but then the positions of the ends remain fixed, so that the effect is all in the coordinate speed. In a local Lorentz frame tied to one mass, the ends do move back and forth, but then the speed of light is invariant.

Exercise (d)

To first order we have

$$R^\mu{}_{\alpha\beta\nu} = \Gamma^\mu{}_{\alpha\nu,\beta} - \Gamma^\mu{}_{\alpha\beta,\nu},$$

$$\Gamma^\mu{}_{\alpha\beta,\nu} = \tfrac{1}{2}\eta^{\mu\sigma}(h_{\sigma\beta,\alpha\nu} + h_{\sigma\alpha,\beta\nu} - h_{\alpha\beta,\sigma\nu}), \quad \text{(i)} \qquad (7.13)$$

$$\Gamma^\mu{}_{\alpha\nu,\beta} = \tfrac{1}{2}\eta^{\mu\sigma}(h_{\sigma\nu,\alpha\beta} + h_{\alpha\sigma,\nu\beta} - h_{\alpha\nu,\sigma\beta}). \quad \text{(ii)} \qquad (7.14)$$

The gauge transformation for a perturbation in linearized theory is

$$h'_{\alpha\beta} = h_{\alpha\beta} - \xi_{\alpha,\beta} - \xi_{\beta,\alpha}. \quad \text{(iii)} \tag{7.15}$$

Substituting (iii) into (i) and (ii), we obtain

$$\text{(i)} = \tfrac{1}{2}\eta^{\mu\sigma}(h'_{\sigma\beta,\alpha\nu} + \xi_{\sigma,\beta\alpha\nu} + h'_{\sigma\alpha,\beta\nu} + \xi_{\sigma,\alpha\beta\nu} - h'_{\alpha\beta,\sigma\nu})$$

$$\text{(ii)} = \tfrac{1}{2}\eta^{\mu\sigma}(h'_{\sigma\nu,\alpha\beta} + \xi_{\sigma,\nu\alpha\beta} + h'_{\sigma\alpha,\beta\nu} + \xi_{\sigma,\alpha\beta\nu} - h'_{\alpha\nu,\sigma\beta}).$$

If we find the difference between the two formulae above we get

$$\underline{R^{\mu}_{\alpha\beta\nu}} = \text{(ii)} - \text{(i)} = \Gamma'^{\mu}_{\alpha\nu,\beta} - \Gamma'^{\mu}_{\alpha\beta,\nu} = \underline{R'^{\mu}_{\alpha\beta\nu}}.$$

Chapter 5

Exercise (e)

The action principle is:

$$\delta I = \int \frac{\delta(R\sqrt{-g})}{\delta g_{\mu\nu}} h_{\mu\nu}\,\mathrm{d}^4 x = -\int G^{\mu\nu}\sqrt{-g}h_{\mu\nu}\,\mathrm{d}^4 x = 0. \quad \text{(i)} \tag{7.16}$$

If we perform an infinitesimal coordinate transformation $x^{\mu} \to x^{\mu} + \xi^{\mu}$ without otherwise varying the metric, then the action I must not change:

$$0 = \delta I = \int G^{\mu\nu}(\xi_{\mu;\nu} + \xi_{\nu;\mu})\sqrt{-g}\,\mathrm{d}^4 x$$

$$= 2\int G^{\mu\nu}\xi_{\mu;\nu}\,\mathrm{d}^4 x.$$

This can be transformed in the following way:

$$\delta I = \int (G^{\mu\nu}\xi_{\mu})_{;\nu}\sqrt{-g}\,\mathrm{d}^4 x - \int (G^{\mu\nu}_{;\nu}\xi_{\mu})\sqrt{-g}\,\mathrm{d}^4 x = 0.$$

The first integral is a divergence and vanishes. The second, because of the arbitrariness of ξ_{μ}, gives the Bianchi's identities:

$$G^{\mu\nu}_{;\nu} = 0.$$

Exercise (f)

The two polarization components are $h^{xx}_{+} = -h^{yy}_{+} = A_{+}e^{-ik(t-z)}$ and $h^{xy}_{\times} = A_{\times}e^{-ik(t-z)}$. The energy flux is the negative of

$$\langle T^{(\text{GW})}_{0z}\rangle = \frac{1}{32\pi}\langle h^{ij}{}_{,0}h_{ij,z}\rangle = -\frac{k^2}{16\pi}(A^2_{+} + A^2_{\times})\langle \sin^2 k(t - z)\rangle$$

$$= -\frac{k^2}{32\pi}(A^2_{+} + A^2_{\times}).$$

PART 2

GRAVITATIONAL-WAVE DETECTORS

Guido Pizzella, Angela Di Virgilio, Peter Bender and Francesco Fucito

Chapter 8

Resonant detectors for gravitational waves and their bandwidth

G Pizzella
University of Rome Tor Vergata and Laboratori Nazionali di
Frascati, PO Box 13, I-00044 Frascati, Italy

The sensitivity of the resonant detectors for gravitational waves (GW) is discussed. It is shown that with a very low-noise electronic amplifier it is possible to obtain a frequency bandwidth up to 50 Hz. Five resonant detectors are presently in operation, with spectral amplitude sensitivity of the order of $\tilde{h} = 3 \times 10^{-22}(1/\sqrt{\text{Hz}})$. Initial results are presented, including the first search for coincidences and a measurement of the GW stochastic background.

8.1 Sensitivity and bandwidth of resonant detectors

The detectors of GW now operating [1–5] use resonant transducers (and therefore there are two resonance modes coupled to the gravitational field) in order to obtain high coupling and high-Q.

However, for a discussion on the detectors' sensitivity and frequency bandwidth it is sufficient to consider the simplest resonant antenna, a cylinder of high-Q material, strongly coupled to a non-resonant transducer followed by a very low-noise electronic amplifier. The equation for the end bar displacement ξ is

$$\ddot{\xi} + 2\beta_1\dot{\xi} + \omega_0^2\xi = \frac{f}{m} \tag{8.1}$$

where f is the applied force, m the oscillator reduced mass (for a cylinder $m = M/2$) and $\beta_1 = \omega_0/2Q$ is the inverse of the decay time of an oscillation due to a delta excitation.

We consider here only the noise which can be easily modelled, the sum of two terms: the thermal (Brownian) noise and the electronic noise. The power spectrum due to the thermal noise is

$$S_f = \frac{2\omega_0}{Q} mkT_e \tag{8.2}$$

where T_e is the equivalent temperature which includes the effect of the backaction from the electronic amplifier.

By referring the noise to the displacement of the bar ends, we obtain the power spectrum of the displacement due to Brownian noise:

$$S_\xi^B = \frac{S_f}{m^2} \frac{1}{(\omega^2 - \omega_0^2)^2 + \frac{\omega^2 \omega_0^2}{Q^2}}. \tag{8.3}$$

From this we can calculate the mean square displacement

$$\bar{\xi^2} = \frac{kT_e}{m\omega_0^2} \tag{8.4}$$

that can also be obtained, as is well known, from the equipartition of the energy.

To this noise we must add the wide-band noise due to the electronic amplifier (the contribution to the narrow-band noise due to the amplifier has already been included in T_e).

For the sake of simplicity we consider an electromechanical transducer that converts the displacement of the detector in a voltage signal

$$V = \alpha \xi \tag{8.5}$$

with the transducer constant α (typically of the order of 10^7 V m^{-1}). Thus, the electronic wide-band power spectrum, S_0, is expressed in units of V^2 Hz^{-1} and the overall noise power spectrum referred to the bar end is given by

$$S_\xi^n = \frac{2kT_e\omega_0}{mQ} \frac{1}{(\omega^2 - \omega_0^2)^2 + \frac{\omega^2 \omega_0^2}{Q^2}} + \frac{S_0}{\alpha^2}. \tag{8.6}$$

We now calculate the signal due to a gravitational wave with amplitude h and with optimum polarization impinging perpendicularly to the bar axis. The bar displacement corresponds [6] to the action of a force

$$f = \frac{2}{\pi^2} mL\ddot{h}. \tag{8.7}$$

The bar end spectral displacement due to a flat spectrum of GW (as for a delta-excitation) is similar to that due to the action of the Brownian force. Therefore, if only the Brownian noise were present, we would have a nearly infinite bandwidth, in terms of the signal-to-noise ratio (SNR). For a GW excitation with power spectrum $S_h(\omega)$, the spectrum of the corresponding bar end displacement is

$$S_\xi = \frac{4L^2\omega^4 S_h}{\pi^4} \frac{1}{(\omega^2 - \omega_0^2)^2 + \frac{\omega^2\omega_0^2}{Q^2}}. \tag{8.8}$$

We can then write the SNR as

$$\text{SNR} = \frac{S_\xi}{S_\xi^n} = \frac{4L^2\omega^4 S_h}{\pi^4 \frac{S_f}{m^2}} \frac{1}{1 + \Gamma(Q^2(1 - \frac{\omega^2}{\omega_0^2})^2 + \frac{\omega^2}{\omega_0^2})} \tag{8.9}$$

where the quantity Γ is defined by [7]

$$\Gamma = \frac{S_0\beta_1}{\alpha^2\bar{\xi}^2} \sim \frac{T_n}{\beta Q T_e} \tag{8.10}$$

T_n is the noise temperature of the electronic amplifier and β indicates the fraction of energy which is transferred from the bar to the transducer. It can readily be seen that $\Gamma \ll 1$.

The GW spectrum that can be detected with SNR = 1 is:

$$S_h(\omega) = \pi^2 \frac{kT_e}{MQv^2} \frac{\omega_0^3}{\omega^3} \frac{1}{\omega} \left(1 + \Gamma\left(Q^2\left(1 - \frac{\omega^2}{\omega_0^2}\right)^2 + \frac{\omega^2}{\omega_0^2}\right)\right) \tag{8.11}$$

where v is the sound velocity in the bar material ($v = 5400$ m s^{-1} in aluminum). For $\omega = \omega_0$ we obtain the highest sensitivity

$$S_h(\omega_0) = \pi^2 \frac{kT_e}{MQv^2} \frac{1}{\omega_0} \tag{8.12}$$

having considered $\Gamma \ll 1$.

Another useful quantity often used is the *spectral amplitude*

$$\tilde{h} = \sqrt{S_h}. \tag{8.13}$$

We remark that the best spectral sensitivity, obtained at the resonance frequency of the detector, only depends, according to equation (8.12), on the temperature T, on mass M and on the quality factor Q of the detector, provided $T = T_e$. Note that this condition is rather different from that required for optimum pulse sensitivity (see later). The bandwidth of the detector is found by imposing that $S_h(\omega)$ is equal to twice the value of $S_h(\omega_0)$. We obtain, in terms of the frequency $f = \omega/2\pi$, that

$$\Delta f = \frac{f_0}{Q} \frac{1}{\sqrt{\Gamma}}. \tag{8.14}$$

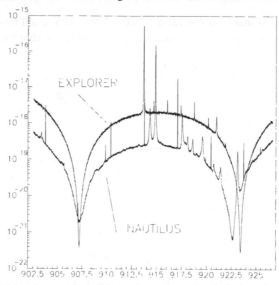

Figure 8.1. Spectral amplitudes $\tilde{h}$ present and planned (see equations (8.13) and (8.11)) using the parameters given in table 8.1 versus frequency (Hz). The presently operating detectors have bandwidth ~ 1 Hz. The bandwidth will be much larger in the near future with improved transducers.

Table 8.1. Bandwith and sensitivity for presently operating detectors and for future detectors with improved transducers.

T (K)	Γ	Q	Δf (Hz)	$\tilde{h}_{min}$ $\left(\frac{1}{\sqrt{Hz}}\right)$
0.1	10^{-6}	8.5×10^5	1.1	5.3×10^{-22}
0.1	10^{-11}	4.2×10^6	70	2.3×10^{-22}

The present detector bandwidths are of the order of 0.5 Hz, but it is expected that the bandwidths will become of the order of 50 Hz, by improving the amplifier noise temperature T_n, the coupling parameter β and the quality factor Q.

In figure 8.1 we show the spectral amplitude $\tilde{h}$ for the present aluminium resonant detectors with mass $M = 2270$ kg operating at temperature $T = 0.1$ K and the target $\tilde{h}$ planned to be reached with improved transducers. The parameters used for calculating the spectral amplitudes are given in table 8.1.

8.2 Sensitivity for various GW signals

Let us consider a signal $s(t)$ in the presence of noise $n(t)$ [8]. The available information is the sum

$$x(t) = s(t) + n(t) \tag{8.15}$$

where $x(t)$ is the measurement at the output of the low-noise amplifier and $n(t)$ is a random process with known properties. Let us start by applying to $x(t)$ a linear filter which must be such to maximize the SNR at a given time t_0 (we emphasize the fact that we search the signal at a given time t_0).

Indicating the impulse response of the filter as $w(t)$ (to be determined) and with $y_s(t) = s(t) * w(t)$ and $y_n(t) = n(t) * w(t)$, respectively, the convolutions of the signal and of the noise, we have

$$\text{SNR} = \frac{|y_s(t_0)|^2}{E[|y_n(t_0)|^2]}. \tag{8.16}$$

The expectation of the noise after the filter, indicating with $N(\omega)$ the power spectrum of the noise $n(t)$, is

$$E[|y_n(t_0)|^2] = \frac{1}{2\pi} \int_{-\infty}^{\infty} N(\omega)|W(\omega)|^2 \, d\omega \tag{8.17}$$

where $W(\omega)$ is the Fourier transform of the unknown $w(t)$.

At $t = t_0$ the output due to $s(t)$ with Fourier transform $S(\omega)$ is given by

$$y_s(t_0) = \frac{1}{2\pi} \int_{-\infty}^{\infty} S(\omega)W(\omega)e^{j\omega t_0} \, d\omega. \tag{8.18}$$

We now apply [8] the Schwartz' inequality to the integral (8.18) and using the identity

$$S(\omega)W(\omega) = \frac{S(\omega)}{\sqrt{N(\omega)}}W(\omega)\sqrt{N(\omega)} \tag{8.19}$$

we obtain

$$\text{SNR} \leq \frac{1}{2\pi} \int_{-\infty}^{\infty} \frac{|S(\omega)|^2}{N(\omega)} \, df. \tag{8.20}$$

It can be shown [8] that the equals sign holds if and only if

$$W(\omega) = \text{constant} \frac{S(\omega)^*}{N(\omega)} e^{-j\omega t_0}. \tag{8.21}$$

Applying this optimum filter to the data we obtain the maximum SNR

$$\text{SNR} = \frac{1}{2\pi} \int_{-\infty}^{\infty} \frac{|S(\omega)|^2}{N(\omega)} \, d\omega \tag{8.22}$$

where $S(\omega)$ and $N(\omega)$ as already specified are, respectively, the Fourier transform of the signal and the power spectrum of the noise at the end of the electronic chain where the measurement $x(t)$ is taken.

Let us apply the above result to the case of measurements $x(t)$ done at the end of a chain of two filters with transfer functions W_a (representing the bar) and W_e (representing the electronics).

Let S_{uu} be the white spectrum of the Brownian noise entering the bar and S_{ee} the white spectrum of the electronics noise. The total noise power spectrum is

$$N(\omega) = S_{uu}|W_a|^2|W_e|^2 + S_{ee}|W_e|^2. \tag{8.23}$$

The Fourier transform of the signal is

$$S(\omega) = S_g(\omega)W_a W_e. \tag{8.24}$$

where S_g is the Fourier transform of the GW signal at the bar entrance.

The optimum filter will have, applying equation (8.21), the transfer function

$$W(\omega) = \frac{S_g^* e^{-j\omega t_0}}{S_{uu}} \frac{1}{W_a W_e} \frac{1}{1 + \frac{\Gamma}{|W_a|^2}} \tag{8.25}$$

where

$$\Gamma = \frac{S_{ee}}{S_{uu}}. \tag{8.26}$$

Using equations (8.23) and (8.24), from equation (8.22) we obtain

$$\text{SNR} = \frac{1}{2\pi S_{uu}} \int_{-\infty}^{\infty} \frac{|S_g(\omega)|^2 \, d\omega}{1 + \frac{\Gamma}{|W_a|^2}}. \tag{8.27}$$

We now apply the above result to a delta GW with Fourier transform S_g independent of ω. The remaining integral of equation (8.27) can be easily solved if we make use of a lock-in device which translates the frequency, bringing the resonant frequency ω_0 to zero. Then the total noise becomes [7]

$$N(\omega) \Rightarrow N(\omega - \omega_0) + N(\omega + \omega_0) \tag{8.28}$$

with the bar transfer function given by

$$W_a = \frac{\beta_1}{\beta_1 + i\omega}. \tag{8.29}$$

We now estimate the signal and noise just after the first transfer function, before the electronic wide-band noise. For the signal due to a delta excitation we have

$$V(t) = \frac{1}{2\pi} \int_{-\infty}^{\infty} \frac{\beta_1}{\beta_1 + i\omega} S_g \, d\omega = S_g \beta_1 e^{-\beta_1 t} = V_s e^{-\beta_1 t} \tag{8.30}$$

where V_s is the maximum signal. For the thermal noise we have

$$V_{nb}^2 = \frac{1}{2\pi} \int_{-\infty}^{\infty} \frac{\beta_1^2}{\beta_1^2 + \omega^2} S_{uu} \, d\omega = \frac{S_{uu}\beta_1}{2} \qquad (8.31)$$

where $V_{nb}^2 = \frac{\alpha^2 kT_e}{m\omega_0^2}$ is the mean square narrow-band noise (see equations (8.4) and (8.5)).

Introducing the signal energy $E_s = \frac{1}{2}m\omega_0^2(\frac{V_s}{\alpha})^2$ we calculate (using the variable $y = \frac{\omega}{\beta_1}$)

$$\text{SNR} = \frac{1}{2}\frac{S_g^2\beta_1}{2\pi S_{uu}} \int_{-\infty}^{\infty} \frac{dy}{1 + \Gamma(1 + y^2)} = \frac{V_s^2}{8V_{nb}^2\sqrt{\Gamma}} = \frac{E_s}{4kT_e\sqrt{\Gamma}}. \qquad (8.32)$$

The factor of $\frac{1}{2}$ has been introduced because we have supposed the signal to be all at a given phase, while we add to the noise in phase the noise in quadrature, thus reducing the SNR by a factor of two. The effective noise temperature is [9]

$$T_{\text{eff}} = 4T_e\sqrt{\Gamma}. \qquad (8.33)$$

For a continuous source we can directly apply equation (8.22). With a total measuring time t_m the continuous source with amplitude h_0 and angular frequency ω_0 appears as a wavepacket with its Fourier transform at ω_0

$$S(\omega_0) = \left(\frac{h_0 t_m}{2}\right)^2 \qquad (8.34)$$

and with bandwidth

$$\delta f = \frac{2}{t_m} \qquad (8.35)$$

which is very small for long observation times. Indicating with $N(\omega_0)$ the power spectrum of the measured noise at the resonance, we obtain the amplitude of the wave that can be observed with SNR = 1:

$$h_0 = \sqrt{\frac{2N(\omega_0)}{t_m}}. \qquad (8.36)$$

A similar result has been obtained in the past [10] using a different procedure.

In practical cases it is often not possible to calculate the Fourier spectrum $N(\omega_0)$ from experimental data over the entire period of measurement t_m, either because the number of steps in the spectrum would be too large for a computer or because the physical conditions change as, for instance, a change in frequency due to the Doppler effect. It is then necessary to divide the period t_m in n sub-periods of length $\Delta t = \frac{t_m}{n}$. In the search for a monochromatic wave we then have to consider two cases: (a) The wave frequency is exactly known. In this case we can combine n Fourier spectra in one unique spectrum taking into consideration also the phase of the signal. The final spectrum has the same characteristics of the spectrum over the entire period t_m and equation (8.36) still applies. (b) The exact frequency is unknown. In this case when we combine the n spectra we lose information on the phase. The result is that the final combined spectrum over the entire period has a larger variance and the left part of equation (8.36) has to be changed to

$$h = \sqrt{\frac{2N(\omega_0)}{\sqrt{t_m \Delta t}}}.$$ (8.37)

We come now to the measurement of the GW stochastic background [11, 12, 16]. Using one detector, the measurement of the noise spectrum corresponding to equation (8.11) only provides an upper limit for the GW stochastic background spectrum, since the noise is not so well known that we can subtract it. The estimation of the GW stochastic background spectrum can be considerably improved by employing two (or more) antennae, whose output signals are cross correlated. Let us consider two antennae, that may in general be different, with transfer functions T_1 and T_2, and displacements ξ_1 and ξ_2: the displacement cross-correlation function

$$R_{\xi_1 \xi_2}(\tau) = \int \xi_1(t + \tau)\xi_2(t)\,dt$$ (8.38)

only depends on the common excitation of the detectors, due to the GW stochastic background spectrum S_{gw} acting on both of them, and is not affected by the noises acting independently on the two detectors.

The Fourier transform of equation (8.38) is the displacement cross spectrum. This spectrum, multiplied by $T_1 T_2 (4L^2/\pi^4)$, is an estimate of the gravitational background S_{gw}. The estimate, obtained over a finite observation time t_m, has a statistical error. It can be shown [11] that

$$\delta S_{gw}(\omega) = S_{gw}(\omega) = \frac{\sqrt{N_{h1}(\omega)N_{h2}(\omega)}}{\sqrt{t_m \delta f}}$$ (8.39)

where t_m is the total measuring time and δf is the frequency step in the power spectrum. From equation (8.11) we have the obvious result that, for resonant detectors, the error is smaller at the resonances. If the resonances of the two detectors coincide the error is even smaller. In practice, it is better to have two detectors with the same resonance and bandwidth. If one bandwidth is smaller, the minimum error occurs in a frequency region overlapping the smallest bandwidth.

From the measured S_{gw} one can calculate the value of the energy density of the stochastic GW referred to as the critical density (the energy density needed for a closed universe). We have

$$\Omega = \frac{4\pi^2}{3}\frac{f^2}{H^2}S_{gw}(f)$$ (8.40)

where H is the Hubble constant.

Table 8.2. Sensitivity of the resonant detectors in operation.

Resonance frequency (Hz)	$\tilde{h} = \sqrt{S_h}$ at resonance $\left(\frac{1}{\sqrt{Hz}}\right)$	Frequency bandwidth δf (Hz)	Minimum h for 1 ms bursts	Minimum h for continuous waves	Minimum Ω
900–700	$3\text{–}20 \times 10^{-22}$	0.5–1	4×10^{-19}	2×10^{-25}	0.1

Table 8.3. Target sensitivity for Auriga and Nautilus.

$\tilde{h} = \sqrt{S_h}$ at resonance $\left(\frac{1}{\sqrt{Hz}}\right)$	Frequency bandwidth δf (Hz)	Minimum h for 1 ms bursts	Minimum h for continuous waves	Minimum Ω
2×10^{-22}	50	3×10^{-21}	2×10^{-26}	10^{-4}

8.3 Recent results obtained with the resonant detectors

The present five cryogenic bars in operation [1–5] (Allegro, Auriga, Explorer, Nautilus and Niobe) have roughly the same experimental sensitivity as given in table 8.2.

Niobe, made with niobium, has a resonance frequency of 700 Hz, the other ones with aluminium, have resonance frequencies near 900 Hz. These minimum values for monochromatic waves and for the quantity Ω have been estimated by considering one year of integration time (for Ω we suppose to use the cross-correlation of two identical antennae).

The burst sensitivity for all bars can be increased by improving the transducer and associated electronics. It has been estimated that these improvements can increase the frequency bandwidth up to 50 Hz.

In addition to increasing the bandwidth, Auriga and Nautilus can improve (see table 8.3) their spectral sensitivity by making full use of their capability to go down in temperature to $T = 0.10$ K. At present the major difficulty is due to excess noise, sometimes of unknown origin, and work is in progress for eliminating this noise.

The search for signals due to GW bursts is done after the raw data have been filtered with optimum filter algorithms [13, 14]. These algorithms may have various expressions but they all have in common an optimum integration time that is roughly the inverse of the detector bandwidth and all give approximately the same value of T_{eff} (all algorithms being optimal filters for short bursts).

The data recorded by the various detectors are now being analysed, searching, in particular, for coincidences above the background. The coincidence

Table 8.4. Results of a coincidence search between data from Explorer and Nautilus. See the text for an explanation.

Number of days	Number of Explorer events	Number of Nautilus events	$\langle n \rangle$	n_c	$P_{poisson}$ (%)	P_{exp} (%)
29.2	8527	5679	11.0	19	1.5	1.44

technique is a powerful mean for reducing the noise. As an example we show in table 8.4 some results obtained recently [15] by searching for coincidences between Explorer and Nautilus during 1995 and 1996.

In the first column we give the number of days when both antennae were operating. The small number of useful days shows that it is difficult to keep a GW antenna in operation continuously with good behaviour. In future it should be possible to increase the useful time to 70% of the total time, considering that some time is always lost for cryogenic maintenance. In the second and third column we show the number of *candidate events*. The candidate events are obtained by introducing a proper threshold on the data filtered with an optimum filter for short burst detection. We notice the large number of candidate events that make it practically impossible, using one detector alone, to search for a particular signal due to a GW. A big improvement is obtained by the comparison of at least two detectors.

In the fourth column we give the expected number of accidental coincidences measured by means of 10 000 shifts of the event times of one detector with respect to the other and using a coincidence window of $w = \pm 0.29$ s (one sampling time for EXPLORER and NAUTILUS). This number of accidentals is small enough to start considering the possibility of searching for a coincidence excess (though, according to astrophysical expectations, this excess should be much smaller than the observed accidentals). In this case the number of coincidences n_c, reported in column five, turns out to be slightly larger than the expected number of accidentals.

Finally, in column six we report the probability calculated with the Poisson formula and in column seven the *experimental probability*, obtained by counting how many times we had a number of accidental coincidences equal or larger than n_c and dividing this number by the number of trials, 10 000. The agreement between theoretical values and experimental values is good, indicating a poissonian statistical behaviour of the data.

A new and interesting result has recently been obtained by cross-correlating the data recorded with EXPLORER and NAUTILUS [17]. The measured spectral amplitudes of Explorer and Nautilus, shown in figure 8.2, were correlated. The result of such cross-correlation at a resonance of 907.2 Hz (the same for the two detectors) has been the determination of an upper limit for Ω that measures the

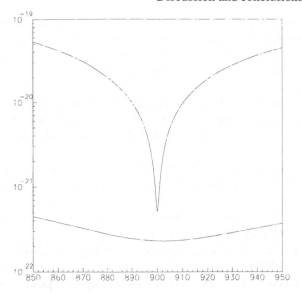

Figure 8.2. Spectral amplitude $\tilde{h}$ of EXPLORER and NAUTILUS versus frequency (Hz).

closure of the universe. It has been found, using equation (8.40), that $\Omega \geq 60$, where a factor of six has been also included for taking into consideration the fact that EXPLORER and NAUTILUS are separated by about 600 km. The value of Ω we have obtained is much larger than the expected value, but we remark that it is the first measurement of this type made with two cryogenic resonant GW detectors.

Finally, we would like to mention that recently, making use of the detector NAUTILUS, we have been able to observe the passage of cosmic rays[1]. The observed events in the bar have amplitude and frequency of occurrence in agreement with the prediction (event energy of the order of 1 mK), showing that NAUTILUS is properly working. In particular, the efficiency of the filter aimed at detecting a small signal embedded into noise is well proven.

8.4   Discussion and conclusions

In previous literature the sensitivity for resonant detectors of gravitational waves has usually been expressed in terms of T_{eff}, the minimum energy delivered by a GW burst that can be detected by the apparatus. However, what the resonant detectors really measure is essentially the Fourier transform (over a certain frequency band) of the GW adimensional amplitude.

For studying the operation of a resonant antenna as a GW detector of stochastic GW we had to deal with noise spectrums. This has made us reconsider

---

[1] This result has been presented at the School by Dr Evan Mauceli.

the sensitivity to bursts and other types of GW in a somewhat different manner, which improved our understanding of the role played by the electromechanical transducer and its associated electronics. The noise spectrum of the apparatus is expressed by equation (8.11), which also directly gives the sensitivity for the GW background.

We notice that the optimum sensitivity (at the resonance) depends essentially on the ratio T_e/MQ and is independent of the transducer. In practice, the transducer and electronics determine only the bandwidth of the apparatus, expressed by equation (8.14).

For the measurement of the GW stochastic background, that should not change drastically in a frequency band of a few hertz, it might therefore be sufficient to make use of a very simple transducer and even small bandwidth electronics. This makes the resonant detectors, in particular, well suited for measuring the GW stochastic background.

The use of a sophisticated transducer, with larger bandwidth, improves the sensitivity to GW bursts. As far as the search for monochromatic waves a larger bandwidth is better, in the sense that it allows a larger frequency region in which to search.

References

[1] Astone P *et al* 1993 *Phys. Rev.* D **47** 362
[2] Mauceli E *et al* 1996 *Phys. Rev.* D **54** 1264
[3] Blair D G *et al* 1995 *Phys. Rev. Lett.* **74** 1908
[4] Astone P *et al* 1997 *Astropart. Phys.* **7** 231–43
[5] Cerdonio M *et al* 1995 Gravitational wave experiments *Proc. First Edoardo Amaldi Conf. (Frascati, 1994)* ed E Coccia, G Pizzella and F Ronga (Singapore: World Scientific)
[6] Pizzella G 1975 *Riv. Nuovo Cimento* **5** 369
[7] Pallottino G V and Pizzella G 1998 *Data Analysis in Astronomy* vol III, ed V Di Gesu *et al* (New York: Plenum) p 361
[8] Papoulis A 1984 *Signal Analysis* (Singapore: McGraw-Hill) pp 325–6
[9] Pizzella G 1979 *Nuovo Cimento* **2** C209
[10] Pallottino G V and Pizzella G 1984 *Nuovo Cimento* **7** C155
[11] Bendat J S and Piersol A G 1966 *Measurement and Analysis of Random Data* (New York: Wiley)
[12] Brustein R, Gasperini M, Giovannini M and Veneziano G 1995 *Phys. Lett.* B **361** 45–51
[13] Astone P, Buttiglione C, Frasca S, Pallottino G V and Pizzella G 1997 *Nuovo Cimento* **20** 9
[14] Astone P, Pallottino G P and Pizzella G 1998 *Gen. Rel. Grav.* **30** 105
[15] Astone P *et al* 1999 *Astropart. Phys.* **10** 83–92
[16] Astone P, Pallottino G V and Pizzella G 1997 *Class. Quantum Grav.* **14** 2019–30
[17] Astone P *et al* 1999 *Astron. Astrophys.* **351** 811–14

Chapter 9

The Earth-based large interferometer Virgo and the Low Frequency Facility

Angela Di Virgilio
INFN Sezione di Pisa, via Vecchia Livornese 1265,
56010 S. Piero a Grado (Pisa), Italy
E-mail: angela.divirgilio@pi.infn.it

The principle of the Earth-based interferometer and the parameters of Virgo, the CNRS-INFN antenna under construction in Cascina near Pisa, are presented. The Super-Attenuator (SA), the suspension ad hoc designed for Virgo, and the Low Frequency Facility, the project devoted to study the SA with a sensitivity close to the Virgo one, are shortly described.

9.1 Introduction

The detection of gravitational waves (GW), produced in astrophysical processes, is one of the most challenging fields of modern physics, aimed at finding gravitational-wave astronomy. Lectures about GW production and expected sources can be found in this book, but let me very briefly remind you of the essentials. General relativity predicts that accelerated matter produces gravitational waves, ripples of spacetime travelling with the speed of light. GW can be defined as a perturbation of the metric, $h_{\alpha\beta}$, and the observational effect of its passage through the Earth is a change among the relative distance of massive bodies. If A and B are the space-state locations of the two testmasses and if we set $AB = (x^\alpha)$, the vector (x^α) obeys the equation of geodesic deviation in weak field:

$$\frac{\mathrm{d}^2 x^\alpha}{\mathrm{d}t^2} = \frac{1}{2}\frac{\partial^2 h_{\alpha\beta}^{\mathrm{TT}}}{\partial t^2}x^\beta \qquad (9.1)$$

where TT denotes the transverse-traceless gauge. The maximum change in the distance AB is then:

$$\delta L = \frac{h}{2} L \qquad (9.2)$$

where h is the dimensionless gravitational-wave amplitude and L is the distance AB at rest. One of the most important gravitational-wave sources is the SuperNovae explosion. In order to detect several SuperNovae events per year, it is necessary to have a sensitivity of h close to 10^{-21} for a millisecond signal; in fact this is the expected amplitude for gravitational waves from SuperNovae coming from the closest cluster of galaxies: Virgo, distant 10–20 Mpc. The idea of using an interferometer to detect gravitational waves has been proposed independently, about 30 years ago, by Weber and two Russian physicists, Gerstenstein and Pustovoit [1] and a lot of work has been done to develop the optical design of the large interferometers presently under construction [2, 3]. Let me mention (see, for example, [4]) the work done in Garching, Glasgow, Caltech and MIT, where interferometers of 30–40 m arms have been developed and are still running with displacement sensitivity equal to the shot noise limit. The power spectrum of the displacement sensitivity obtained so far is of the order of 10^{-19} m/$\sqrt{\text{Hz}}$. Following the above equation it is straightforward to see that a few kilometre armlength is necessary in order to reach the sensitivity of $h = 10^{-21}$ for the millisecond signal. Several antennae are under construction [5]: two antennae 4 km arms in the USA (LIGO), one 0.6 km in Germany (GEO600), one in Japan 0.3 km (TAMA), in Italy 3 km (Virgo) and one is planned with 3 km arms in Australia (ACIGA). Most of them (TAMA, LIGO and GEO) have been constructed and are making working together parts of the interferometer. TAMA is already operational, and work is concentrated in order to reach the planned sensitivity; moreover Japan already plans to build a longer one. The above interferometers, all Earth based, will cover the 10–10 000 Hz window, where events from SuperNovae explosions, coalescing binaries and pulsars events are expected. Using the signals of the resonant bars, and the other kind of detectors already operational, in the near future gravitational-wave astronomy is expected to become a reality. Furthermore, a big space antenna is planned to be launched in this decade, it is called LISA [5], and will cover the frequency spectrum 10^{-4}–10^{-1} Hz. In the following the essential characteristics of the Virgo antenna [6,7], the Virgo SuperAttenuator suspension, the Low Frequency Facility [9] and the R and D experiment of the Virgo project, are described.

9.1.1 Interferometer principles and Virgo parameters

Figure 9.1 shows a simple Michelson interferometer: the passage of a gravitational wave changes the phase shift between the two outgoing beams interfering on the beam splitter. For the sake of simplicity it is assumed that the wave is optimally polarized and travels perpendicularly to the interferometer

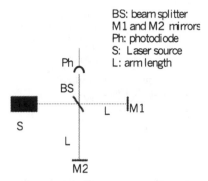

BS: beam splitter
M1 and M2 mirrors
Ph: photodiode
S: Laser source
L: arm length

Figure 9.1. The simple Michelson.

plane, moreover the Michelson is assumed ideal [10, 11] (contrast $C = 1$):

$$\delta\phi = \pi + \alpha + \Phi_{gw}. \tag{9.3}$$

Here α is the offset between the two interferometer arms and Φ_{gw} is the effect of the gravitational wave. The transmitted outgoing power P_{out} has a simple relation with the input power P_{in}.

$$P_{out} = P_{in} \sin^2 \frac{\alpha + \Phi_{gw}}{2} \tag{9.4}$$

for $\Phi_{gw} \ll 1$

$$P_{out} \simeq P_{in} \left[\sin^2 \frac{\alpha}{2} + \frac{1}{2} \sin\alpha \Phi_{gw} \right]. \tag{9.5}$$

Equation (9.5) clearly states that P_{out} is proportional to Φ_{gw}, the gravitational-wave signal, and the fundamental limit of the measurement comes from the fluctuation of the incoming power, i.e. the fluctuation of the number of photons. For a coherent state of light, the number of photons fluctuates with Poissonian statistics, and it can be shown that the signal-to-noise ratio (SNR), when $C = 1$, is:

$$SNR = \sqrt{\frac{P_{in}\Delta t}{\hbar\omega}} \cos a \Phi_{gw} \tag{9.6}$$

where ω is the circular frequency of the laser light, Δt is the time of the measure and $\hbar$ is the Planck constant. The SNR is maximum when $\alpha = 0$ and the interferometer is tuned to a dark fringe. The minimum phase shift detectable, per unit time, is evaluated assuming SNR = 1

$$\Phi_{gw}^{min} = \sqrt{\frac{\hbar\omega}{P_{in}}} \tag{9.7}$$

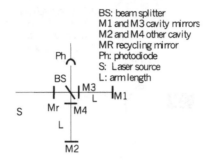

Figure 9.2. The interferometer with Fabry–Perot cavity and power recycling.

and it is called the shot noise limit. The contrast $C < 1$ does not change this result. The large interferometer design has been optimized in order to enhance the sensitivity: inside the arm there is a resonant Fabry–Perot cavity; this is equivalent to increasing the interferometer armlength, by a coefficient proportional to the finesse of the Fabry–Perot cavity. Moreover, as shown in equation (9.5), the SNR increases with the square root of the power and, since the output is at the dark fringe, the unused light, reflected back from the interferometer, is re-injected with a technique called recycling. Figure 9.2 shows the basic scheme of most of the interferometers under construction [3].

   The other fundamental noise of the interferometer is the so called 'thermal noise' [11, 12]. The Virgo test masses are in thermal equilibrium with the environment. An average energy of kT, where k is the Boltzmann constant and T is the temperature, is associated with each degree of freedom, giving origin, through the 'fluctuation dissipation theorem', to position fluctuations of the macroscopic coordinates. Using the equation of motion and the fluctuation-dissipation theorem, the displacement noise is estimated. There are two main dissipation mechanisms: the structural and the viscous damping. It is assumed that mechanisms of viscous damping are pretty much reduced, and only the structural one survives. Internal friction in materials is described by adding an imaginary term to the elastic constant, and the equation of motion for a spring becomes:

$$m\ddot{x} + K(1 + i\Phi)x = 0 \qquad (9.8)$$

where x is the coordinate of interest, m is the mass associated with the oscillator, K is the elastic constant and Φ takes care of the dissipation and is related to the usual Q of the oscillator.

   As most of the interferometers under construction, Virgo [6, 7] will be a recycled interferometer, with resonant Fabry–Perot cavity in each arm. The arm length is 3 km, the recycling factor 40 and the finesse of the cavity 40; the light source is a 25 W Nd:YAG laser ($\lambda = 1.06 \ \mu$m). The Virgo Project is under construction in Italy at Cascina, near Pisa, by CNRS (France) and INFN (Italy). The participating laboratories are: ESPCI-Paris, Nice, LAL-

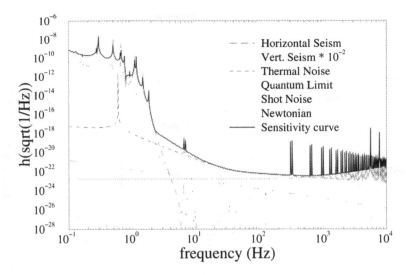

Figure 9.3. Virgo sensitivity curves.

Orsay, LAPP-Annecy, INPN-Lyon for CNRS and Firenze, Frascati, Napoli, Perugia, Pisa, Roma for INFN. The aim is to reach, in a useful detection bandwidth between a few hertz up to a few kilohertz, a sensitivity which allows us to detect gravitational waves emitted by coalescing binary stars, gravitational collapses, spinning neutron stars or constituting the stochastic background of the gravitational radiation. The Virgo sensitivity curve is shown in figure 9.3, as a function of frequency; the vertical scale shows the linear spectral density of the minimum detectable dimensionless gravitational amplitude: at low frequency there is the pendulum thermal noise, then the mirror substrate thermal noise and at high frequency the shot noise of the laser light, the interferometer sensitivity decreases at high frequency for the high storage time of the 3 km Fabry–Perot cavity. Figure 9.3 shows the thermal and shot noise and the other important noise of the Virgo antenna.

The laser has to be frequency stabilized at the level of 10^{-1} Hz/$\sqrt{}$(Hz) in the frequency band of interest [8], the mirror losses have to be below 1 ppm in order to obtain the required recycling factor and cavity finesse; moreover, power stabilization is required, and a mode cleaner 144 m long, with finesse 1000, is necessary in order to have a good quality TM_{00} mode injected into the interferometer, and to contribute to the amplitude stability, since such a long mode cleaner has a pole at 10 Hz. All the optical components will be suspended in vacuum by anti-seismic suspensions called Super Attenuators (SA); the optical path will be completely under vacuum. The vacuum system will consist of: two orthogonal vacuum tubes, 3 km long, 1.2 m in diameter, containing the arm optical paths; one vacuum tube, 144 m long, 0.3 m in diameter for the mode cleaner

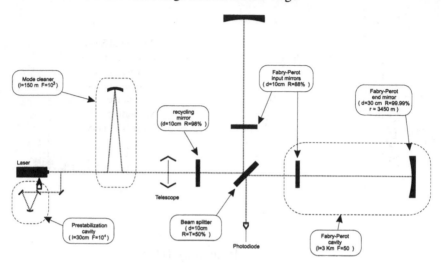

Figure 9.4. Scheme of the Virgo interferometer.

cavity; ten vertical vacuum towers, up to 11 m high, 2 m in diameter, containing the different SAs. Figure 9.4 shows the Virgo scheme. The construction of Virgo is planned in two main steps: the realization of the central interferometer by late 2000, and the realization of the 3 km arm full interferometer by late 2002.

9.2   The SA suspension and requirements on the control

Roughly speaking, above a few hertz, the seismic noise is a function of the frequency ν and goes as ν^{-2}. In our site the measured spectrum is about 10^{-6}– $10^{-7}/\nu^2$ m/$\sqrt{\text{Hz}}$, in all degrees of freedom; this implies a factor of at least 10^{10} of the noise reduction at 10 Hz. An harmonic oscillator, with resonant frequency ν_0 acts as a low pass filter if N oscillators are connected in cascade, and for frequencies above the resonant frequencies, the displacement noise of the attachment point is transmitted to the test-mass with a transfer function going as $\sim 1/\nu^{2N}$. Each SA is a multistage pendulum [13], acting as a multipendulum in all six degrees of freedom. The basic tool is a piece called a mechanical filter. It is essentially a rigid steel cylinder (70 cm in diameter, 18 cm in height and with a mass of about 130 kg), with a thin steel wire. In this way the horizontal oscillator in two degrees of freedom and the torsion spring are made. The attachment points of the two wires are connected close to the centre of mass of the filter: in this way the pitch spring is formed. The vertical string is a mechanical one, a triangular curled piece of a special steel (MARAGING). The SA blades are designed to hold about 50 kg, at rest they are bent and when loaded act as a spring and become flat. Each filter of the SA is equipped with a number of blades from 12–4, in function of the load. Each filter of the SA is composed of two main

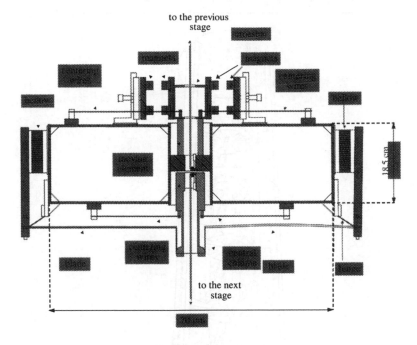

Figure 9.5. The filter.

parts: the vessel, roughly speaking a cylinder with a hole in the centre and
movable part, composed of a piece called 'cross' attached to a tube called 'centi
column', rigidly connected together. The 'central column' is free to move aloi
the vertical inside the vessel, for about a centimetre. The upper wire is attach
to the vessel, while the lower is attached to the column. The two pieces ε
constrained to move along the vertical by means of centring wires. The ba
of each blade is rigidly attached to the outer diameter of the vessel, while the ti
are connected to the central column. When the filter is loaded the blades act a:
spring, and are free to move for several millimetres, with a resonant frequen
above 1 Hz. In order to decrease the vertical resonant frequency a magnei
antispring acts between the vessel and the 'cross', in parallel with the blad(
It is composed of an array of permanent magnets attached to the 'cross', plac
in front of an equal array of magnets attached to the vessel; the two array ε
ad hoc built in order to make repulsive force, proportional to the vertical offs(
for a few millimetres. The effective elastic constant is the difference between t
blade spring and the magnetic antispring, and each filter is tuned to oscillate
∼400 mHz. In figure 9.3 with the other noise sources, the seismic motion of t
Virgo test masses is shown [14], it is possible to see that it dominates below 2 F
and the resonance peaks of the SA are visible.

Ground motion can be of the order of hundreds of micrometres, due to tic
deformation of the Earth's crust and thermal effects on the surface. In order

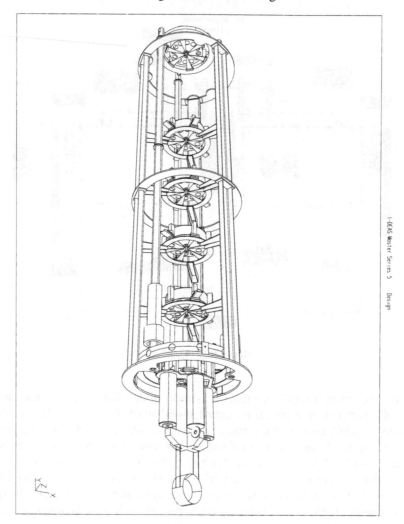

re 9.6. The complete SA suspension, some details of the last stage control are shown.

the interferometer running and compensate such a large drift, the SA is pped with a special stage called an inverted pendulum (IP). It is composed ree aluminium legs, about 6 m tall. When it is loaded with the 1.2 ton of the it becomes a very soft spring, with about 30 mHz frequency. With magnets op of the IP and coils attached to the ground, forces are produced, and it is ible to displace to SA by a few centimetres. It is necessary not to introduce nic noise through the forces to be applied to the test mirror in order to keep nterferometer in its working position. This is an important task of the SA, all the control forces are done by means of coils and magnets acting between different stages of the SA. The last part of the SA is composed of steering filters, which support four coils, the marionetta, a cross shaped piece of steel, which supports four magnets, positioned in the form of four coils, see figure 9.6. In this way the mirror can translate along the beam direction and along the vertical, rotate around the vertical axis and around the horizontal axis perpendicular to the beam. The test-mass (the interferometer mirror) and the reference mass are attached as pendulums to the marionetta [15]. Magnets and coils produce forces between the reference mass and the mirror.

The control has two main parts: the damping and the mirror control. The damping is a local control, which reduces the large motion of the SA, due to the resonance frequencies below 1 Hz: on top of the inverted pendulum feedbacks are produced using accelerometer signals and electromagnetic actuators [16]. The other control uses the interferometer signal itself and it is aimed to keep aligned the whole Virgo interferometer in its working point. The actuators, couples of magnet-coil, are divided per frequency band: below 100 mHz the feedback goes to the IP actuators, in the intermediate band to the marionetta, and above to the reference mass. This kind of control, called hierarchical is necessary for the huge displacement range required and the necessity not to induce electronic noise at the mirror level.

9.3   A few words about the Low Frequency Facility

Much work is being done around interferometers at the moment, but the planned sensitivity is not enough to guarantee the detection of events. A large community [17] is working hard to improve the sensitivity of the antennae. LIGO plans to make an upgrade in five years from now. The effort is mainly concentrated on improving the suspension and keeping as low as possible the thermal noise of pendulum and mirror substrate. With the aim of reducing thermal noise, several methods have been proposed, like cryogenic cooling, choice of low loss material for the test mass, as zaffire or silicon, and novel cancellation techniques. Figure 9.3 shows that, below 600 Hz, Virgo is entirely dominated by the thermal noise of the suspension. The improvement below 600 Hz implies changes in the suspension. In order to demonstrate that a new choice, once applied to the antenna, makes a real improvement in the sensitivity of the whole antenna it is necessary to make tests in conditions as close as possible to the real one. The Low Frequency Facility (LFF) [9, 19], has been conceived in order to measure the noise of test masses attached to the SA, with a sensitivity at 10 Hz close to the Virgo one: 10^{-18} m/$\sqrt{\text{Hz}}$. The goal is to take a measurement of the thermal noise of the Virgo suspension, while the system is under active control of the low-frequency motion. In this way, it will be possible to characterize the suspension in an environment close to the real one.

The use of a Fabry–Perot cavity as a displacement transducer relies on the excellent frequency stabilization of the laser beam. A frequency stabilization, for a Nd-YAG, of the order of 10^{-1}–10^{-2} Hz/$\sqrt{\text{Hz}}$ at 10 Hz has been obtained, for example by the Nice-Virgo team [8]. A copy of this circuit provides the frequency stabilization for the LFF. It is already operational and has obtained a closed loop behaviour similar to the Virgo prestabilization circuit. The technique to reduce the influence of power fluctuation is to extract signals from the Fabry–Perot with the well known Pound–Drever–Hall, where the injected light is phase modulated at high frequency (usually around 10 MHz, where the laser is shot noise limited). The signal is extracted in reflection, the noise is due to the shot noise, and the SNR for the cavity detuning δv is:

$$\text{SNR} = \frac{P_{\text{in}} 8 J_0(m) J_1(m) F (1 - \frac{A_{\text{loss}} F}{\pi}) \frac{2L}{c} \delta v}{2\sqrt{\eta P_{\text{DC}} h v_0}}. \tag{9.9}$$

where P_{in} represents the incident power on the cavity, A_{loss} the total loss in the cavity during one round trip, P_{DC} is the DC term of the power reflected from the cavity, $\sqrt{2\eta P_{\text{DC}} h v}$ is the shot noise expression with η the quantum efficiency of the photodiode, h is the Planck constant, v_0 is the laser frequency, L is the cavity length, F is the finesse, c is the speed of light and J_0 and J_1 are Bessel functions, functions of the modulation depth m. The LFF cavity parameters are: Finesse 3000, $P_0 = 30$ mW, cavity length $L = 1$ cm. With these parameters, the sensitivity at 10 Hz is limited by the frequency fluctuation, and a power spectrum density of 10^{-1} Hz/$\sqrt{\text{Hz}}$ is necessary in order to reach the displacement sensitivity of 10^{-18} m/$\sqrt{\text{Hz}}$ at 10 Hz. A prototype of the SA, called R and D SA, has been assembled in Pisa, and has tested the assembly procedure and developed the control loop. LFF will suspend to the R and D SA two mirrors, in a way to create a very high finesse Fabry–Perot; of whose element, one is the standard Virgo mirror, with a mass above 20 kg, while the other [18], usually called (AX), will be different. In the early phase of the experiment the auxiliary mirror will be a standard small mirror; the auxiliary mirror will be loaded with an extra mass, in order to reduce the displacement noise due to low-frequency radiation pressure. The Fabry–Perot cavity will act as a displacement transducer and allow the measurement of the combined thermal noise of the two mirrors, and related suspension. Figure 9.7 shows the experimental apparatus: the R and D SA holding the two mirrors of the cavity and the table with the optical circuit, which will inject light into the cavity.

9.4   Conclusion

Several gravitational-wave interferometers will be operational in a few years, and a lot of effort is being concentrated in building and making them work. Expectations are great around such experiments, since they can give essential information on general relativity, fundamental physics and astrophysics. The Virgo antenna will be operational in 2–3 years from now. A large study is going on around the SA suspension. A special prototype called R and D SA is under

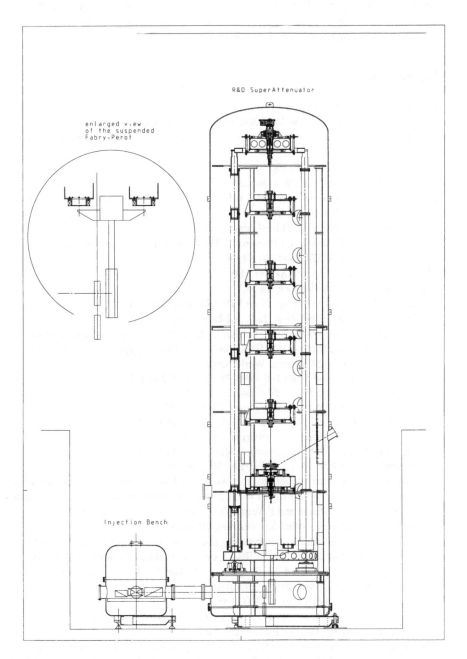

Figure 9.7. The R and D SA and the injection table of the LFF, inside vacuum tanks; the enlarged picture shows the two mirrors forming a Fabry–Perot cavity.

test in Pisa. The R and D SA is part of the LFF experiment, aimed to measure thermal noise of the test mass suspended to the R and D SA, and to become a test area for future improved antenna suspensions.

References

[1] Weber J 1960 *Phys. Rev.* **117** 306
    Gerstenstein and Pustovoit 1963 *Sov. Phys.–JETP* **16** 433
[2] Blair D G (ed) 1991 *The Detection of Gravitational Waves* (Cambridge: Cambridge University Press)
[3] Giazotto A *Phys. Rev.* **182** 365
[4] Shoemaker D *et al* 1988 *Phys. Rev.* **38** 423
[5] See articles in: 1999 *Gravitational Waves, 3rd Edoardo Amaldi Conf. (Caltec, July)* (Singapore: World Scientific)
[6] Brillet A *et al* 1992 VIRGO final conceptual design
    Gamaitoni L, Kovalik J and Punturo M 1997 The VIRGO sensitivity curve *VIRGO Internal Note* NOT-PER-1390-84, 30-3
[7] 1997 *Virgo—Final Design Report* May
[8] Bondu F *et al* 1996 *Opt. Lett.* **14** 582
[9] Di Virgilio A *et al* 1997 *Virgo Note* VIR-NOT-PIS-8000-101
    Bernardini M *et al* 1998 *Phys. Lett.* A **243** 187–94
[10] Hello P 1998 *Progress in Optics* vol XXXVIII, ed E Wolf (Amsterdam: Elsevier)
[11] See, for example, Saulson P R 1994 *Fundamentals of Interferometric Gravitational Wave Detectors* (Singapore: World Scientific)
[12] Bondu F, Hello P and Vinet J-Y 1998 *Phys. Lett.* A **246** 227–36
    Levin Yu 1998 *Phys. Rev.* D **57** 659–63
[13] Braccini S *et al* 1996 *Rev. Sci. Instrum.* **67** 2899–902
    Beccaria M *et al* 1997 *Nucl. Instrum. Methods* A **394** 397–408
    Braccini S *et al* 1997 *Rev. Sci. Instrum.* **68** 3904–6
    De Salvo R *et al* 1999 *Nucl. Instrum. Methods* A **420** 316–35
    Losurdo G *et al* 1999 *Rev. Sci. Instrum.* **70** 2507–15
[14] Vicerè A 2000 Introduction to the mechanical simulation of seismic isolation systems *Experimantal Physics of Gravitational Waves* ed M Barcone, G Calamai, M Mazzoni, R Stanga and F Vetrano (Singapore: World Scientific)
[15] Bernardini A *et al* 1999 *Rev. Sci. Instrum.* **70** 3463–71
[16] Losurdo G 2000 Active controls in interferometric detectors of gravitational waves: inertial damping of VIRGO super attenuators *Experimantal Physics of Gravitational Waves* ed M Barcone, G Calamai, M Mazzoni, R Stanga and F Vetrano (Singapore: World Scientific)
    Losurdo G 2000 *Preprint* gr-qc/00020006
[17] Contribution to the *Aspen Winter Conf. on Gravitational Waves, and their Detection* can be found on: http://www.ligo.caltech.edu/
[18] Cella G *et al* 1999 Suspension for the Low Frequency Facility *Phys. Lett.* A accepted
[19] Benabid F *et al* 2000 The Low Frequency Facility, R&D experiment of the Virgo Project *J. Opt.* B accepted (special issue)

Chapter 10

LISA: A proposed joint ESA–NASA gravitational-wave mission

Peter L Bender, on behalf of the LISA Study and Mission Definition Teams
JILA, National Institute of Standards and Technology and University of Colorado, Boulder, Colorado 80309-0440, USA

10.1 Description of the LISA mission

10.1.1 Introduction

The evidence for supermassive black holes at the centres of quasars and active galactic nuclei (AGNs) has been strong but not conclusive for several decades. Recently, the evidence for massive black holes in one particular AGN, in our own Galaxy, and in the Local Group Galaxy M32 has become extremely convincing. Thus, questions concerning gravitational waves generated by the interaction of massive black holes with smaller compact objects and with each other have become of strong interest. Signals from such sources also are likely to provide the strongest possible tests of general relativity.

Plans are being developed in both Europe and the USA for flying a dedicated gravitational wave mission called the Laser Interferometer Space Antenna (LISA). The antenna will measure gravitational waves in the frequency range from roughly 1 μHz to 1 Hz, and thus will strongly complement the results expected from ground-based detectors. The primary objectives are to obtain unique new information about massive black holes throughout the universe, and to map the metric around massive black holes (MBHs) with much higher accuracy than otherwise would be possible. Other important objectives include studies of resolved signals from thousands of compact binary star systems in our Galaxy, and looking for a possible gravitational-wave cosmic background at millihertz frequencies.

The possibility of making sensitive gravitational-wave measurements at low frequencies by laser interferometry between freely floating test masses in widely separated spacecraft appears to have been first suggested in print about 1972 [1,2]. More extensive discussions started in 1974. An initial proposal similar to that for the LISA mission was presented at the *Second International Conference on Precision Measurement and Fundamental Constants* in 1981 [3], and at the *ESA Colloquium on Kilometric Optical Arrays in Space* in 1984 [4]. Work on the concept was supported initially by the National Bureau of Standards, and later in the USA by NASA. However, the concept became much better defined and widely known in the 1993–1994 period, when more extensive studies were carried out in Europe under ESA support. Since then, further studies have been carried out by both ESA and NASA of a proposed joint ESA–NASA mission, that could fly as early as 2010 if all goes well.

The currently proposed mission is described in section 1 of this chapter. This includes the overall antenna and spacecraft design, the optics and interferometry system, the free mass sensors, the required micronewton thrusters for the spacecraft, and the mission scenario. The emphasis will be on aspects of the antenna that are quite different from those for ground-based detectors. Section 2 describes the main scientific results that seem likely to be obtained by LISA. This includes unique new information on three major astrophysical questions concerning MBHs and a nearly ideal test of general relativity, as well as the detection of thousands of compact binaries in our Galaxy. In addition, some speculations will be given on possible future prospects for gravitational-wave observations in space after the LISA mission.

Much more information on most of the above subjects can be found in 'Laser Interferometer Space Antenna', the *Proceedings of the Second International LISA Symposium* [5], in the *LISA Pre-Phase A Study* [6], and in a special issue of *Classical and Quantum Gravity* [7], which is the *Proceedings of the First International LISA Symposium*. More recent information from the 1999–2000 ESA Industrial Study of the LISA mission is being provided in the report of that study and in several papers in the *Proceedings of the Third International LISA Symposium* (Albert Einstein Institute, Potsdam, 11–14 July 2000).

10.1.2 Overall antenna and spacecraft design

The basic geometry is shown schematically in figure 10.1. Three spacecraft form an equilateral triangle 5 000 000 km on a side, and laser beams are sent both ways along each side of the triangle. A Y-shaped thermal shield inside each spacecraft contains the sensitive parts of the scientific payload, consisting of two separate optical assemblies mounted in the two top arms of the Y, so that they are aimed along the two adjacent sides of the triangle. The spacecraft instrumentation and a cover over the sunward side of the spacecraft are not shown.

Each spacecraft is in a one year period solar orbit with an eccentricity e of about 0.01 and an orbit inclination to the ecliptic of $3^{0.5} \times e$ [8]. By choosing

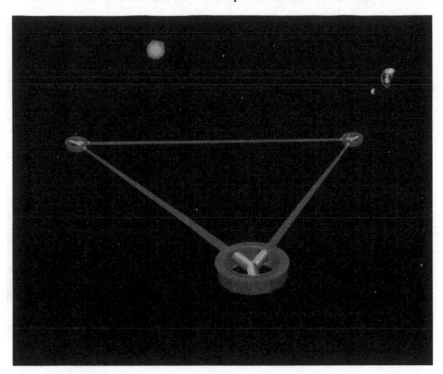

Figure 10.1. Basic geometry of the LISA antenna. The sides of the triangle are 5000 000 km long.

the phasing and the orientations of the orbits properly, the three spacecraft will form the desired nearly equilateral triangle, which is tipped at $60°$ to the ecliptic (figure 10.2). The plane of the triangle will precess around the pole of the ecliptic once per year, and the triangle will rotate in that plane at the same rate. The centre of the triangle is chosen to be about 50 000 000 km ($20°$) behind the Earth. The arm lengths for the triangle will stay constant to about 1% for a number of years. Locating the triangle $60°$ from the Earth would keep the arm lengths more constant, but at the expense of more propulsion required to reach the desired orbits and more telemetry capability to send the data back.

The layout of one of the two optical assemblies in each spacecraft is shown in figure 10.3 (see [9]). Each optical assembly contains an optical bench, a transmit/receive telescope, and a low-power electronics package. Each of these three subassemblies is mounted by low-thermal-conductivity struts from a stiffened support cylinder that forms the outside of the optical assembly. Because the distances between the spacecraft will change by up to 1% during the year, the angles between the sides will also change by roughly one degree. Thus, it is necessary to provide some adjustment for the angle between the axes of the two

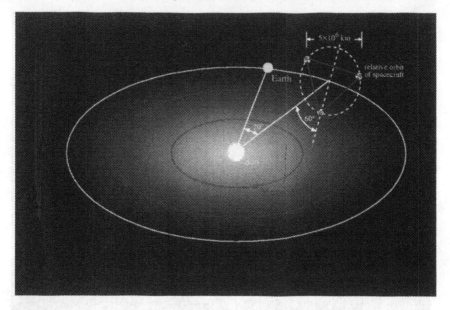

Figure 10.2. Location of the LISA antenna, 50 000 000 km behind the Earth in orbit around the Sun. The plane of the antenna is tipped at 60° to the ecliptic.

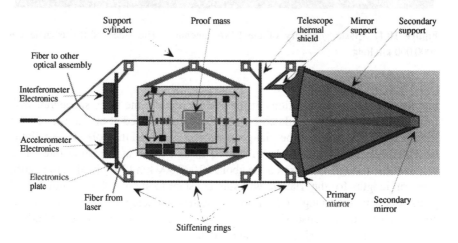

Figure 10.3. Layout of one of the two optical assemblies in each spacecraft. The main components are the optical bench at the centre and the transmit/receive telescope at one end.

optical assemblies in each spacecraft. This is accomplished by supporting the lower part of each optical assembly by a flexure from a common block near the base of the Y-shaped shield, and the upper part by two adjustable displacement

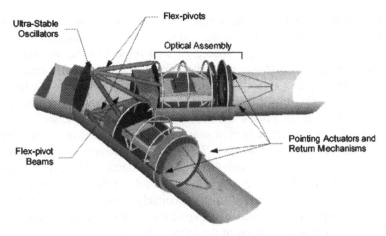

Figure 10.4. Mounting of optical assemblies inside main thermal shield. Each optical assembly must rotate slowly and smoothly during the year over about a one degree range.

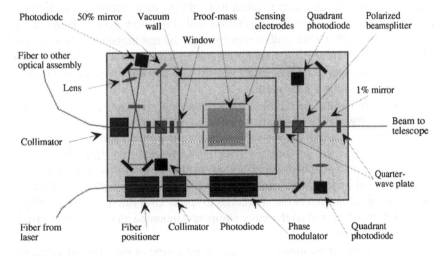

Figure 10.5. Optical bench for one optical assembly. Light from the laser is mainly sent out to the telescope, with a little going to the photodiode. Received light from the other end of the arm passes through the beamsplitter to the test mass, and then goes to the photodiode. Changes in the phase of the beat signal determine the changes in the arm length.

mechanisms from an arm of the Y, as shown in figure 10.4 (see [10]). The mechanisms provide a very smooth one degree change in the angle between the arms with mainly a one year period.

At the centre of each optical bench, shown in figure 10.5 (see [9]), is a freely floating test mass that is protected as much as possible from sources of spurious accelerations. The basic measurements made by the LISA antenna are the changes with time in the distances between test masses in the different spacecraft. Capacitive sensors on the inside of a housing surrounding each test mass sense the position of the spacecraft with respect to the test mass. This position information is fed via a proportional control servosystem to micronewton thrusters that keep the spacecraft essentially fixed with respect to the test mass. Each test mass plus the housing around it and the associated electronics will be referred to as a 'free mass sensor'. The names 'inertial sensor', 'drag-free sensor', and 'disturbance reduction sensor' also have been used frequently, but each choice has some drawbacks. 'Inertial sensor' causes confusion with the gyroscopic sensors used in inertial navigation systems. 'Drag-free sensor' or 'disturbance reduction sensor' emphasizes the role of the sensor in permitting external forces on the spacecraft to be accurately compensated for, but do not indicate the importance of keeping the spurious accelerations of the test mass far below the residual accelerations of the spacecraft.

At frequencies below roughly 3 mHz, the threshold sensitivity of the LISA antenna will be determined mainly by the spurious accelerations of the test masses, despite the care that is taken to minimize such disturbances. However, at higher frequencies the main limitation will come from how well the relative changes in the 5000 000 km distances between the test masses can be measured. The optical interferometry system for measuring these distances will be described in the next section. The free mass sensors and the way they are used will be discussed in detail in a later section.

In describing the operation of the gravitational-wave antenna, it is useful to distinguish between the 'main spacecraft' and the 'instrument package'. Here the instrument package will mean the Y-shaped thermal shield and everything inside it. In addition, it will include the lasers and a thermal radiation plate they are located on that is attached to the bottom of the Y. With this arrangement, the heat generated by the lasers can be radiated by the radiation plate out through a large hole in the bottom of the spacecraft to space. Everything else, including the spacecraft structure and all the other equipment mounted on it, will be referred to as the main spacecraft.

Probably the most important factor in the design of the LISA mission, after minimizing spurious forces on the test masses, is the need to keep the temperature distribution throughout each main spacecraft and instrument package as constant as possible. This is essential for two main reasons. One is that mass displacements due to local temperature changes within the desired measurement band, roughly 10^{-6} Hz to 1 Hz, will change the gravitational force on the test mass and look like a real signal [10, 11]. The other is that a variation in the temperature difference between the two sides of the housing in a free mass sensor will cause a difference in the thermal radiation pressure force on the two sides of the test mass, and thus give a spurious acceleration.

The thermal design of the LISA spacecraft can be understood on the basis of the following rough model. The main spacecraft will have some thermal time constant for responding to changes in the incident solar flux or in the electrical power dissipated in its equipment. The Y-shaped thermal shield is mounted from the main spacecraft by low-thermal-conductivity flexural strips or stressed fibreglass bands, and forms a second passive thermal isolation stage. The third stage is the support cylinder around each optical assembly and the fourth stage is the optical bench that contains the free mass sensor. Since low-thermal-conductivity supports are used between each successive stage, and with low infrared emissivity coatings on as much of the surfaces as possible, quite long thermal time constants can be achieved for each stage.

With the above type of multistage passive thermal isolation approach, quite low levels of temperature fluctuations can be achieved at frequencies of 0.1 mHz and higher. But there are practical limits to how long the various thermal time constants can be made. To achieve high thermal stability at lower frequencies, some active thermal control will be needed. Even a fairly crude servosystem for the temperature of the Y-shaped thermal shield would help considerably, and some active control of the temperature of the main spacecraft probably will be desirable also. Such active temperature control systems may be added to the baseline mission design later to meet a goal of improved antenna sensitivity at frequencies below 0.1 mHz, but are not included in the present mission requirements.

A fundamental feature of the mission design is the high stability of the solar energy dissipation in the spacecraft. In an ecliptic-based reference frame that rotates once per year about the pole of the ecliptic, the three spacecraft form a nearly equilateral triangle in a plane that is tipped at 60° to the ecliptic, as discussed earlier. Thus, the direction to the Sun makes a nearly constant 30° angle with the normal to the sunward side of each spacecraft, where the solar cells providing power for the spacecraft are located. Because of the stable solar illumination geometry, only fluctuations in the intrinsic solar intensity and in the electrical power dissipation in the various parts of the spacecraft produce significant temperature variations. Measures such as keeping the microwave transmitters operating at the same power level whether data is being transmitted or not are taken to reduce thermal changes. Also, the power needed in the optical assemblies and particularly on the optical benches is kept as low as possible, and is highly regulated.

10.1.3 Optics and interferometry system

The present parameters assumed for the LISA optical system are the use of 1 W of output power from cw NdYAG lasers operating at 1.064 μm wavelength, and of 30 cm diameter telescopes to transmit and receive the laser beams between different spacecraft. Perhaps the most impressive thing about the mission is that these fairly modest parameter values lead to roughly 2×10^8 photoelectrons s^{-1} being detected at the far end of a 5 000 000 km arm, which would correspond to 2×10^{-21} precision in measuring changes in the arm length in one second if there were no other sources of error. Since almost all expected types of LISA gravitational-wave sources can be observed for a year or more, the potential measurement accuracy clearly is extremely high.

Each optical assembly has two lasers for redundancy, and a switching mechanism for switching between them. Each laser is capable of 2 W output over a mean lifetime of several years, but is operated at 1 W to extend the lifetime by a substantial factor. A schematic drawing of the optical bench for one optical assembly was shown earlier in figure 10.5. Light from the active laser is taken by a single mode fibre onto the optical bench, and then formed into a beam of roughly 5 mm diameter. A few per cent of the beam is split off for other purposes, and then a beamsplitter sends almost all of the power out to the telescope, where it is transmitted to the matching optical assembly on the other end of the arm. The return beam from the laser in the distant optical assembly is sent by the telescope to the test mass, where it is reflected from the face of the test mass, and then beat against a small amount of power from the local laser.

The beat signals detected in the six optical assemblies are the main data used to make the gravitational-wave measurements. In the present baseline scheme, one laser in spacecraft SC-1 is considered the master laser and locked to a stable reference cavity on its optical bench. The other laser in SC-1 is phase locked to the first with a small offset frequency of perhaps a couple of kilohertz. The two laser beams are transmitted along the adjacent arms of the triangle to SC-2 and SC-3, where they are reflected from the test masses in the receiving optical assemblies. The local lasers are then phase locked to the received beams, also with small frequency offsets, and their beams are sent back to SC-1. Finally, the second lasers in SC-2 and SC-3 are offset locked to the first lasers, their beams are sent over the third arm of the triangle, and the beats between them are recorded on both ends of the side, after reflection from the test masses.

The frequencies of the beat signals can range from well under 1 MHz to roughly 15 MHz, depending on the exact orbits of the spacecraft, or rather the orbits of the freely floating test masses in them. In one possible measurement scheme, the signals are beat down to a convenient frequency range for precise phase measurements, such as perhaps 10 kHz, using outputs from frequency synthesizers. The drive frequencies for the synthesizers come from ultra-stable crystal oscillators (USOs) on each spacecraft. In a second measurement scheme, the signals are first sampled at a high rate such as 40 MHz, and the results are analysed in the software. In this case, the sampling times are controlled by the USOs.

If only the signals from the two arms adjacent to SC-1 are considered, the antenna is like a single groundbased detector, with changes in the arm length difference being the quantity of interest. However, it does not appear practical to keep the difference in arm length nearly constant, since this would require applying quite large forces to the test masses to overcome the changing gravitational forces on them due to the Sun and to the planets. If such large forces were applied, it would be very difficult to avoid noise in those forces within the frequency band of interest from giving undesired changes that could not be distinguished from real signals. To avoid this problem, the test masses are left nearly free, except for small differential applied forces on the two test masses in a given spacecraft that are necessary to compensate for the difference in spurious forces on them at dc and at frequencies below the useful measurement band.

Because of the up to 1% difference in arm lengths expected, the phase noise in the master laser has to be corrected for too high accuracy. The original suggestion that this was possible was made by Faller in the abstract for a 1981 conference [3]. The basic idea is to use the apparent changes in the sum of the lengths of the two arms to estimate the laser phase noise, and then to apply the corresponding correction to the measured difference in the arm lengths. Approximate algorithms for doing this were published by Giampieri *et al* in 1996 [12]. More recently, Armstrong, Tinto and Estabrook [13, 14] have given what are believed to be rigorously correct algorithims, and it is planned to use these during the LISA mission.

So far, only the use of the signals from the two detectors on SC-1 has been discussed. In addition, the signal from one of the detectors on each of SC-2 and SC-3 is used to phase lock a laser to the received laser beam from SC-1. So the signals measured on SC-2 and SC-3 from the beams sent between them can be combined to give the changes in length of the third arm. As discussed by Cutler [15], the length of the third arm minus the average of the lengths of the other two is an observable that gives the other polarization from the one determined by the difference in lengths of the first two. Thus, having the measurements over the third arm gives a valuable addition to the scientific information that can be obtained from a mission like LISA.

In the recent papers of Armstrong, Tinto and Estabrook [13, 14], the analysis is done in a way that assumes all the lasers are running independently, presumably stabilized by locking to their own stable reference cavities, but not locked to each other. This appears to put some additional requirements on the mission measurement system, but they point out that there is additional information in the resulting signals at the shorter gravitational wavelengths that is not available with the current baseline measurement system. It appears that this may be equivalent to the information that would be obtained from running LISA as a Sagnac interferometer, but this possibility has not yet been considered in detail.

In addition to providing the second polarization if all is going well, having the capability of making measurements over the third arm has another very important benefit for the LISA mission. With two fully functional optical assemblies on each of the three spacecraft, if something in one of the six were not operating properly, the antenna would still provide two-arm gravitational-wave data with the full planned sensitivity. Even if a second optical assembly were out of operation, provided the two were not on the same spacecraft, the two-arm data could still be obtained. Thus, the third arm capability provides an essential level of redundancy for the mission.

In addition to the shot noise in the photocurrents from the received laser beams, other noise sources in measuring changes in the arm length differences also have to be considered. One such source is fluctuations in the attitude of each spacecraft. If the transmitted beam gave a perfectly spherical wavefront in the far field, attitude changes for the transmitting spacecraft would not give phase changes in the received signal at the distant spacecraft. However, the combination of diffraction due to the finite size transmitting aperture plus imperfections in the optical system cause the wavefronts at the distant spacecraft to be somewhat distorted. Thus, the attitude of the spacecraft and any other sources of jitter in the pointing of the transmitted beam have to be controlled quite closely.

There are two methods under consideration for measuring jitter in the spacecraft attitude. In one, perhaps 10% of the received light from the distant spacecraft is picked off and focused on a CCD array or quadrant diode via a fairly long effective focal length optical system. Changes in attitude then give changes in the position of the focal spot on the detector, which are used in a servo system to control the spacecraft attitude, along with the similar signals from the second optical assembly on the spacecraft. In the second approach, all of the light goes to the main detector for measuring changes in the arm length, but the detector is replaced by a quadrant device. If the spacecraft tips slightly with respect to the received wavefronts, the differences in phase of the four detected signals will change, and these changes are used as the inputs to the attitude control system.

Even if the attitude jitter is made small, there still is a need to set the mean beam pointing direction carefully. Diffraction plus a defocus of the transmitted beam would result in the phase of the received wavefront varying only quadratically with the angular offset from the optical axis of the transmitting system. However, non-axisymmetric defects in the transmitted wavefronts can make the change in the received phase vary linearly with the attitude change, even on what otherwise would have been the optical axis. To avoid this increased sensitivity to beam pointing, the pointing along each axis of each optical assembly is modulated at a known frequency, and the resulting apparent changes in the arm length differences at the modulation frequencies and their second harmonics are detected. This information permits the outputs at the modulation frequencies to be minimized by small offsets in the dc pointing directions, which corresponds to having just the quadratic variations in the received phase due to attitude jitter.

As a measure of the remaining requirement on pointing jitter, a convenient test case is to assume only astigmatic error in the transmitted wavefronts with an error of a tenth of a wavelength rms. For this example, if the distance error due to pointing is allowed to be equal to that from shot noise, the errors in the dc pointing offset and in the pointing jitter can be as large as 10 milliarcsec and 4 milliarcsec/rtHz, respectively. (Here and later, /rtHz stands for 'per square root Hz', and the given error or noise is the spectral amplitude of the error, which is the square root of the power spectral density.) These error allocations are each about three times larger than those given in section 3.1.8 of [6], since the error allocation in that case was assumed to be only 10% of the error due to shot noise.

Further information on the optical path error allocation budget for LISA is given in sections 4.2.1 and 4.2.2 of [6]. The total error allocation for the measurement of the difference in the round-trip path lengths for two arms of the antenna is 40 nm/rtHz. With only shot noise considered, the corresponding value would be 22 nm/rtHz, and with beam pointing jitter included also, the value becomes 30 nm/rtHz. Other error sources considered explicitly are as follows: residual laser phase noise after the correction procedure discussed earlier is applied; noise in the ultra-stable oscillators after a similar correction procedure; noise in the laser phase measurements and phase locks; and scattered light effects. The 40 nm/rtHz total error allocation for the difference in round-trip paths for two arms corresponds to 20 nm/rtHz for determining the difference in lengths for the two arms.

10.1.4 Free mass sensors

Historically, the first free mass sensor was developed jointly by the Johns Hopkins University Applied Physics Laboratory and by Stanford University, and was flown on the TRIAD satellite in 1972 [16]. A spherical test mass was contained in a spherical cavity with three opposing pairs of capacitive electrodes on the inside of the housing for sensing the relative position of the test mass. Whenever the atmospheric drag caused the housing to move a few millimetres with respect to the test mass, pulsed thrusters on the satellite fired to keep it centred on average on the test mass. Care was taken to keep forces on the test mass other than those due to external gravitational fields as small as possible. Thus, the orbit of the satellite was nearly drag-free, and was much more predictable than normal. The theory of such drag-free systems had been given earlier in 1964 by Lange [17].

There is a close connection between free mass sensors and the high-performance force-rebalance accelerometers developed over the last three decades by the Office National d'Etudes et de Recherches Aerospatiales (ONERA) in Paris for flight on various missions. The basic approach in the accelerometers is to measure the position of a free or nearly free test mass inside a housing by means of capacitive electrodes on the inside of the housing. Forces are then applied to the test mass by means of voltages on the electrodes to keep the test mass centred in the housing. The required voltages are measures of the accelerations along the three perpendicular axes.

The first such accelerometer designed and built by ONERA was flown from 1975 to 1979 on the French CASTOR-D5B satellite. It had a spherical test mass. However, later ONERA-designed accelerometers have used test masses in the form of rectangular parallelepipeds. The first such design, called the GRADIO accelerometer, was for possible use in a proposed gravity gradiometer mission (ARISTOTLES) to map the Earth's gravity field, and has been the basis for a number of later designs.

The test mass for the GRADIO accelerometer is 4 cm by 4 cm by 1 cm in dimensions. The material used for the test mass is a Pt–Rh alloy with a density of about 20 g cm^{-3}. The housing is made of ultra-low expansion glass (ULE), with a gold coating on the inside that is carefully patterned to form the capacitive plate electrodes. The gaps between the plates are recessed and kept very small to reduce the charge buildup on the underlying material. The position sensing is done with 100 kHz transformer bridge circuits. Extra pairs of plates are included to permit relative rotation of the test mass with respect to the housing to be sensed and controlled.

The gaps between the capacitive plates and the test mass is 30 μm for the large square faces and 300 μm for the smaller rectangular ones. With the test mass horizontal, the factor four smaller vertical dimension and the 30 μm gap permit the test mass to be electrostatically suspended for laboratory tests. The entire accelerometer is contained in a vacuum enclosure. Extensive tests have been carried out at ONERA with two GRADIO accelerometers mounted on a single pendular support platform in a special sub-basement room to reduce the effects of tilts in the floor. The results for the differential horizontal accelerations have been very encouraging, but are still several orders of magnitude worse than the theoretical sensitivity of roughly 10^{-12} m s^{-2}/rtHz from a few millihertz to 1 Hz in a quiet and constant temperature zero-g environment.

A later but lower sensitivity version of the GRADIO accelerometer has been flown a few times on the space shuttle, and other versions will fly soon on other missions. These include the German–French CHAMP mission scheduled for launch in 2000, a USA–German mission called GRACE scheduled for a 2001 launch, and a later ESA mission called GOCE. GRACE is a satellite-to-satellite tracking mission that will map time variations in the Earth's gravity field over a five-year period with high accuracy, and an accelerometer on each satellite will monitor drag and other non-gravitational forces on the satellites. GOCE is a gravity gradiometer mission to map very short wavelength spatial structure in the gravity field, using the differential accelerations between pairs of accelerometers in the same spacecraft.

For the free mass sensors for the LISA mission, the requirements are quite different from those for accelerometers. The purpose of the accelerometers is to measure the non-gravitational accelerations of the spacecraft to which they are attached, and thus the readout sensitivity has to be high. For the GRADIO accelerometer, with 300 μm gaps along the two most sensitive axes, the position measurement sensitivity is 6×10^{-12} m/rtHz at frequencies above 5 mHz. However, the small gaps lead to increases in some of the time-varying spurious accelerations of the test mass, as discussed later. For LISA, it turns out that the necessary sensitivity of the capacitive position measurements is only about 10^{-9} m/rtHz, so gaps of 2 mm or larger and different values of some of the other measurement parameters can be used to reduce the spurious accelerations.

An initial ONERA study of a proposed free mass sensor for LISA, called CAESAR, was published by Touboul, Rodrigues and Le Clerc in 1996 [18]. The test mass is a cube 4 cm on a side, made of a 10% Pt–90% Au alloy to achieve low magnetic susceptibility and a density of 20 g cm^{-3} (see figures 10.6 and 10.7). The gaps assumed were 4 mm along the most sensitive axis and 1 mm for the other two axes. An error budget for the sensor was developed, based on the high thermal stability expected for the optical bench during the LISA mission.

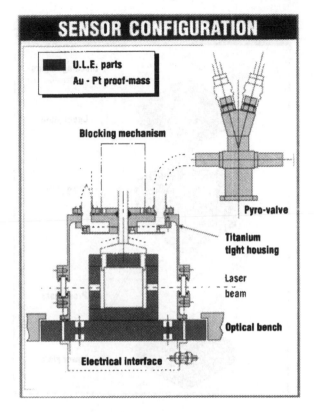

Figure 10.6. Early ONERA design of LISA free mass sensor called CAESAR. Up to four capacitor plates on each face of the housing around the test mass allow the relative position and angular orientation to be measured accurately. Electrical forces also can be applied, if desired.

After completion of the initial CAESAR design, additional studies of free mass sensor designs and of the LISA requirements have been carried out, influenced strongly by the CAESAR design. A number of the factors influencing the design were discussed by Speake [19] and by Vitale and Speake [20]. A design with a considerably different geometry for the capacitor plates was investigated by Josselin, Rodrigues and Touboul [21], and has since been constructed and tested in the laboratory and the LISA free mass sensor requirements were simplified and somewhat relaxed [6].

One current version of the error allocation budget for the free mass sensor is described in section 4.2.3 of [6]. The total acceleration error allowed for one sensor is 3×10^{-15} m s^{-2}/rtHz over the frequency range 0.1–30 mHz. The six largest allocations to individual error sources are as follows: thermal distortion of the main spacecraft; temperature difference variations across the

128   *LISA: A proposed joint ESA–NASA gravitational-wave mission*

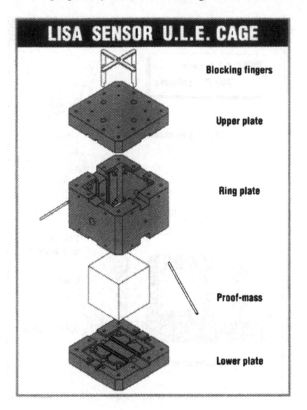

Figure 10.7. Ultra-low expansion glass housing for the proof (test) mass in the early CAESAR design.

test mass housing; electrical force on the charge on the test mass; Lorentz force on the charged test mass from the fluctuating interplanetary magnetic field; residual gas impacts on the test mass; and fluctuating forces due to electrical field dissipation in the test mass housing. These sources are each allocated an error of 1×10^{-15} m s^{-2}/rtHz.

In view of the performance level desired for the LISA free mass sensors, it is planned to have a technology demonstration flight for them at the earliest possible date. Efforts to arrange for such a flight currently are under way both in Europe and the USA. Two sensors would be flown on a thermally isolated optical bench on a small spacecraft, with an interferometer to measure changes in the separation of the two test masses. To keep the rate of charging of the test masses comparable with what it would be well away from the Earth, and to reduce the effect of the Earth's gravity gradient, a perigee altitude of 10 000 km or higher is desired. To reduce the cost, a tentative goal of demonstrating only 3×10^{-14} m s^{-2}/rtHz performance from 1–10 mHz currently is assumed. It is believed that cautious engineering extrapolation from tests at this level plus some in-flight tests with intentionally increased disturbances will provide a sound basis for proceeding with the LISA mission.

10.1.5 Micronewton thrusters

The main non-gravitational force on a LISA spacecraft is expected to be from the solar radiation pressure and to have a magnitude of about 20 μN. If not compensated for, it would cause an acceleration of the spacecraft of roughly 10^{-7} m s^{-2}. Since the test masses are shielded from this source of acceleration, it is necessary to apply force to the spacecraft to keep it from moving with respect to them. The spectral amplitude of the fractional fluctuations in the solar pressure force [22] over the 0.1–10 mHz frequency range is approximately

$$1.3 \times 10^{-3} \, (f/1 \text{ mHz})^{-1/3}.$$

Comparable fluctuations in force but with a more reddened spectrum and a much lower dc level will be present from the solar wind.

The type of thrusters that are planned for use on the LISA mission are field emission electric propulsion (FEEP) thrusters. They operate by accelerating ions through a potential drop of 5–10 kV and ejecting them to provide thrust. The ejection velocity is roughly 60–100 km s^{-1}, corresponding to a specific impulse of 6000 to 10 000 s. In view of this high specific impulse and the low thrust level needed, the fuel required per thruster for a ten year extended mission lifetime is only a few grams.

Historically, most of the development of FEEP thrusters has been based on the use of Cs ions. This work has been done mainly at the European Space Research and Technology Centre in Noordwijk, The Netherlands and at Centrospazio in Pisa, Italy (see section 7.3 of [6]).

A schematic drawing of a thruster is shown in figure 10.8. In each thruster, liquid Cs metal at a temperature somewhat above the melting point of 29 °C is contained in a small reservoir. It is drawn by capillary forces through a narrow channel between two polished metal plates spaced 1 or 2 μm apart. The accelerating voltage is applied by a plate with a slot in it, located at the outer edge of the channel.

The high field at the surface of the Cs metal causes an instability, and Taylor cones roughly a micrometre in diameter and a few micrometres apart form on the surface. The tips of these cones are very sharp, and the high field around the tips causes Cs ions to be drawn out by field emission and accelerated away in a beam perhaps 30° wide. The one substantial drawback to the use of Cs ions is that any water vapour that is present will react with the Cs to form CsOH. Thus, the thrusters are kept in vacuum containers roughly 5 cm in dimensions until in space and ready for use. A spring-loaded cover with an O-ring seal is then released.

An alternative to Cs is the use of In. Field emission of In ions for the neutralization of positive charge on spacecraft has been pioneered by the Austrian

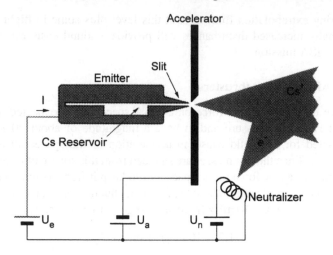

Figure 10.8. Schematic drawing of cesium Field Emission Electric Propulsion (FEEP) thruster. The thrust can be controlled by changing the accelerator voltage smoothly. A set of six or more thrusters is used to control the spacecraft position and attitude.

Research Centre Seibersdorf, and such devices have been flown on several missions. The geometry used consists of a sharpened tungsten needle with a tip a few micrometres in diameter mounted in the centre of a heated In reservoir. Capillary forces bring the In to the tip, and again an accelerator plate applies a high voltage. A single Taylor cone is formed, and the In ions are produced by field emission from the sharp tip of the cone.

If such devices are used as thrusters, there is a limit to the thrust achievable per tungsten needle. If too much current is drawn, small droplets form and give much less thrust per unit mass than the ions do. However, it has been demonstrated that a number of needles can be operated in parallel to meet the quite low thrust requirements of the LISA mission during its operational phase. The results of research on such thrusters have been reported by the Seibersdorf group [23]. The possibility of using In instead of Cs with the channel capillary flow approach is now being investigated by Centrospazio, but no decision has been made on the overall merits of the two materials with this approach.

The advantages of FEEP thrusters, in addition to their high specific impulse, is that they can run continuously and be controlled easily by small changes in the acceleration voltage. Work on their further development and testing is being supported by ESA because of their potential for use on a number of future missions. If In ions are used, the thrusters have to run hotter because of the 156 °C melting temperature for In. However, the power required is still low, and does not appear to be a problem. To control both attitude and translation and allow sufficient redundancy, probably between 12 and 24 individual thrusters pointing in different directions will be included per spacecraft.

10.1.6   Mission scenario

A single launch with the Delta II launch vehicle is assumed in current studies of the LISA mission [7]. Each of the three spacecraft has a thin cylindrical propulsion module attached to it. The three composite vehicles are stacked on top of each other for launch, and then separate from the launch vehicle and from each other a short time later. Since the spacecraft will be separated by 5000 000 km and have different orbit planes in their final configuration, they travel separately and each has its own onboard guidance system.

It is currently planned to use solar electric propulsion for the propulsion modules. The top area of the main spacecraft is covered with solar cells, and enough power is generated to provide 15 or 20 mN of thrust with either Hall effect or another type of Xe ion thruster. The time necessary for the three spacecraft to reach their proper locations and velocities about $20°$ behind the Earth in orbit about the Sun is 14 or 15 months.

After the nominal orbits have been achieved, the spacecraft will be tracked for a week or two with NASA's Deep Space Tracking Network. Then, any desired minor corrections to the orbits can be made. The next step is separation of the propulsion modules from the final spacecraft. After that, the orbits will be almost completely gravitational, with each spacecraft servocontrolled to follow either the average of the two test masses inside it or just one of the two. The difference of the spurious accelerations of the two test masses will be roughly 10^{-10} m s^{-2} or less, and can be corrected for by applying weak and stable electrical forces to the test masses at frequencies below the measurement range via the capacitor plates.

The next step is for each spacecraft to acquire the other two optically. Star trackers aboard each spacecraft give the attitude to a few arcseconds, and the transmitted laser beams are defocused by roughly a factor ten so that they can be detected on the other end of an arm even with relatively poor beam pointing. The received laser beam normally is bright enough to look like about a magnitude-3 star, so there will be a substantial brightness level even with some defocusing. If necessary, an angular scan pattern on the transmitted beam could be used to find the beam, since the stability of the star tracker will be considerably better than its absolute accuracy.

As soon as a receiving spacecraft finds the beam, the output from the CCD array or quadrant diode in its angle tracker can be used to lock the attitude to the direction to the distant spacecraft, and the defocus in the transmitted beam can be removed to give a high S/N ratio. The local laser then is turned on, and its frequency adjusted to give a fairly low beat frequency with the received beam. The resulting heterodyne signal back at the original transmitting end provides a higher S/N ratio, and makes detecting the defocused return beam easier. After the beam pointing signals are locked in and the focus is corrected on both ends of each arm, the system is ready for operation.

Even with careful phase locking of the master laser to a stable cavity, or possibly the locking of more lasers, the phase noise will still be significant up to frequencies of a few hundred hertz to 1 kHz. Also, the amplitude of the phase noise at frequencies near 1 Hz and below will be large. To avoid aliasing the higher frequency noise into the band of interest, below perhaps 3 Hz, it may be desirable to make the phase measurements at a rate approaching a kilohertz, and then filter the data in software. For low-frequency noise, it is planned to combine all the signals at a single spacecraft and perform the algorithm for correcting for laser phase noise there. The resulting signals then can be compressed before being sent to the ground.

The number and amplitude of mechanical motions aboard the spacecraft are kept as small as possible to reduce gravitational attraction changes and excitation of vibrations. Roughly 30 cm diameter X-band antennae probably will be used to send data down to 34 m Deep Space Network antennae on the ground, and will have to be repointed about once a week. Step changes in the angle between the axes of the two optical assemblies probably will be made at the same time with coarse adjustment mechanisms, but smooth adjustments over roughly a range of 5 arcmin are required between the steps. It is expected that offsets in the position, velocity, and dc acceleration of the test masses will have to be solved for at the times of the steps.

The nominal mission lifetime after the antenna geometry is established probably will be three years. However, there are no expendables required except for the In or Cs fuel for the micronewton thrusters, and that will be made adequate for a ten year or longer lifetime. Thus, a rolling three year approval period hopefully can be established, provided that the antenna continues to operate properly.

10.2   Expected gravitational-wave results from LISA

10.2.1   LISA sensitivity and galactic sources

The two types of noise sources for the LISA antenna have been discussed earlier. The one that dominates at frequencies below about 3 mHz is spurious accelerations of the test masses, and the level of error allocated for frequencies f between 0.1 and 30 mHz is 3×10^{-15} m s^{-2}/rtHz for each free mass sensor. Considering only two arms of the antenna for simplicity, and adding the errors from the four sensors quadratically, the resulting noise level for the difference in (one-way) length of the two arms is

$$\Delta(L2 - L1) = 1.520 \times 10^{-16} \, (1 \text{ Hz}/f)^2 \text{ m/rtHz}.$$

This should be combined quadratically with the error allocation of 2×10^{-11} m/rtHz for measuring the difference in the distances between the test masses along the two arms.

For low frequencies, where the wavelength is long compared with the arm length, the response of the antenna to a gravitational wave [24] can be given

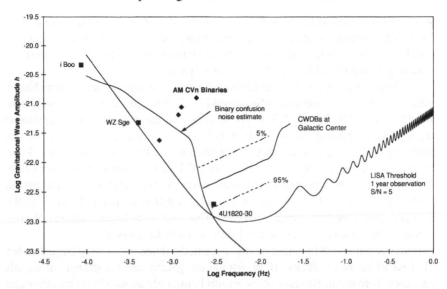

Figure 10.9. LISA sensitivity and galactic sources. The instrumental and binary confusion threshold sensitivity curves are shown for one year of observations and an S/N ratio of 5. Estimated signal strengths and frequencies for galactic binaries are indicated also.

simply for the optimum direction of propagation and polarization of the wave. The change in $L2 - L1$ is just the amplitude h of the wave times the cosine of the angle between the two arms. Averaging over the direction of propagation and the polarization gives a factor $5^{0.5}$ lower signal amplitude. However, at higher frequencies, the signal is reduced by a factor that depends in a complicated way on the direction of propagation and the polarization. The transfer function giving the rms value of this reduction factor, and including the cosine of the angle between the two arms, has been given by Schilling [25]. It was used in calculating the LISA threshold sensitivity, as shown in figure 10.9 and discussed below. Essentially the same curve has been calculated independently by Armstrong, Tinto and Estabrook [13], but for a slightly different choice of arm length.

For almost all of the gravitational-wave signals LISA will see, the frequency, amplitude and polarization of the wave will be very stable, and will not change substantially over a few years of observation. Thus, it is desirable to plot the LISA threshold sensitivity as a curve such that there is a good chance of seeing a source in a reasonable observing time if its rms amplitude lies above the curve. To accomplish this, the threshold sensitivity curves in figures 10.9 and 10.11 are plotted as the rms amplitude of a stable signal needed in order to have a S/N ratio of five for one year of observation, when averaged over source direction and polarization. The S/N ratio of five is chosen because almost all the sources will be unknown from optical observations, and roughly this S/N ratio is needed to determine the reality of sources with unknown frequencies and with only statistical information on their directions in the sky. When averaged over a year, the LISA antenna response is remarkably independent of the source direction in the sky [26], so no allowance for the ecliptic latitude or longitude of the source is needed in using the LISA threshold sensitivity curve.

The dominant type of signal LISA will see is gravitational waves from binary stars systems in our galaxy. Mironowskii [27] pointed out in 1966 that there would be roughly 10^8 W UMa type binaries in the galaxy giving substantial signals. These binaries consist of two main sequence stars so close together that their Roche lobes are in contact, and material can flow between them. Their frequencies lie in a fairly narrow band near 0.1 mHz. With so many sources at low frequencies, it is clear that even many years of observations would not permit individual signals to be resolved, except for a few that happen to be close to the Sun and thus have high signal strengths. There are even larger numbers of non-contact binaries composed of normal stars at lower frequencies.

In 1984 Iben [28] pointed out the expected existence of a very large number of close white dwarf binaries (CWDBs) in our galaxy, and the strength of signals expected from them. Because of their radii being only about 10 000 km, they can give gravitational-wave signals at frequencies up to roughly 30 mHz. However, they had not been observed directly until the last few years, and their space density in the galaxy is only poorly known. On the other hand, since the orbits of the interesting CWDBs evolve mainly by gravitational radiation, the number per hertz will decrease as the 11/3 power of the frequency, and thus the frequency above which most of them can be resolved is only fairly weakly dependent on the space density.

Recent observations by Marsh and collaborators [29–32] have found a number of examples of CWDBs, so their abundance must be moderately close to the astrophysical estimates. However, the criteria used in selecting their targets of observation make it difficult to derive a space density. Based on the best astrophysical studies available at present [33,34], it is expected that most CWDBs will be resolved at frequencies above roughly 3 mHz, even for those near the galactic centre. The signals are strong enough so that the frequencies of and directions to a few thousand such sources will be determined by LISA.

The mean strength of signals from CWDBs at the galactic center as a function of frequency is shown in figure 10.9 as a solid curve. It hooks sharply upward at about 15 mHz because the majority of the CWDBs below that frequency contain either one or two He white dwarfs, and coalesce just above 15 mHz. This leaves only binaries containing mainly white dwarfs composed almost completely of carbon and/or oxygen, which are called CO white dwarfs. The CO white dwarfs are roughly a factor two more massive and are smaller in radius, so they can reach somewhat higher frequencies before coalescing.

Above and below the curve for the galactic centre are dot-dashed curves labelled 5% and 95%. The 5% curve shows the mean strength of the signals from CWDBs at a distance from the Earth such that 5% of all the ones in the galaxy are closer, and the definition for the 95% curve is similar. Thus roughly 90%

of the CWDBs in the galaxy will give signal strengths in the region between the two curves. Since this region is substantially above the LISA threshold sensitivity curve for S/N = 5, most of these sources will have S/N ratios of 10–50.

Several other types of binaries also can contribute signals above the instrumental sensitivity level for LISA [35–43]. When all of these types are considered together, there is some frequency at which the expected number of galactic sources per frequency bin for one year of observations drops below a value of very roughly 1/3. Below that frequency, very little information can be obtained about the scientifically very important extragalactic sources, unless they are stronger than the sum of signals from the galactic binaries. Above that frequency, some information is lost because of having to fit out the galactic signals, but some survives [44–46]. The effective binary noise level then drops by a factor of about 3–10, at which point it is coming from all the binaries in other galaxies throughout most of the universe.

A simple calculation shows that the extragalactic contribution from each shell around our galaxy of a given thickness will give the same average contribution out until redshifts comparable with one, provided the effects of galaxy evolution are not too large [47]. A recent estimate of the effective noise level introduced by all of the galactic and extragalactic binary star systems, but not allowing for evolution [46], is shown in figure 10.9. It has been adjusted to correspond to S/N = 5. This confusion noise curve lies above the instrumental sensitivity curve for frequencies from about 0.1–3 mHz, and has a major effect on determining whether LISA can reliably detect particular sources from one year of observations. This affects some galactic sources, as well as extragalactic ones.

A few individual known sources also are shown on figure 10.9. Since the frequencies, as well as directions in the sky, are known for most of these sources, the S/N ratio needed to detect them is only about two. Thus, the ones somewhat below the threshold sensitivity or confusion noise curves probably are detectable from one to two years of observations. i Boo is a non-contact binary composed of normal stars, and WZ Sge is a very short period dwarf nova binary. The four Am CVn binaries shown are each believed to consist of an extremely low mass helium dwarf losing mass to a CO dwarf primary [39]. They also are called helium cataclysmics, and effects due to such binaries and their possible progenitors have been included in the recent confusion noise estimates [46]. The final known binary shown is 4U 1820-30, an x-ray binary believed to contain a neutron star or stellar mass black hole.

Before discussing the important questions about massive black holes and about relativity that LISA hopefully will contribute unique new information on, it is useful to say something about how the data will be analysed [48]. First, any large and easily recognizable signals will be fitted and removed from the data. Next, a convenient data set such as a one year data record will be analysed carefully to determine as many as possible of the galactic binary signals. This search will be easy for frequencies above perhaps 10 mHz, where the signals are quite well separated. However, the sidebands on the signals due to rotation and motion of the antenna will overlap strongly as the frequency decreases to near 3 mHz, and a simultaneous fit of all the signals in each roughly 100 or 200 cycle/year band is likely to be necessary in order to fit the signals well. At still lower frequencies, only the strongest sources whose signals rise substantially above the large number from near the galactic center will be detectable. All of the identified sources will then be removed from the data record, before looking for the swept frequency signals expected from extragalactic sources involving massive black holes.

10.2.2   Origin of massive black holes

An important astrophysical question is how the seed black holes that later grew to be the massive and supermassive black holes observed today were formed. To aid in identifying different mass ranges for black holes, we will refer to those from roughly 1.5 to 30 solar mass ($M_\odot$) that are thought to be capable of being formed by the evolution of very massive stars as stellar mass black holes, larger ones up to about $3 \times 10^7 M_\odot$ as massive black holes (MBHs), and still larger ones as supermassive black holes.

As is well known, supermassive black holes were invoked first to provide the energy source necessary to explain quasars. However, more recent optical and other observations have provided strong evidence for the existence of MBHs in AGNs much closer to the Earth [49]. In one case, the evidence is from observations of OH masers in Keplarian orbits around the MBH, and is essentially conclusive [50, 51].

For some normal galaxies, the evidence also is very strong. One case comes from extremely high-resolution infrared observations of stars moving around a $2.6 \pm 0.2 \times 10^6 M_\odot$ object at the centre of our galaxy, that really can only be a MBH [52, 53]. Two other cases come from optical observations of the motions of stars or gas around the centres of the M31 (Andromeda) and M32 galaxies in the Local Group at distances of about 0.6 Mparsec. The evidence for stellar mass black holes comes from observations of x-ray binaries.

For MBH masses of $10^6 M_\odot$ or less, the only fairly convincing observations so far are two OH maser observations indicating about $10^6 M_\odot$ central objects. Except for a few more observations of this kind, it is not apparent whether electromagnetic observations are likely to tell us much about the fraction of normal galaxies containing MBHs that are this low in mass. Because of this, the information potentially available from gravitational wave observations with LISA concerning the existence and masses of MBHs in other normal galaxies and how they formed is likely to be quite valuable.

There are two main types of theories concerning how seed black holes were formed. In one, collisions of stellar mass objects in dense galactic nuclei led to the formation of higher mass objects, and these sank down toward the centre (mass segregation), where their collision rates were enhanced. If the mass got to be a few hundred times the solar mass and an object was not already a black hole, it would evolve quickly to form one. When the largest objects got to be roughly a thousand solar mass in size, they would then be able to continue growing fairly rapidly by absorption of gas in the galactic nucleus and by tidal disruption of stars. At some point, the largest black hole would grow enough faster than the others that it would swallow up the ones of comparable size and become the seed for growth of a perhaps $10^5 M_\odot$ or larger MBH.

The alternate type of theory involves the evolution of a dense cloud of gas and dust to the point where it becomes optically thick, and radiation pressure plus magnetic fields can prevent further fragmentation of the cloud to form stars. At that point, if energy and angular momentum can be dissipated fairly rapidly, there are two options. In one, a supermassive star possibly 10^5–$10^6 M_\odot$ in size is formed, and quickly evolves to the point of relativistic collapse and forms a MBH. In the other option, the cloud can become dense enough to reach the point of relativistic instability and collapse directly to a MBH without going through the supermassive star stage. There also is the possibility of a relativistic star cluster becoming unstable and collapsing to a MBH.

For the collisional growth scenario, quite detailed calculations starting from $1 M_\odot$ stars were carried out by Quinlan and Shapiro [54] (see this 1990 paper for earlier references). They found that roughly $100 M_\odot$ objects could form in a few times 10^8 years, starting from plausible conditions in a dense galactic nucleus, and including the effects of mass segregation. However, it was not possible at that time to follow the process further.

In an alternate approach, Lee [55] started from assuming that 1% of the mass in a dense galactic nucleus was in the form of $7 M_\odot$ black holes that resulted from evolution of stars at the high end of the initial mass function (i.e. initial mass distribution). The rest of the material was in the form of $0.7 M_\odot$ normal stars. Dynamical friction led to segregation of the black holes to the core, and core collapse among the black holes occurred on a time scale much shorter than for a single component cluster. For rms stellar velocities above 100 km s^{-1} and for plausible densities in the nucleus, it was shown that many black hole binaries formed and merged to produce $14 M_\odot$ black holes within about two billion years. The process was not followed further, but it seems likely that most of the black holes would have merged rapidly to form a substantial sized seed MBH.

The main objection raised to the collisional growth scenario is that it seems difficult to produce the seed black holes and have them grow much further to the supermassive black hole size before the appearance of quasars as early as a redshift of four [56–58]. Instead, it was suggested that the inefficiency of star formation would leave most of the material in a dense cloud in the form of gas and dust. The cloud would cool and condense toward the centre until angular momentum support became important. Gravitational instabilities and other effects would help to remove energy and angular momentum and permit the density to become high enough for collapse to a supermassive star or directly to a MBH perhaps $10^5 M_\odot$ or larger in size. A related argument made is that a self-gravitating gaseous object of more than $10^8 M_\odot$ does not appear to have any stationary non-relativistic equilibrium state that can be supported for very long.

Under the cloud collapse scenario, an important question is whether signals are likely to be produced that LISA could see. For example, if a supermassive star forms and then evolves to the relativistic instability, the final collapse to a MBH could be slow enough that most of the gravitational-wave radiation would be at such low frequencies that LISA would have poor sensitivity. Also, if the collapse were nearly spherically symmetric, the radiative efficiency would be poor. However, recent fully relativistic calculations by Baumgarte and Shapiro [59] of the evolution of a rotating supermassive star up to the onset of collapse provide some basis for a more optimistic view. The later evolution can be determined reliably only by a numerical, three-dimensional hydrodynamics simulation in general relativity. However, estimates of what will happen indicate that most of the mass will go into a MBH, and that a bar instability which radiates efficiently at frequencies observable by LISA may form.

Even if the supermassive black holes in quasars at high redshifts are formed initially by cloud collapse, it still seems quite possible that the collisional growth scenario may contribute substantially to the formation of seed black holes for more modest sized MBHs, like the one in our galaxy. Under the collisional growth scenario, if a number of $500 M_\odot$ seed black holes are formed before runaway growth occurs and the largest has already swallowed the others, then the coalescence of two of these seeds could be seen by LISA even at a substantial redshift. For such a coalescence at $z = 5$, it can be shown that the signal strength as a function of frequency during the last year before coalescence, for a circular orbit, would stay just about at the LISA threshold sensitivity curve level, so the event would be detectable with S/N = 5.

A remaining question, even if many intermediate sized MBHs are produced by the collisional growth of seeds, is whether the time of runaway growth of the largest seed black hole would be delayed to high enough mass for LISA to observe the coalescences. As pointed out by Lee [60], the calculations of Quinlan and Shapiro are based on the Fokker–Planck approach, and that approach does not allow for the proper statistical treatment of a runaway instability. This is true for the calculations of Lee [55] also. In considering this issue, it should be noted that only the chirp mass for the binary is important for determining the signal strength and frequency as a function of time. Thus the coalescence of a $100 M_\odot$ black hole with a $4000 M_\odot$ one would be as observable as for two $500 M_\odot$ black holes. Further work on the runaway growth question, as well as on the overall collisional growth and cloud collapse scenarios, certainly would be valuable.

10.2.3 Massive black holes in normal galaxies

Another important astrophysical question concerns the abundance of intermediate size MBHs of roughly $10^5–10^6 M_\odot$. From observations based almost entirely on galaxies containing larger MBHs and SMBHs, various authors have estimated that the mass of the central object is about 500 times smaller than the mass of the spheroid (central bulge) of the galaxy. An even tighter relationship to the velocity dispersion in the spheroid has been reported recently [61]. However, the reasons for these relationships are not yet known [56,62–64]. It appears likely that LISA data can address whether a relation something like this extends to galaxies with smaller spheroids, which constitute the majority of all galaxies.

MBHs in galactic centres are expected to usually have an increased density of stars around them in the region where the potential of the MBH dominates that of the galaxy. This density cusp is usually taken to be a power law cusp, with a $-7/4$ power dependence on radius if the distribution of stellar motions is relaxed and a $-3/2$ power for some unrelaxed cusps. It is generally expected that there will be large numbers of compact stars, i.e. white dwarfs and neutron stars, in the cusp. Occasionally, one of them that is on a nearly radial orbit and passes close to the MBH may be deflected enough by the other stars so that it comes within five or so gravitational radii of the MBH and loses significant amounts of energy and angular momentum by gravitational radiation. If so, and if further deflections by the other stars are not important, the compact star orbit will continue to shrink gradually until coalescence with the MBH occurs.

Unfortunately, in almost all cases for white dwarfs and neutron stars, the above gradual approach scenario is interferred with by interactions with other stars. Hils and Bender [65] have simulated what happens for a particular model which assumes $1 M_\odot$ for both the compact stars and the normal stars in the cusp. After the first pass near the MBH, the orbit of the compact star is modified by interactions with the other stars more rapidly than by the gravitational radiation, unless the compact star is bound very tightly to the MBH initially. Thus the compact star usually will plunge rapidly into the MBH, or its point of closest approach will wander far enough away to not give appreciable interaction. A rapid plunge will not provide enough integration time for detection. In the remaining favourable cases, the signal typically will be observable by LISA from a one year data record starting up to roughly 100 years before coalescence.

Despite a loss of several orders of magnitude in the event rate due to plunging, the study by Hils and Bender [65] gave some hope of LISA seeing such signals. However, studies by Sigurdsson and Rees [66] and by Sigurdsson [67], plus an unpublished extension of the above study by Hils and Bender, indicated that the prospects were considerably better for observing gradual approaches to coalescence with galactic centre MBHs for roughly 5 or $10 M_\odot$ black holes. The effect of mass segregation was included. Such events would be detectable at a redshift of $z = 1$ during the last year before coalescence for MBH masses from 5×10^4 to $2 \times 10^6 M_\odot$.

Estimates of the event rate are certainly model dependent, but still offer encouragment that multiple signals of this kind will be observed. As an example, results obtained by Hils and Bender for one particular model are shown in figure 10.10. About 1% of the mass in the cusp was assumed to be in $7 M_\odot$ black holes, and mass segregation was included. For each factor two range in the mass of the central MBH about a nominal value M, and for each factor two

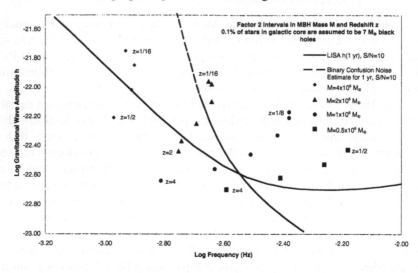

Figure 10.10. Expected signals from BH–MBH binaries. The instrumental and confusion noise thresholds are shown for one year of observations, but for a S/N ratio of 10. The different symbols correspond to different MBH masses. For a given mass, the individual points correspond to factor of two different values of the redshift z.

range in the redshift z, the signal strength and frequency of the strongest expected signal was calculated. The different symbols correspond to the different MBH masses, ranging from 5×10^5 to $4 \times 10^6 M_\odot$, and the highest and lowest values of z for each M are labelled. It should be noted that, only in this figure, the threshold S/N ratio is taken to be ten rather than five. This is because the orbits in this case generally will be quite complex because of relativistic effects, as discussed later, and thus require a higher S/N ratio for detection. With improved understanding of the conditions in the cusps, the distribution of MBH masses for such events may give information that cannot be obtained in other ways on the demographics of intermediate mass black holes in galactic centres throughout much of the universe.

A recent paper by Miralda-Escuda and Gould [68] looks specifically at the stellar mass black holes in the cusp around the MBH in the centre of our galaxy. With assumed masses of $0.7 M_\odot$ for stars in the cusp and $7 M_\odot$ for the black holes, and with 1% of the mass in the black holes, they show that the black holes will rapidly sink down closer to the center. With this model, they estimate that about 24 000 black holes would be concentrated within 0.7 parsec of the MBH, and calculate a lifetime for loss of the black holes to the MBH of about 3×10^{10} years. If a substantial fraction of the black holes approach coalescence gradually, and with many galaxies like ours in the universe, the possible LISA event rate from this study also may be considerable. On the other hand, the authors suggest that most of the stars in the inner few parsecs of the cusp would be forced out to larger distances by interaction with the black holes. If so, the estimated observable event rate for white dwarfs and neutron stars from [65] could be drastically reduced.

10.2.4 Structure formation and massive black hole coalescence

The third of the important astrophysical questions that LISA has a good chance of giving new information on concerns the development of structure in the universe. It is currently believed that objects considerably smaller than galaxies formed first, and that continuing sequences of interactions and mergers led to the present distribution of galaxy types and sizes, and to galaxy clusters and superclusters. If MBHs were already present in pre-galactic structures that merged, this could lead to the MBHs sinking down to the centre of the new structure by dynamical friction and getting close enough together to coalesce due to gravitational radiation before the next merger [58, 69].

The above scenario is quite attractive, and could give a high event rate for LISA. If so, valuable new constraints on the merger process and on the conditions after mergers would be obtained. However, there are a number of issues that affect the event rate that have to be considered.

One question is how effective dynamical friction will be in bringing MBHs of a given size to the centre of the new structure in less than the time between mergers. For galaxies like ours, the time required is less than the Hubble time for MBHs of roughly $3 \times 10^6 M_\odot$ or larger (see, e.g., Zhu and Ostriker [70]). However, the stars in the cusp around a MBH before merger will stay with it for some time, and will affect the dynamics. Thus, whether the mergers of pre-galactic structures in galaxies like ours with fairly modest spheroid and MBH masses are likely to have produced coalescences of perhaps 10^5 or $10^6 M_\odot$ MBHs appears to be an open one.

There also is a question about whether the MBHs will modify the star distribution near the centre of the new structure enough so that the dynamical friction will be decreased considerably. Calculations indicate that this may occur when the MBHs are fairly close together, but not yet close enough to coalesce by gravitational radiation in less than a Hubble time (see, e.g., Makino [71]. On the other hand, conditions such as Brownian motion of the MBH and tri-axiality in the structure may affect the results, and there also may be complex enough motions going on soon after a merger to change the conclusions. Thus the effectiveness of this 'hang-up' in the coalescence process is not known.

Another question arises if another merger occurs before the MBHs from an earlier merger have coalesced. In principle, one, two or all three of the MBHs could be thrown out of the new structure by the slingshot effect. However, the new merger could even have a beneficial effect, if the resulting disturbances in the central star density overcame the possible hang-up for the earlier two MBHs and allowed them to coalesce.

The signals arising from MBH coalescences after mergers of galaxies or of pre-galactic structures would be very large. Figure 10.11 shows several cases

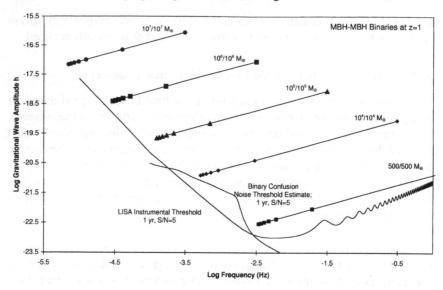

Figure 10.11. Strain amplitudes during the last year before MBH–MBH coalescence. The curves are for MBH–MBH binaries at $z = 1$ with circular orbits, and for one year of observations with S/N = 5. For a given mass combination, the first five points from the left are for 1.0, 0.8, 0.6, 0.4, and 0.2 months before coalescence. The last two points are for 0.5 week before and for very roughly the last stable orbit for Schwarzschild MBHs.

of the signal strength as a function of frequency during the last year before coalescence for events at a redshift of $z = 1$ and for circular orbits. The square symbols show the time at 0.2 year intervals, so the first is one year before, the second is 0.8 year before, etc. The last symbol is shifted slightly to be 0.5 week before coalescence, instead of at that time. The cases shown are for equal MBH masses, and only the spiral-in part of the event before the last stable orbit is reached is considered. However, it should be remembered that the detectability of the signal for the spiral-in phase depends mainly on the chirp mass, so the curves given can be used to estimate conditions under which unequal mass coalescences would be observable also. The case of possible coalescence of $500 M_\odot$ seed black holes during the early growth of MBHs that was discussed earlier is shown for $z = 1$ for comparison.

Since the LISA threshold sensitivity curve and the confusion noise estimate are defined on the assumption that the frequency and signal strength are fixed for the one year period of the observation, and the MBH–MBH coalescence signals change dramatically during the year, it is necessary to integrate the square of the amplitude S/N ratio during the year and then take the square root to obtain the effective S/N ratio. The way in which the effective S/N ratio builds up during the year, with the LISA instrumental noise and the confusion noise combined quadratically, is shown in figure 10.12. What is shown for each of the cases from

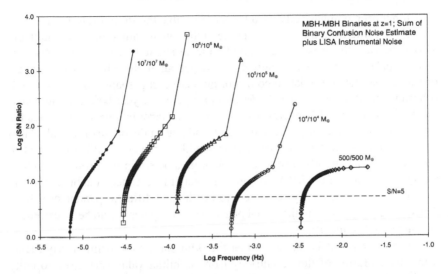

Figure 10.12. Cumulative weekly S/N ratios during the last year before MBH–MBH coalescence at $z = 1$. Both the instrumental noise and the estimated binary confusion noise are included.

figure 10.11 is the accumulated S/N ratio after the first week, the second week, etc, up until the end of the year. The large jump in the last week or so, in all but the $500M_\odot$ case, is due to the frequency of the signal having swept up to where the LISA sensitivity is much higher.

It is clear by extrapolating roughly from figures 10.11 and 10.12 that MBH–MBH coalescences resulting from structure mergers could be seen clearly even at redshifts of 10 for $10^7 M_\odot$ MBHs and 20 or more for less massive ones. Thus, any such events at any plausible occurrence time will be observable. Particularly because of the large uncertainties concerning what happens for roughly $10^6 M_\odot$ or smaller MBHs after mergers of structures, the event rate for LISA apparently could range from less than one per decade to quite a large value.

10.2.5   Fundamental physics tests with LISA

Of similar importance to the astrophysical questions about MBHs discussed in the last three sections is the fundamental physics question of whether Einstein's general relativity theory is correct. Very strong tests of the theory can be provided either by MBH–MBH coalescences with high S/N ratio, if they are observed, or by highly unequal mass coalescences of 5 or $10M_\odot$ black holes with MBHs in galactic centres. Only the latter case will be discussed here, since the much larger number of orbital periods observable and the substantial orbit eccentricity expected are likely to be more important for testing relativity than a higher S/N ratio.

It has been emphasized by Cutler *et al* [72] that the strongest test of a theory is likely to come from the observed phase of the signal, rather than the amplitude. This is because the cross-correlation of a theoretical template with the observed signal will be reduced substantially if the two get out of phase by even half a cycle or less during the entire data record. For our case of a perhaps $10M_\odot$ black hole orbiting around a 10^5 or $10^6 M_\odot$ MBH during the last year before coalescence, there will be roughly 10^5 cycles. Even a very small error in the metric will lead to a continuously increasing error in the orbit, and thus in the phase of the calculated signal. While some type of metric error could, in principle, affect the orbit in the same way as slightly different values of parameters in the problem, such as the spin and mass of the MBH, it seems unlikely that a conceptual breakdown in the theory would be nearly equivalent to just having different parameter values. Thus, correctly predicting the signal phase over 10^5 cycles would be an extremely strong test of the theory.

For this case, it is important that the orbit started out being nearly radial. The later evolution of the eccentricity from its initial value very close to unity can be calculated from the rates of loss of energy and angular momentum due to gravitational radiation. The results obtained by Tanaka *et al* for a Schwarzschild MBH [73] show for a typical case that the orbit never becomes circular, but instead remains substantially eccentric (0.5 ?) up until the final plunge begins. However, rigorous calculations will be required in order to fit the data well [74].

During the last year, the periapsis distance changes only very slowly, but the apoapsis distance decreases substantially. At periapsis, the speed is about half that of light, so the dynamics are highly non-Newtonian. In fact, the precession of periapsis during one radial motion period can be about a whole cycle. With relativistic beaming of the gravitational radiation and variation of the strength of the radiation with time during a radial period, the amplitude observed in a given direction can vary in a quite complex way. For an orbit plane that is not perpendicular to the spin axis for a rapidly rotating MBH, rapid Lense–Thirring precession also will be present. In view of the complexity of the relativistic motion and the large number of cycles over which the phase of the signal can be followed, such a signal would give a nearly ideal test of the predictions of general relativity.

It is of course necessary to be able to detect the signal in order to test the theory. It was mentioned earlier that a S/N ratio of about ten would be needed in order to detect such a complex signal. But there also is the question of how difficult the search for such a signal would be. A very crude estimate of the number of templates needed for a brute force search with one year of data gives, within a couple of orders of magnitude, something like 10^{18}. Even with rapid advances in computing power, such a search probably would not be possible. Some improvement could be made by a hierarchical search strategy, but it is not clear that this approach would be sufficient. On the other hand, more powerful search algorithms such as 'genetic algorithms' and 'stimulated annealing' have been developed and demonstrated for substantially more challenging search problems. Consideration of such algorithms for use with LISA data is just starting, but it currently appears unlikely that the difficulty of the search problem will be a real limitation in the use of the LISA data, even for the case of highly unequal mass binaries. For all other expected types of LISA sources, the search problem is much easier than for ground-based detectors because of the roughly 10^4 times lower number of data points that have to be handled for a year of observations.

The other important question is whether the ability to calculate theoretical templates will have improved enough by the time LISA data is available to carry out a thorough test of general relativity. For comparable mass MBHs, there is a very long way to go to accomplish this. However, for the highly unequal mass case, the chances for fairly rapid progress seem much better. It is hoped that numerical methods can be developed that start from the test mass approximation, and converge moderately well. Still, since small changes in the initial conditions can lead to very large changes in the motion a year later, it is important that this problem as well as the search problem for the highly unequal mass case be pursued vigorously in the next few years.

For either detailed tests of general relativity with LISA or for studies of astrophysical questions concerning MBHs, the first requirement is that some signals involving MBHs be seen. While it seems likely that several of the types of MBH sources discussed earlier will be observed, this certainly is not guaranteed. Thus, the situation is somewhat like that for ground-based observations, where the detection of signals within the next decade seems quite likely, but is not certain. Still, a reasonable prediction is that both ground-based detectors in the next decade and LISA not too much later will detect the desired types of signals, and extremely strong tests of general relativity will be among the most important scientific results.

Another fundamental physics test that LISA may contribute to concerns the possible existence of a primordial gravitational-wave background [75, 76]. Standard inflation theory predicts a nearly scale-invariant spectrum, in which the spectral amplitude for the gravitational waves falls off at about the -1.5 power of the frequency. However, the observed COBE microwave background spectrum is believed to be determined mainly by the large-scale density fluctuations at the time of decoupling, and with a scale invariant spectrum would predict a very low amplitude in the frequency range of LISA and of ground-based detectors.

There are theories that could give a non-scale-invariant spectrum and detectable amplitudes for a cosmic background at LISA or ground-based frequencies. LISA could detect an isotropic cosmic background near 10 mHz if its energy density were roughly 10^{-11} of the closure density. A candidate for giving a detectable background is a phase transition at about the electro-weak energy scale, but it would have to be strongly first order in order to give enough amplitude. But this is not currently thought to be likely. Another possibility is associated with suggested effects of extra dimensions, which may have become very small at a time which would produce a peak in the gravitational-wave spectrum near the

LISA frequency band. In any case, LISA will improve the limits on a possible cosmic background in its frequency band, but the chances of seeing anything appear to be very uncertain.

10.2.6 Future prospects

Provided that signals involving massive black holes are indeed seen by LISA, it is likely that there will be an opportunity for a later advanced gravitational-wave mission. Depending on what is seen, there are several directions in which major instrumental improvements could be made. If improvements in sensitivity at frequencies above 3 mHz are of most interest, this can be achieved by increasing the telescope size and the laser power, provided that comparable reductions in other noise sources such as beam pointing jitter and phase measurement errors also can be made. Shortening the antenna arm length also would be desirable. If improvements below about 0.1 mHz are the main objective, then improved free mass sensors and longer arm lengths would be needed.

As an example of a mission to achieve strongly increased high-frequency sensitivity, a goal of using 10 W lasers and 1 m diameter telescopes might be chosen, along with a reduction in the antenna arm length to 50 000 km. The resulting factor of about 10^7 increase in the received power would reduce the shot noise contribution to measuring changes in the arm length difference from about 1×10^{-11} m/rtHz to 3×10^{-15} m/rtHz, but the signal strength for frequencies below about 1 Hz would be reduced also because of the reduced arm lengths. If other errors such as from beam pointing jitter and from the phase measurements only double the distance measurement error, the level of the LISA threshold sensitivity curve can be reduced by a factor ranging from about 20 at 10 mHz to 3000 at 1 Hz, and then retaining that value at higher frequencies. However, with the same assumptions about the confusion noise level due to extragalactic CWDBs as used for figures 10.8–10.11, and plausible estimates for extragalactic neutron star and black hole binaries, the overall sensitivity will be limited more by the confusion noise than the instrumental noise up to at least 100 mHz.

One strategy that is not planned for LISA but probably would be used in a later high-frequency mission with shorter arm lengths is to apply enough force to each test mass to keep the rates of change of the arm lengths constant. This is because the force required is about a factor 6000 less for 50 000 km arms, or roughly 4×10^{-10} m s^{-2}. Also, the main emphasis is on the noise level above 3 mHz, where keeping fluctuations in the applied voltages low is easier than at lower frequencies. This would make the phase measurements very much simpler than they would be with substantial Doppler shifts.

Quite a few types of measurements could be improved considerably with the above sensitivity. For example, gradual coalescences of $10 M_\odot$ black holes with MBHs with masses down to $10^3 M_\odot$ could be observed virtually anywhere in the universe, if they occur. And tests of general relativity for coalescences with highly unequal masses would have considerably higher S/N ratios, as well as large numbers of cycles. In addition, sensitivity to events involved in the initial formation of intermediate mass black holes would be improved.

For improved measurements at low frequencies, below about 0.1 mHz, the most important goals are reduced noise levels from the free mass sensors and longer arm lengths. Moderate goals might be a factor ten reduction in the free mass sensor noise level plus a factor three increase in the arm length. Basically the same geometry could be used as for LISA, but the antenna probably would have to be considerably farther from the Earth to keep the arm length changes from being too large. This antenna would make it possible to observe MBH–MBH binaries with both masses above about $10^5 M_\odot$ much longer before coalescence, and thus increase the number of such events observed by a large factor.

A much more challenging goal would be to try for roughly 1 AU arm lengths and a factor 100 reduction in the free mass sensor noise. This would permit improved measurements mainly at frequencies below about 0.02 mHz, assuming that the confusion noise level is nearly constant below this frequency, as given in Hils, Bender and Webbink [39]. This would further push back the time before coalescence when MBH–MBH binaries could be observed, and probably tell us considerably more about such coalescences after mergers during the process of galaxy formation.

One possible geometry would be to locate spacecraft near the $L4$ and $L5$ points of the Earth–Sun system, $60°$ in front of and behind the Earth, and near either the $L1$ or $L2$ point, within about 1500 000 km of the Earth. In this case the third arm would be about $3^{0.5}$ times the length of the other two. Because the $L1$ and $L2$ points are unstable, putting the third spacecraft near the $L3$ point on the opposite side of the Sun from the Earth and various other possibilities would be considered. One disadvantage of such geometries is that the antenna is in the ecliptic plane, and the sensitivity to the ecliptic latitude of sources at low latitudes is reduced. However, the extra impulse required to go considerably out of the ecliptic is large.

It is interesting to note that the requirement on the accuracy of distance measurements between the test masses for frequencies below 0.1 mHz are much less severe than those for LISA, even for roughly 1 AU arm lengths. The distance measurement error would have to be about 1×10^{-8} m/rtHz to equal the limitation due to the confusion noise at 0.1 mHz, and more than a factor $(0.1 \text{ mHz}/f)^2$ larger at a lower frequency f. Thus the required combination of laser power and telescope size will be affected more by considerations such as staying sufficiently above possible noise in the scattered light level rather than by the desire to keep the shot noise level low. In any case, 10 W of laser power and 1 m diameter telescopes would at least be adequate, since it would give the same received power as for LISA.

Achieving a factor 100 lower free mass sensor noise than for LISA would be a major challenge. It might be desirable, for example, to consider having each free mass sensor in a small separate slave spacecraft containing as little other equipment as possible in order to minimize disturbances. Several such slave spacecraft could be included if they could be made much smaller than the main spacecraft. With the increased tolerance on the distance measurement uncertainty below 0.1 mHz, a small local diode laser system could track changes in the internal geometry of each combination of a spacecraft plus its slaves. However, this approach would help only if the main noise sources in each free mass sensor were independent. This would not be the case if common fluctuations in the solar intensity contributed substantially to the noise.

In view of the apparent advantages of improving the sensitivity at both high and low frequencies after the LISA mission, it may be desirable to consider flying two separate antennae, with one at least somewhat like the high frequency antenna discussed earlier and the other like the moderately improved low frequency antenna. However, questions such as this can be addressed much more easily after some data from LISA have been received.

Acknowledgments

It is a pleasure to thank all of the people who have been involved in the various studies of LISA, including, in particular, Karsten Danzmann, Jim Hough, Bernard Schutz, Stefano Vitale and the rest of our European colleagues. In the USA, Rai Weiss and Ron Drever and later Mark Vincent were very helpful in getting the discussions started; Kip Thorne and Jan Hall have provided constant information and encouragement throughout; and Bill Folkner, Sterl Phinney, Doug Richstone, Ron Webbink, John Armstrong, Ron Hellings and other members of the LISA Mission Definition Team have been essential to the progress that has been achieved. The 1999–2000 ESA Industrial Study has added a number of other people who have made important contributions to future progress. At JILA, essential ideas and long-term creative efforts from Tuck Stebbins, Dieter Hils and Jim Faller have been responsible for much of our present understanding of the opportunity presented by the LISA mission.

References

[1] Press W H and Thorne K S 1972 *Ann. Rev. Astron. Astrophys.* **10** 335–71
[2] Moss G E, Miller L R and Forward R L 1971 *Appl. Opt.* **10** 2495
[3] Faller J E and Bender P L 1984 *Precision Measurements and Fundamental Constants II (NBS Spec. Pub. 617)* (Washington, DC: US Govt Printing Office) pp 689–90
[4] Faller J E, Bender P L, Hall J L, Hils D and Vincent M A 1985 *Proc. Colloq. on Kilometric Optical Arrays in Space, Cargese (Corsica), ESA Report SP-226* pp 157–63
[5] Laser Interferometer Space Antenna 1998 *Proc. 2nd Int. LISA Symp. (AIP Conf. Proc. 456)* ed W M Folkner (Woodbury, NY: American Institute of Physics) pp 3–239
[6] LISA 1998 *Pre-Phase A Report* MPQ-233, 2nd edn, ed K V Danzmann (Garching: Max-Planck Institute for Quantum Optics) pp 1–191
[7] 1997 *Class. Quantum Grav.* **14** 1397–585

[8] Folkner W M 1998 Laser Interferometer Space Antenna *Proc. 2nd Int. LISA Symp. (AIP Conf. Proc. 456)* ed W M Folkner (Woodbury, NY: American Institute of Physics) pp 11–16

[9] Stebbins R T 1998 Laser Interferometer Space Antenna *Proc. 2nd Int. LISA Symp. (AIP Conf. Proc. 456)* ed W M Folkner (Woodbury, NY: American Institute of Physics) pp 17–23

[10] Whalley M S, Turner R F and Sandford M C W 1998 Laser Interferometer Space Antenna *Proc. 2nd Int. LISA Symp. (AIP Conf. Proc. 456)* ed W M Folkner (Woodbury, NY: American Institute of Physics) pp 24–30

[11] Peskett S, Kent B, Whalley M and Sandford M 1998 Laser Interferometer Space Antenna *Proc. 2nd Int. LISA Symp. (AIP Conf. Proc. 456)* ed W M Folkner (Woodbury, NY: American Institute of Physics) pp 31–8

[12] Giampieri G, Hellings R W, Tinto M and Faller J E 1996 *Opt. Commun.* **123** 669–78

[13] Armstrong J W, Estabrook F B and Tinto M 1999 *Ap. J.* **527** 814–26

[14] Estabrook F B, Tinto M and Armstrong J W 1999 *Ap. J.* **527** 814

[15] Cutler C 1998 *Phys. Rev.* D **70** 89–102

[16] Staff of Space Science Department, Johns Hopkins University Applied Physics Laboratory and Staff of Guidance and Control Laboratory, Stanford University 1974 *AIAA J. Spacecraft* **11** 637–44

[17] Lange B 1964 *AIAA J.* **2** 1590–606

[18] Touboul P, Rodrigues M and Le Clerc G M 1996 *Class. Quantum Grav.* **13** A259–70 See also section 3.3 of LISA 1996 *Pre-Phase A Report* MPQ-208, 1st edn, ed K V Danzmann (Garching: Max-Planck Institute for Quantum Optics) pp 60–75

[19] Speake C C 1996 *Class. Quantum Grav.* **13** A291

[20] Vitale S and Speake C C 1998 Laser Interferometer Space Antenna *Proc. 2nd Int. LISA Symp. (AIP Conf. Proc. 456)* ed W M Folkner (Woodbury, NY: American Institute of Physics) pp 172–7

[21] Josselin V, Rodrigues M and Touboul P 1998 Laser Interferometer Space Antenna *Proc. 2nd Int. LISA Symp. (AIP Conf. Proc. 456)* ed W M Folkner (Woodbury, NY: American Institute of Physics) pp 236–9

[22] Wollard M 1984 Short-period oscillations in the total solar irradiance *Thesis* University of California, San Diego, CA

[23] Fehringer M, Ruedenauer F and Steiger W 1998 Laser Interferometer Space Antenna *Proc. 2nd Int. LISA Symp. (AIP Conf. Proc. 456)* ed W M Folkner (Woodbury, NY: American Institute of Physics) pp 207–13

[24] Thorne K 1987 *300 Years of Gravitation* ed S W Hawking and W Israel (Cambridge: Cambridge University Press) pp 330–458

[25] Schilling R 1997 *Class. Quantum Grav.* **14** 1513–19

[26] Giampieri G 1997 *Mon. Not. R. Astron. Soc.* **289** 185–95

[27] Mironowskii V N 1966 *Sov. Astron.* **9** 752–5

[28] Iben I 1984 *Paper Presented at Workshop on Grav. Waves and Axions* (Santa Barbara, CA: Institute for Theoretical Physics)

[29] Marsh T R 1995 *Mon. Not. R. Astron. Soc.* **275** L1–5

[30] Marsh T R, Dhillon V S and Duck S R 1995 *Mon. Not. R. Astron. Soc.* **275** 828–40

[31] Moran C, Marsh T R and Bragaglia A 1997 *Mon. Not. R. Astron. Soc.* **288** 538–44

[32] Maxted P F L and Marsh T R 1999 *Mon. Not. R. Astron. Soc.* **307** 122–30

[33] Webbink R F and Han Z 1998 Laser Interferometer Space Antenna *Proc. 2nd Int. LISA Symp. (AIP Conf. Proc. 456)* ed W M Folkner (Woodbury, NY: American

Institute of Physics) pp 61–7

[34] Nelemans G, Verbunt F, Yungelson L R and Portegies Zwart S F 2000 *Astron. Astrophys.* **360** 1011–18 (astro-ph/0006216)

[35] Hils D, Bender P L, Faller J E and Webbink R F 1986 *11th Int. Conf. On General Relativity and Gravitation: Abstracts of Cont. Papers* vol 2 (University of Stockholm) p 509

[36] Lipunov V M, Postnov K A and Prokhorov M E 1987 *Astron. Astrophys.* **176** L1–4

[37] Lipunov V M and Postnov K A 1987 *Sov. Astron.* **31** 228–30

[38] Evans C R, Iben I and Smarr L 1987 *Ap. J.* **323** 129–39

[39] Hils D, Bender P L and Webbink R F 1990 *Ap. J.* **360** 75–94

[40] Warner B 1995 *Astrophys. Space Sci.* **225** 249–70

[41] Verbunt F 1997 *Class. Quantum Grav.* **14** 1417–23

[42] Verbunt F 1998 *Proc. 2nd Amaldi Conf. on Gravitational Waves* ed E Coccia, G Pizella and G Veneziano (Singapore: World Scientific) pp 50–61

[43] Nelemans G, Portegies Zwart S F and Verbunt F 2000 Gravitational waves and experimental relativity *Proc. 36th Rencontres de Moriond (Les Arc, Jan. 1999)* (Hanoi: World Publishers) pp 119–24

[44] Bender P L and Hils D 1997 *Class. Quantum Grav.* **14** 1439–44

[45] Hils D 1998 Laser Interferometer Space Antenna *Proc. 2nd Int. LISA Symp. (AIP Conf. Proc. 456)* ed W M Folkner (Woodbury, NY: American Institute of Physics) pp 68–78

[46] Hils D and Bender P L 2000 *Ap. J.* **537** 334–41

[47] Kosenko D I and Postnov K A 1998 *Astron. Astrophys.* **336** 786–90

[48] Stebbins R T, Bender P L and Folkner W M 1997 *Class. Quantum Grav.* **14** 1499–505

[49] Richstone D *et al* 1998 *Nature* **395** A14–19

[50] Miyoshi M *et al* 1995 *Nature* **373** 127–9

[51] Maoz E 1998 *Ap. J. Lett.* **494** L181–4

[52] Ghez A M, Klein B L, Morris M and Becklin E E 1998 *Ap. J.* **509** 678–86

[53] Genzel R, Pichon C, Eckart A, Gerhard O E and Ott T 2000 *Mon. Not. R. Astron. Soc.* **317** 348–74 (astro-ph/0001428)

[54] Quinlan G D and Shapiro S L 1990 *Ap. J.* **356** 483–500

[55] Lee H M 1995 *Mon. Not. R. Astron. Soc.* **272** 605–17

[56] Haehnelt M G and Rees M J 1993 *Mon. Not. R. Astron. Soc.* **263** 168–78

[57] Rees M J 1997 *Rev. Mod. Astron.* **10** 179–89

[58] Haehnelt M G 1998 Laser Interferometer Space Antenna *Proc. 2nd Int. LISA Symp. (AIP Conf. Proc. 456)* ed W M Folkner (Woodbury, NY: American Institute of Physics) pp 45–9

[59] Baumgarte T and Shapiro S L 1999 *Ap. J.* **526** 941–52

[60] Lee M H 1993 *Ap. J.* **418** 147–62

[61] Gebhardt K *et al* 2000 *Ap. J.* in press (astro-ph/0006289)

[62] Haiman Z and Loeb A 1998 *Ap. J.* **503** 505–17

[63] Silk J and Rees M J 1998 *Astron. Astrophys.* **331** L1–4

[64] Kauffmann G and Haehnelt M 2000 *Mon. Not. R. Astron. Soc.* **311** 576–88

[65] Hils D and Bender P L 1995 *Ap. J. Lett.* **445** L7–10

[66] Sigurdsson S and Rees M J 1997 *Mon. Not. R. Astron Soc.* **284** 318–26

[67] Sigurdsson S 1998 Laser Interferometer Space Antenna *Proc. 2nd Int. LISA Symp. (AIP Conf. Proc. 456)* ed W M Folkner (Woodbury, NY: American Institute of Physics) pp 53–6

[68] Miralda-Escuda J and Gould A 2000 *Preprint* astro-ph/0003269
[69] Haehnelt M G 1994 *Mon. Not. R. Astron. Soc.* **269** 199–208
[70] Xu G and Ostriker J P 1994 *Ap. J.* **437** 184–93
[71] Makino J 1997 *Ap. J.* **478** 58–65
[72] Cutler C *et al* 1993 *Phys. Rev. Lett.* **70** 2984–7
[73] Tanaka T, Shibata M, Sasaki M, Tagoshi H and Nakamura T 1993 *Prog. Theor. Phys.* **90** 65–83
[74] Shibata M 1994 *Phys. Rev.* D **50** 6297–311
[75] Allen B 1996 *Les Houches School on Astrophysical Sources of Gravitational Waves* ed J-A Marck and J-P Lasota (Cambridge: Cambridge University Press)
[76] Hogan C J 1998 Laser Interferometer Space Antenna *Proc. 2nd Int. LISA Symp. (AIP Conf. Proc. 456)* ed W M Folkner (Woodbury, NY: American Institute of Physics) pp 79–86

Chapter 11

Detection of scalar gravitational waves

Francesco Fucito
INFN, sez. di Roma 2, Via della Ricerca Scientifica, 00133 Rome,
Italy
E-mail: Fucito@roma2.infn.it

In this talk I review recent progress in the detection of scalar gravitational waves. Furthermore, in the framework of the Jordan–Brans–Dicke theory, I compute the signal-to-noise ratio for a resonant mass detector of spherical shape and for binary sources and collapsing stars. Finally, I compare these results with those obtained from laser interferometers and from Einsteinian gravity.

11.1   Introduction

The efforts aimed at the detection of gravitational waves (GW) started more than a quarter of a century ago and have been, so far, unsuccessful [1,2]. Resonant bars have proved their reliability, being capable of continous data gathering for long periods of time [3,4]. Their energy sensitivity has improved by more than four orders of magnitude since Weber's pioneering experiment. However, a further improvement is still necessary to achieve successful detection. While further developments of bar detectors are underway, two new generations of Earth-based experiments have been proposed: detectors based on large laser interferometers are already under construction [5] and resonant detectors of spherical shape are under study [2].

In this lecture I report on a series of papers [6] in which the opportunity of introducing resonant mass detectors of spherical shape was studied. As a general motivation for their study, spherical detectors have the advantage over bar-shaped detectors of a larger degree of symmetry. This translates into the possibility of building detectors of greater mass and consequently of higher cross section.

Besides this obvious observation, the higher degree of symmetry enjoyed by the spherical shape puts such a detector in the unique position of being able to detect GWs with a spin content different from two. This is a means of testing non-Einsteinian theories of gravity.

I would now like to remind the reader of the very special position of Einstein's general relativity (GR) among the possible gravitational theories. Theories of gravitation, in fact, can be divided into two families: metric and non-metric theories [7]. The former can be defined to be all theories obeying the following three postulates:

- spacetime is endowed with a metric;
- the world lines of test particles are geodesic of the above-mentioned metric;
- in local free-falling frames, the non-gravitational laws of physics are those of special relativity.

It is an obvious consequence of these postulates that a metric theory obeys the principle of equivalence. More succinctly a theory is said to be metric if the action of gravitation on the matter sector is due exclusively to the metric tensor. GR is the most famous example of a metric theory. Kaluza–Klein-type theories, also belong to this class along with the Brans–Dicke theory. Different representatives of this class differ by their equations of motion which in turn can be deduced from a Lagrangian principle. Since there seems to be no compelling experimental or theoretical reasons to introduce non-Einsteinian or non-metric theories, they are sometimes considered a curiosity. This point should perhaps be reconsidered. String theories are, in fact, the most serious candidate for a theory of quantum gravity, the standard cosmological model has been emended with the introduction of inflation and even the introduction of a cosmological constant (which seems to be needed to explain recent cosmological data) could imply the existence of other gravitationally coupled fields. In all of the above cited cases we are forced to introduce fields which are non-metrically coupled in the sense explained above.

In the first section of this lecture we will explain that a spherical detector is able to detect any spin component of an impinging GW. Moreover, its vibrational eigenvalues can be divided into two sets called spheroidal and toroidal. Only the first set couples to the metric. This leads to the opportunity of using such a detector as a veto for non-Einsteinian theories. In the second section we take as a model the Jordan–Brans–Dicke (JBD), in which along with the metric we also have a scalar field which is metrically coupled. We are then able to study the signal-to-noise ratio for sources such as binary systems and collapsing stars and compare the strength of the scalar signal with respect to the tensor one. Finally, in the third and last section we repeat this computation in the case of the hollow sphere which seems to be the detector which is most likely to be built.

11.2 Testing theories of gravity

11.2.1 Free vibrations of an elastic sphere

Before discussing the interaction with an external GW field, let us consider the basic equations governing the free vibrations of a perfectly homogeneous, isotropic sphere of radius R, made of a material having density ρ and Lamé coefficients λ and μ [8].

Following the notation of [9], let x_i, $i = 1, 2, 3$ be the equilibrium position of the element of the elastic sphere and x'_i be the deformed position, then $u_i = x'_i - x_i$ is the displacement vector. Such vector is assumed small, so that the linear theory of elasticity is applicable. The strain tensor is defined as $u_{ij} = (1/2)(u_{i,j} + u_{j,i})$ and is related to the stress tensor by $\sigma_{ij} = \delta_{ij}\lambda u_{ll} + 2\mu u_{ij}$. The equations of motion of the free vibrating sphere are thus [8]

$$\rho \frac{\partial^2 u_i}{\partial t^2} = \frac{\partial}{\partial x^j}(\delta_{ij}\lambda u_{ll} + 2\mu u_{ij}) \qquad (11.1)$$

with the boundary condition:

$$n_j \sigma_{ij} = 0 \qquad (11.2)$$

at $r = R$ where $n_i \equiv x_i/r$ is the unit normal. These conditions simply state that the surface of the sphere is free to vibrate. The displacement u_i is a time-dependent vector, whose time dependence can be factorized as $u_i(\vec{x}, t) = u_i(\vec{x})\exp(i\omega t)$, where ω is the frequency. The equations of motion then become:

$$\mu\nabla^2 u_i + (\lambda + \mu)\nabla_i(\nabla_j u_j) = -\omega^2\rho u_i. \qquad (11.3)$$

Their solutions can be expressed as a sum of a longitudinal and two transverse vectors [10]:

$$\vec{u}(\vec{x}) = C_0\vec{\nabla}\phi(\vec{x}) + C_1\vec{L}\chi(\vec{x}) + C_2\vec{\nabla} \times \vec{L}\chi(\vec{x}) \qquad (11.4)$$

where C_0, C_1, C_2 are dimensioned constants and $\vec{L} \equiv \vec{x} \times \vec{\nabla}$ is the angular momentum operator. Regularity at $r = 0$ restricts the scalar functions ϕ and χ to be expressed as $\phi(r, \theta, \varphi) \equiv j_l(qr)Y_{lm}(\theta, \varphi)$ and $\chi(r, \theta, \varphi) \equiv j_l(kr)Y_{lm}(\theta, \varphi)$. $Y_{lm}(\theta, \varphi)$ are the spherical harmonics and j_l the spherical Bessel functions [11]:

$$j_l(x) = \left(\frac{1}{x}\frac{d}{dx}\right)^l \left(\frac{\sin x}{x}\right) \qquad (11.5)$$

$q^2 \equiv \rho\omega^2/(\lambda + 2\mu)$ and $k^2 \equiv \rho\omega^2/\mu$ are the longitudinal and transverse wavevectors, respectively.

Imposing the boundary conditions (11.2) at $r = R$ yields two families of solutions:

- *Toroidal* modes: these are obtained by setting $C_0 = C_2 = 0$, and $C_1 \neq 0$. In this case the displacements in (11.4) can be written in terms of the basis:

$$\vec{\psi}_{nlm}^T (r, \theta, \varphi) = T_{nl}(r) \vec{L} Y_{lm}(\theta, \varphi) \tag{11.6}$$

with $T_{nl}(r)$ proportional to $j_l(k_{nl}r)$. The eigenfrequencies are determined by the boundary conditions (11.2) which read [10]

$$f_1(kR) = 0 \tag{11.7}$$

where

$$f_1(z) \equiv \frac{\mathrm{d}}{\mathrm{d}z}\left[\frac{j_l(z)}{z}\right]. \tag{11.8}$$

- *Spheroidal* modes: these are obtained by setting $C_1 = 0$, $C_0 \neq 0$ and $C_2 \neq 0$. The displacements of (11.4) can be expanded in the basis

$$\vec{\psi}_{nlm}^S (\vec{x}) = A_{nl}(r) Y_{lm}(\theta, \varphi)\vec{n} - B_{nl}(r)\vec{n} \times \vec{L} Y_{lm}(\theta, \varphi) \tag{11.9}$$

where $A_{nl}(r)$ and $B_{nl}(r)$ are dimensionless radial eigenfunctions [9], which can be expressed in terms of the spherical Bessel functions and their derivatives. The eigenfrequencies are determined by the boundary conditions (11.2) which read [9]

$$\det\begin{pmatrix} f_2(qR) - \frac{\lambda}{2\mu}q^2R^2 f_0(qR) & l(l+1)f_1(kR) \\ f_1(qR) & \frac{1}{2}f_2(kR) + [\frac{l(l+1)}{2} - 1]f_0(kR) \end{pmatrix} = 0 \tag{11.10}$$

where

$$f_0(z) \equiv \frac{j_l(z)}{z^2}, \quad f_2(z) \equiv \frac{\mathrm{d}^2}{\mathrm{d}z^2}j_l(z). \tag{11.11}$$

The eigenfrequencies can be determined numerically for both toroidal and spheroidal vibrations. Each mode of order l is $(2l + 1)$-fold degenerate. The eigenfrequency values can be obtained from

$$\omega_{nl} = \sqrt{\frac{\mu}{\rho}} \frac{(kR)_{nl}}{R}. \tag{11.12}$$

11.2.2 Interaction of a metric GW with the sphere vibrational modes

The detector is assumed to be non-relativistic (with sound velocity $v_s \ll c$ and radius $R \ll \lambda$ the GW wavelength) and endowed with a high quality factor ($Q_{nl} = \omega_{nl}\tau_{nl} \gg 1$, where τ_{nl} is the decay time of the mode nl). The displacement $\vec{u}$ of a point in the detector can be decomposed in normal modes as:

$$\vec{u}(\vec{x}, t) = \sum_N A_N(t)\vec{\psi}_N(\vec{x}) \tag{11.13}$$

where N collectively denotes the set of quantum numbers identifying the mode. The basic equation governing the response of the detector is [12]

$$\ddot{A}_N(t) + \tau_N^{-1}\dot{A}_N(t) + \omega_N^2 A_N(t) = f_N(t). \tag{11.14}$$

We assume that the gravitational interaction obeys the principle of equivalence which has been experimentally supported to high accuracy. In terms of the so-called electric components of the Riemann tensor $E_{ij} \equiv R_{0i0j}$, the driving force $f_N(t)$ is then given by [13]

$$f_N(t) = -M^{-1}E_{ij}(t)\int \psi_N^{i*}(\vec{x})x^j\rho\,\mathrm{d}^3x \tag{11.15}$$

where M is the sphere mass and we consider the density ρ as a constant. In any metric theory of gravity E_{ij} is a (3×3) symmetric tensor, which depends on time, but not on spatial coordinates.

Let us now investigate which sphere eigenmodes can be excited by a metric GW, i.e. which sets of quantum numbers N give a non-zero driving force.

(a) Toroidal modes

The eigenmode vector, ψ_{nlm}^{T} can be expressed as in equation (11.6). Up to an adimensional normalization constant C, the driving force is

$$f_N^{(\mathrm{T})}(t) = -\,\mathrm{e}^{-i\omega_N t}\frac{3C}{4\pi R^3}\int_0^R \mathrm{d}r\,r^3 j_l(k_{nl}^{(\mathrm{T})}r)\int_0^\pi \mathrm{d}\theta\,\sin\theta$$

$$\times \int_0^{2\pi}\mathrm{d}\phi\left\{\frac{E_{yy}-E_{xx}}{2}\left(\sin\theta\sin 2\phi\frac{\partial Y_{lm}^*}{\partial\theta} + \cos\theta\cos 2\phi\frac{\partial Y_{lm}^*}{\partial\phi}\right)\right.$$

$$+ E_{xy}\left(\sin\theta\cos 2\phi\frac{\partial Y_{lm}^*}{\partial\theta} - \cos\theta\sin 2\phi\frac{\partial Y_{lm}^*}{\partial\phi}\right)$$

$$+ E_{xz}\left[-\sin\phi\cos\theta\frac{\partial Y_{lm}^*}{\partial\theta} + (\sin\theta\cos\phi - \frac{\cos^2\theta}{\sin\theta}\cos\phi)\frac{\partial Y_{lm}^*}{\partial\phi}\right]$$

$$+ E_{yz}\left[\cos\phi\cos\theta\frac{\partial Y_{lm}^*}{\partial\theta} + (\sin\theta\sin\phi - \frac{\cos^2\theta}{\sin\theta}\sin\phi)\frac{\partial Y_{lm}^*}{\partial\phi}\right]$$

$$+ \left(E_{zz} - \frac{E_{xx}+E_{yy}}{2}\right)\cos\theta\frac{\partial Y_{lm}^*}{\partial\phi}\right\}. \tag{11.16}$$

Using the equations

$$\frac{\partial Y_{lm}^*}{\partial\theta} = (-)^m\left[\frac{2l+1}{4\pi}\frac{(l-m)!}{(l+m)!}\right]^{\frac{1}{2}}\frac{\partial P_l^m(\cos\theta)}{\partial\theta}\mathrm{e}^{-im\phi} \tag{11.17}$$

and

$$\frac{\partial Y_{lm}^*}{\partial\phi} = -im(-)^m\left[\frac{2l+1}{4\pi}\frac{(l-m)!}{(l+m)!}\right]^{\frac{1}{2}}P_l^m(\cos\theta)\mathrm{e}^{-im\phi} \tag{11.18}$$

the integration over ϕ can be performed. Equation (11.16) then contains integrals over θ of the form:

$$\int_0^\pi \left[(\sin^2 \theta - \cos^2 \theta) P_l^{\pm 1}(\cos \theta) - \sin \theta \cos \theta \frac{\partial P_l^{\pm 1}(\cos \theta)}{\partial \theta} \right] d\theta \qquad (11.19)$$

and

$$\int_0^\pi \left[2 \sin \theta \cos \theta P_l^{\pm 2}(\cos \theta) + \sin^2 \theta \frac{\partial P_l^{\pm 2}(\cos \theta)}{\partial \theta} \right] d\theta. \qquad (11.20)$$

After integration by parts, the derivative terms in equations (11.19) and (11.20) exactly cancel the non-derivative ones. The remaining boundary terms vanish too, thanks to the periodicity of the trigonometric functions and to the regularity of the associated Legendre polynomials. The vanishing of these integrals has a profound physical consequence. It means that in any metric theory of gravity the toroidal modes of the sphere cannot be excited by GW and can thus be used as a veto in the detection.

(b) Spheroidal modes

The forcing term is given by:

$$f_N^{(S)}(t) = - M^{-1} E_{ij}(t) \int x^j \left(\frac{x^i}{r} A_N(r) Y_{lm}(\theta, \varphi) \right.$$

$$\left. - B_N(r) e^{ink} \frac{x_n}{r} L_k Y_{lm}(\theta, \varphi) \right) \rho \, d^3x. \qquad (11.21)$$

One is thus led to compute integrals of the following types

$$\int x^j x^i Y_{lm}(\theta, \varphi) \, d^3x \qquad (11.22)$$

and

$$\int x^j x^i L_k Y_{lm}(\theta, \varphi) \, d^3x \qquad (11.23)$$

Since the product $x^i x^j$ can be expressed in terms of the spherical harmonics with $l = 0, 2$ and the angular momentum operator does not change the value of l, one immediately concludes that in any metric theory of gravity only the $l = 0, 2$ spheroidal modes of the sphere can be excited. At the lowest level there are a total of five plus one independent spheroidal modes that can be used for GW detection and study.

11.2.3 Measurements of the sphere vibrations and wave polarization states

From the analysis of the spheroidal modes active for metric GW, we now want to infer the field content of the theory. For this purpose it is convenient to express the Riemann tensor in a null (Newman–Penrose) tetrad basis [7].

To lowest non-trivial order in the perturbation the six independent 'electric' components of the Riemann tensor may be expressed in terms of the Newman–Penrose (NP) parameters as

$$E_{ij} = \begin{pmatrix} -\operatorname{Re}\Psi_4 - \Phi_{22} & \operatorname{Im}\Psi_4 & -2\sqrt{2}\operatorname{Re}\Psi_3 \\ \operatorname{Im}\Psi_4 & \operatorname{Re}\Psi_4 - \Phi_{22} & 2\sqrt{2}\operatorname{Im}\Psi_3 \\ -2\sqrt{2}\operatorname{Re}\Psi_3 & 2\sqrt{2}\operatorname{Im}\Psi_3 & -6\Psi_2 \end{pmatrix}. \tag{11.24}$$

The NP parameters allow the identification of the spin content of the metric theory responsible for the generation of the wave [7]. The classification can be summarized in order of increasing complexity as follows:

- General relativity (spin 2): $\Psi_4 \neq 0$ while $\Psi_2 = \Psi_3 = \Phi_{22} = 0$.
- Tensor–scalar theories (spin 2 and 0): $\Psi_4 \neq 0$, $\Psi_3 = 0$, $\Psi_2 \neq 0$ and/or $\Phi_{22} \neq 0$ (e.g. Brans–Dicke theory with $\Psi_4 \neq 0$, $\Psi_2 = 0$, $\Psi_3 = 0$ and $\Phi_{22} \neq 0$).
- Tensor–vector theories (spin 2 and 1): $\Psi_4 \neq 0$, $\Psi_3 \neq 0$, $\Phi_{22} = \Psi_2 = 0$.
- Most general metric theory (spin 2, 1 and 0): $\Psi_4 \neq 0$, $\Psi_2 \neq 0$, $\Psi_3 \neq 0$ and $\Phi_{22} \neq 0$, (e.g. Kaluza–Klein theories with $\Psi_4 \neq 0$, $\Psi_3 \neq 0$, $\Phi_{22} \neq 0$ while $\Psi_2 = 0$).

In equation (11.24), we have assumed that the wave comes from a localized source with wavevector $\vec{k}$ parallel to the z-axis of the detector frame. In this case the NP parameters (and thus the wave polarization states) can be uniquely determined by monitoring the six lowest spheroidal modes. If the direction of the incoming wave is not known two more unknowns appear in the problem, i.e. the two angles of rotation of the detector frame needed to align $\vec{k}$ along the z-axis. In order to dispose of this problem one can envisage the possibility of combining the pieces of information from an array of detectors [14]. We restrict our attention to the simplest case in which the source direction is known.

In order to infer the value of the NP parameters from the measurements of the excited vibrational modes of the sphere, we decompose E_{ij} in terms of spherical harmonics. In fact, the experimental measurements give the vibrational amplitudes of the sphere modes which are also naturally expanded in the above basis. The use of the same basis makes the connection between the NP parameters and the measured amplitudes straightforward. In formulae

$$E_{ij}(t) = \sum_{l,m} c_{l,m}(t) S_{ij}^{(l,m)} \tag{11.25}$$

where $S_{ij}^{(0,0)} \equiv \delta_{ij}/\sqrt{4\pi}$ (with δ_{ij} the Kronecker symbol) and $S_{ij}^{(2,m)}$ ($m = -2,\ldots,2$) are five linearly independent symmetric and traceless matrices such as

$$S_{ij}^{(l,m)} n^i n^j = Y_{lm}, \quad l = 0,2. \tag{11.26}$$

The vector n_i in equations (11.26) has been defined after equation (11.2).

Taking the scalar product we find

$$c_{0,0}(t) = \frac{4\pi}{3} S_{ij}^{0,0} E_{ij}(t)$$

$$c_{2,m}(t) = \frac{8\pi}{15} S_{ij}^{2,m} E_{ij}(t) \tag{11.27}$$

and then for the NP parameters

$$\Phi_{22} = \sqrt{\frac{5}{16\pi}} c_{2,0}(t) - \sqrt{\frac{1}{4\pi}} c_{0,0}(t), \quad \Psi_2 = -\frac{1}{12}\sqrt{\frac{5}{\pi}} c_{2,0}(t) - \frac{1}{12}\sqrt{\frac{1}{\pi}} c_{0,0}(t)$$

$$\operatorname{Re}\Psi_4 = -\sqrt{\frac{15}{32\pi}}[c_{2,2} + c_{2,-2}], \quad \operatorname{Im}\Psi_4 = -i\sqrt{\frac{15}{32\pi}}[c_{2,2} - c_{2,-2}]$$

$$\operatorname{Re}\Psi_3 = \frac{1}{16}\sqrt{\frac{15}{\pi}}[c_{2,1} - c_{2,-1}], \quad \operatorname{Im}\Psi_3 = \frac{i}{16}\sqrt{\frac{15}{\pi}}[c_{2,1} + c_{2,-1}]. \tag{11.28}$$

Equations (11.28) relate the measurable quantities $c_{l,m}$ with the GW polarization states, described by the NP parameters. Equations (11.28) can be put in correspondence with the output of experimental measurements if $c_{l,m}$ are substituted with their Fourier components at the quadrupole and monopole resonant frequencies which, for the sake of simplicity, we collectively denote by ω_0. $c_{l,m}(\omega_0)$ can be determined in the following way: once the Fourier amplitudes $A_N(\omega_0)$ are measured, by Fourier transforming (11.14) and (11.15) we get the Riemann amplitudes $E_{ij}(\omega_0)$ which, using (11.27), yield the desired result.

In order to determine the $A_N(\omega_0)$ amplitudes from a given GW signal the following two conditions must be fulfilled:

- the vibrational states of the five-fold degenerate quadrupole and monopole modes must be determined. The quadrupole modes can be studied by properly combining the outputs of a set of at least five motion sensors placed in independent positions on the sphere surface. Explicit formulas for practical and elegant configurations of the motion sensors have been reported by various authors [15, 16]. The vibrational state of the monopole mode is provided directly by the output of any of the above-mentioned motion sensors. If resonant motion sensors are used, since the quadrupole and monopole states resonate at different frequencies, a sixth sensor is needed.
- The spectrum of the GW signal must be sufficiently broadband to overlap with the antenna quadrupole and monopole frequencies.

11.3 Gravitational wave radiation in the JBD theory

In this section we analyse the signal emitted by a compact binary system in the JBD theory. We compute the scalar and tensor components of the power radiated by the source and study the scalar waveform. Eventually we consider the detectability of the scalar component of the radiation by interferometers and resonant-mass detectors.

11.3.1   Scalar and Tensor GWs in the JBD Theory

In the Jordan–Fierz frame, in which the scalar field mixes with the metric but decouples from matter, the action reads [17]

$$
\begin{aligned}
S &= S_{\text{grav}}[\phi, g_{\mu\nu}] + S_{\text{m}}[\psi_{\text{m}}, g_{\mu\nu}] \\
&= \frac{c^3}{16\pi} \int d^4x \, \sqrt{-g} \left[\phi R - \frac{\omega_{\text{BD}}}{\phi} g^{\mu\nu} \partial_\mu \phi \partial_\nu \phi \right] + \frac{1}{c} \int d^4x \, L_{\text{m}}[\psi_{\text{m}}, g_{\mu\nu}]
\end{aligned}
$$

$$(11.29)$$

where ω_{BD} is a dimensionless constant, whose lower bound is fixed to be $\omega_{\text{BD}} \approx 600$ by experimental data [18], $g_{\mu\nu}$ is the metric tensor, ϕ is a scalar field and ψ_{m} collectively denotes the matter fields of the theory.

As a preliminary analysis, we perform a weak-field approximation around the background given by a Minkowskian metric and a constant expectation value for the scalar field

$$
\begin{aligned}
g_{\mu\nu} &= \eta_{\mu\nu} + h_{\mu\nu} \\
\varphi &= \varphi_0 + \xi.
\end{aligned}
$$

$$(11.30)$$

The standard parametrization $\varphi_0 = 2(\omega_{\text{BD}}+2)/G(2\omega_{\text{BD}}+3)$, with G the Newton constant, reproduces GR in the limit $\omega_{\text{BD}} \to \infty$, which implies $\varphi_0 \to 1/G$. Defining the new field

$$
\theta_{\mu\nu} = h_{\mu\nu} - \frac{1}{2} \eta_{\mu\nu} h - \eta_{\mu\nu} \frac{\xi}{\varphi_0}
$$

$$(11.31)$$

where h is the trace of the fluctuation $h_{\mu\nu}$, and choosing the gauge

$$
\partial_\mu \theta^{\mu\nu} = 0
$$

$$(11.32)$$

one can write the field equations in the following form

$$
\partial_\alpha \partial^\alpha \theta_{\mu\nu} = -\frac{16\pi}{\varphi_0} \tau_{\mu\nu}
$$

$$(11.33)$$

$$
\partial_\alpha \partial^\alpha \xi = \frac{8\pi}{2\omega_{\text{BD}} + 3} S
$$

$$(11.34)$$

where

$$
\tau_{\mu\nu} = \frac{1}{\varphi_0} (T_{\mu\nu} + t_{\mu\nu})
$$

$$(11.35)$$

$$
S = -\frac{T}{2(2\omega_{\text{BD}} + 3)} \left(1 - \frac{1}{2}\theta - 2\frac{\xi}{\varphi_0} \right) - \frac{1}{16\pi} \left[\frac{1}{2}\partial_\alpha (\theta \partial^\alpha \xi) + \frac{2}{\varphi_0}\partial_a(\xi \partial^\alpha \xi) \right].
$$

$$(11.36)$$

In equation (11.35), $T_{\mu\nu}$ is the matter stress–energy tensor and $t_{\mu\nu}$ is the gravitational stress–energy pseudotensor, that is a function of quadratic order in the weak gravitational fields $\theta_{\mu\nu}$ and ξ. The reason why we have written the field equations at the quadratic order in $\theta_{\mu\nu}$ and ξ is that in this way, as we will see later, the expressions for $\theta_{\mu\nu}$ and ξ include all the terms of order $(v/c)^2$, where v is the typical velocity of the source (Newtonian approximation).

Let us now compute τ^{00} and S at the order $(v/c)^2$. Introducing the Newtonian potential U produced by the rest-mass density ρ

$$U(\vec{x}, t) = \int \frac{\rho(\vec{x}', t)}{|\vec{x} - \vec{x}'|} \, d^3 x' \tag{11.37}$$

the total pressure p and the specific energy density Π (that is the ratio of energy density to rest-mass density) we get (for a more detailed derivation, see [7]):

$$\tau^{00} = \frac{1}{\varphi_0} \rho, \tag{11.38}$$

$$S \simeq -\frac{T}{2(2\omega_{BD} + 3)} \left(1 - \frac{1}{2}\theta - 2\frac{\xi}{\varphi_0} \right)$$
$$= \frac{\rho}{2(2\omega_{BD} + 3)} \left(1 + \Pi - 3\frac{p}{\rho} + \frac{2\omega_{BD} + 1}{\omega_{BD} + 2} U \right). \tag{11.39}$$

Far from the source, equations (11.33) and (11.34) admit wave-like solutions, which are superpositions of terms of the form

$$\theta_{\mu\nu}(x) = A_{\mu\nu}(\vec{x}, \omega) \exp(ik^\alpha x_\alpha) + \text{c.c.} \tag{11.40}$$
$$\xi(x) = B(\vec{x}, \omega) \exp(ik^\alpha x_\alpha) + \text{c.c.} \tag{11.41}$$

Without affecting the gauge condition (11.32), one can impose $h = -2\xi/\varphi_0$ (so that $\theta_{\mu\nu} = h_{\mu\nu}$). Gauging away the superflous components, one can write the amplitude $A_{\mu\nu}$ in terms of the three degrees of freedom corresponding to states with helicities ± 2 and 0 [19]. For a wave travelling in the z-direction, one thus obtains

$$A_{\mu\nu} = \begin{pmatrix} 0 & 0 & 0 & 0 \\ 0 & e_{11} - b & e_{12} & 0 \\ 0 & e_{12} & -e_{11} - b & 0 \\ 0 & 0 & 0 & 0 \end{pmatrix}, \tag{11.42}$$

where $b = B/\varphi_0$.

11.3.2 Power emitted in GWs

The power emitted by a source in GWs depends on the stress–energy pseudotensor $t^{\mu\nu}$ according to the following expression

$$P_{GW} = r^2 \int \Phi \, d\Omega = r^2 \int \langle t^{0k} \rangle \hat{x}_k \, d\Omega \tag{11.43}$$

where r is the radius of a sphere which contains the source, Ω is the solid angle, Φ is the energy flux and the symbol $\langle \cdots \rangle$ implies an average over a region of size much larger than the wavelength of the GW. At the quadratic order in the weak fields we find

$$\langle t_{0z} \rangle = -\hat{z} \frac{\varphi_0 c^4}{32\pi} \left[\frac{4(\omega_{BD} + 1)}{\varphi_0^2} \langle (\partial_0 \xi)(\partial_0 \xi) \rangle + \langle (\partial_0 h_{\alpha\beta})(\partial_0 h^{\alpha\beta}) \rangle \right]. \qquad (11.44)$$

Substituting (11.40) and (11.41) into (11.44), one gets

$$\langle t_{0z} \rangle = -\hat{z} \frac{\varphi_0 c^4 \omega^2}{16\pi} \left[\frac{2(2\omega_{BD} + 3)}{\varphi_0^2} |B|^2 + A^{\alpha\beta*} A_{\alpha\beta} - \frac{1}{2} |A^\alpha{}_\alpha|^2 \right], \qquad (11.45)$$

and using (11.42)

$$\langle t_{0z} \rangle = -\hat{z} \frac{\varphi_0 c^4 \omega^2}{8\pi} \left[|e_{11}|^2 + |e_{12}|^2 + (2\omega_{BD} + 3)|b|^2 \right]. \qquad (11.46)$$

From (11.46) we see that the purely scalar contribution, associated with b and the traceless tensorial contribution, associated with $e_{\mu\nu}$, are completely decoupled and can thus be treated independently.

11.3.3 Power emitted in scalar GWs

We now rewrite the scalar wave solution (11.41) in the following way

$$\xi(\vec{x}, t) = \xi(\vec{x}, \omega) e^{-i\omega t} + \text{c.c.} \qquad (11.47)$$

In vacuo, the spatial part of the previous solution (11.47) satisfies the Helmholtz equation

$$(\nabla^2 + \omega^2)\xi(\vec{x}, \omega) = 0. \qquad (11.48)$$

The solution of (11.48) can be written as

$$\xi(\vec{x}, \omega) = \sum_{jm} X_{jm} h_j^{(1)}(\omega r) Y_{jm}(\theta, \varphi) \qquad (11.49)$$

where $h_j^{(1)}(x)$ are the spherical Hankel functions of the first kind, r is the distance of the source from the observer, $Y_{jm}(\theta, \varphi)$ are the scalar spherical harmonics and the coefficients X_{jm} give the amplitudes of the various multipoles which are present in the scalar radiation field. Solving the inhomogeneous wave equation (11.34), we find

$$X_{jm} = 16\pi i\omega \int_V j_l(\omega r') Y_{lm}^*(\theta, \varphi) S(\vec{x}, \omega) \, dV \qquad (11.50)$$

where $j_l(x)$ are the spherical Bessel functions and r' is a radial coordinate which assumes its values in the volume V occupied by the source.

Substituting (11.44) in (11.43), considering the expressions (11.47) and (11.49) and averaging over time, one finally obtains

$$P_{\text{scal}} = \frac{(2\omega_{\text{BD}} + 3)c^4}{8\pi\varphi_0} \sum_{jm} |X_{jm}|^2. \tag{11.51}$$

To compute the power radiated in scalar GWs, one has to determine the coefficients X_{jm}, defined in (11.50). The detailed calculations can be found in appendix A of the third reference in [6], while here we only give the final results. Introducing the reduced mass of the binary system $\mu = m_1 m_2/m$ and the gravitational self-energy for the body a (with $a = 1, 2$)

$$\Omega_a = -\frac{1}{2} \int_{V_a} \frac{\rho(\vec{x})\rho(\vec{x}')}{|\vec{x} - \vec{x}'|} \, d^3x \, d^3x' \tag{11.52}$$

one can write the Fourier components with frequency $n\omega_0$ in the Newtonian approximation

$$(X_{00})_n = -\frac{16\sqrt{2\pi}}{3} \frac{i\omega_0\varphi_0}{\omega_{\text{BD}} + 2} \frac{m\mu}{a} n J_n(ne) \tag{11.53}$$

$$(X_{1\pm1})_n = -\sqrt{\frac{2\pi}{3}} \frac{2i\omega_0{}^2\varphi_0}{\omega_{\text{BD}} + 2} \left(\frac{\Omega_2}{m_2} - \frac{\Omega_1}{m_1}\right)\mu a$$

$$\times \left[\pm J_n'(ne) - \frac{1}{e}(1 - e^2)^{1/2} J_n(ne)\right] \tag{11.54}$$

$$(X_{20})_n = \frac{2}{3}\sqrt{\frac{\pi}{5}} \frac{i\omega_0{}^3\varphi_0}{\omega_{\text{BD}} + 2}\mu a^2 n J_n(ne) \tag{11.55}$$

$$(X_{2\pm2})_n = \mp 2\sqrt{\frac{\pi}{30}} \frac{i\omega_0{}^3\varphi_0}{\omega_{\text{BD}} + 2}\mu a^2$$

$$\times \frac{1}{n}\{(e^2 - 2)J_n(ne)/(ne^2) + 2(1 - e^2)J_n'(ne)/e$$

$$\mp 2(1 - e^2)^{1/2}[(1 - e^2)J_n(ne)/e^2 - J_n'(ne)/(ne)]\}. \tag{11.56}$$

Substituting these expressions in (11.51), leads to the power radiated in scalar GWs in the nth harmonic

$$(P_{\text{scal}})_n = P_n^{j=0} + P_n^{j=1} + P_n^{j=2} \tag{11.57}$$

where the monopole, dipole and quadrupole terms are, respectively,

$$P_n^{j=0} = \frac{64}{9(\omega_{\text{BD}} + 2)} \frac{m^3\mu^2 G^4}{a^5 c^5} n^2 J_n^2(ne)$$

$$= \frac{64}{9(\omega_{\text{BD}} + 2)} \frac{m^3\mu^2 G^4}{a^5 c^5} m(n; e) \tag{11.58}$$

$$P_n^{j=1} = \frac{4}{3(\omega_{BD}+2)} \frac{m^2\mu^2 G^3}{a^4 c^3} \left(\frac{\Omega_2}{m_2} - \frac{\Omega_1}{m_1}\right)^2$$

$$\times n^2 \left[J_n'^2(ne) + \frac{1}{e^2}(1-e^2)J_n^2(ne)\right]$$

$$= \frac{4}{3(\omega_{BD}+2)} \frac{m^2\mu^2 G^3}{a^4 c^3} \left(\frac{\Omega_2}{m_2} - \frac{\Omega_1}{m_1}\right)^2 d(n;e) \qquad (11.59)$$

$$P_n^{j=2} = \frac{8}{15(\omega_{BD}+2)} \frac{m^3\mu^2 G^4}{a^5 c^5} g(n;e). \qquad (11.60)$$

The total power radiated in scalar GWs by a binary system is the sum of three terms

$$P_{\text{scal}} = P^{j=0} + P^{j=1} + P^{j=2} \qquad (11.61)$$

where

$$P^{j=0} = \frac{16}{9(\omega_{BD}+2)} \frac{G^4}{c^5} \frac{m_1^2 m_2^2 m}{a^5} \frac{e^2}{(1-e^2)^{7/2}} \left(1 + \frac{e^2}{4}\right) \qquad (11.62)$$

$$P^{j=1} = \frac{2}{\omega_{BD}+2} \left(\frac{\Omega_2}{m_2} - \frac{\Omega_1}{m_1}\right)^2 \frac{G^3}{c^3} \frac{m_1^2 m_2^2}{a^4} \frac{1}{(1-e^2)^{5/2}} \left(1 + \frac{e^2}{2}\right) \qquad (11.63)$$

$$P^{j=2} = \frac{8}{15(\omega_{BD}+2)} \frac{G^4}{c^5} \frac{m_1^2 m_2^2 m}{a^5} \frac{1}{(1-e^2)^{7/2}} \left(1 + \frac{73}{24}e^2 + \frac{37}{96}e^4\right). \qquad (11.64)$$

Note that $P^{j=0}$, $P^{j=1}$, $P^{j=2}$ all go to zero in the limit $\omega_{BD} \to \infty$.

11.3.4   Scalar GWs

We now give the explicit form of the scalar GWs radiated by a binary system. To this end, note that the major semi-axis, a, is related to the total energy, E, of the system through the following equation

$$a = -\frac{Gm_1 m_2}{2E}. \qquad (11.65)$$

Let us consider the case of a circular orbit, remembering that in the last phase of evolution of a binary system this condition is usually satisfied. Furthermore we will also assume $m_1 = m_2$. With these positions only the quadrupole term, (11.60), of the gravitational radiation is different from zero. The total power radiated in GWs, averaged over time, is then given by (11.62)–(11.64)

$$P = \frac{8}{15(\omega_{BD}+2)} \frac{G^4}{c^5} \frac{m_1^2 m_2^2 m}{d^5} [6(2\omega_{BD}+3)+1] \qquad (11.66)$$

where d is the relative distance between the two stars. The time variation of d in one orbital period is

$$\dot{d} = -\frac{Gm_1 m_2}{2E^2} P \qquad (11.67)$$

Finally, substituting (11.65), (11.66) in (11.67) and integrating over time, one obtains

$$d = 2 \left(\frac{2}{15} \frac{12\omega_{BD} + 19}{\omega_{BD} + 2} \frac{G^3 m_1 m_2 m}{c^5} \right)^{1/4} \tau^4 \qquad (11.68)$$

where we have defined $\tau = t_c - t$, t_c being the time of the collapse between the two bodies.

From (11.49), (11.53)–(11.56) one can deduce the form of the scalar field (see appendix B of the third reference in [6] for details) which, for equal masses, is

$$\xi(t) = -\frac{2\mu}{r(2\omega_{BD} + 3)} \left[v^2 + \frac{m}{d} - (\hat{n} \cdot \vec{v})^2 + \frac{m}{d^3} (\hat{n} \cdot \vec{d}) \right] \qquad (11.69)$$

where r is the distance of the source from the observer, and $\hat{n}$ is the versor of the line of sight from the observer to the binary system centre of mass. Indicating with γ the inclination angle, that is the angle between the orbital plane and the reference plane (defined to be a plane perpendicular to the line of sight), and with ψ the true anomaly, that is the angle between d and the x-axis in the orbital plane x–y, yields $\hat{n} \cdot \vec{d} = d \sin \gamma \sin \psi$. Then, from (11.69) one obtains

$$\xi(t) = \frac{2G\mu m}{(2\omega_{BD} + 3)c^4 dr} \sin^2 \gamma \cos(2\psi(t)) \qquad (11.70)$$

which can also be written as

$$\xi(\tau) = \xi_0(\tau) \sin(\chi(\tau) + \bar{\chi}) \qquad (11.71)$$

where $\bar{\chi}$ is an arbitrary phase and the amplitude $\xi_0(\tau)$ is given by

$$\xi_0(\tau) = \frac{2G\mu m}{(2\omega_{BD} + 3)c^4 dr} \sin^2 \gamma$$

$$= \frac{1}{2(2\omega_{BD} + 3)r} \left(\frac{\omega_{BD} + 2}{12\omega_{BD} + 19} \right)^{1/4} \left(\frac{15G}{2c^{11}} \right)^{1/4} \frac{M_c^{5/4}}{\tau^{1/4}} \sin^2 \gamma. \quad (11.72)$$

In the last expression, we have introduced the definition of the chirp mass $M_c = (m_1 m_2)^{3/5} / m^{1/5}$.

11.3.5 Detectability of the scalar GWs

Let us now study the interaction of the scalar GWs, a spherical GW detector.

As usual, we characterize the sensitivity of the detector by the spectral density of strain $S_h(f)$ [Hz]$^{-1}$. The optimum performance of a detector is obtained by filtering the output with a filter matched to the signal. The energy signal-to-noise ratio (SNR) of the filter output is given by the well known formula:

$$\mathrm{SNR} = \int_{-\infty}^{+\infty} \frac{|H(f)|^2}{S_h(f)} \, \mathrm{d}f \qquad (11.73)$$

where $H(f)$ is the Fourier transform of the scalar gravitational waveform $h_s(t) = G\xi_0(t)$.

We must now take into account the astrophysical restrictions on the validity of the waveform (11.71) which is obtained in the Newtonian approximation for point-like masses. In the following, we will take the point of view that this approximation breaks down when there are five cycles remaining to collapse [20,21].

The five-cycles limit will be used to restrict the range of M_c over which our analysis will be performed. From (11.68), one can obtain

$$
\omega_g(\tau) = 2\omega_0 = 2\sqrt{\frac{Gm}{d^3}}
$$
$$
= 2\left(\frac{15c^5}{64G^{5/3}}\right)^{3/8}\left(\frac{\omega_{BD} + 2}{12\omega_{BD} + 19}\right)^{3/8}\frac{1}{M_c^{5/8}}\tau^{3/8}. \tag{11.74}
$$

Integrating (11.74) yields the amount of phase until coalescence

$$
\chi(\tau) = \frac{16}{5}\left(\frac{15c^5}{64G^{5/3}}\right)^{3/8}\left(\frac{\omega_{BD} + 2}{12\omega_{BD} + 19}\right)^{3/8}\left(\frac{\tau}{M_c}\right)^{5/8}. \tag{11.75}
$$

Setting (11.75) equal to the limit period, $T_{5\ cycles} = 5(2\pi)$, solving for τ and using (11.74) leads to

$$
\omega_{5\ cycles} = 2\pi(6870\ \text{Hz})\left(\frac{\omega_{BD} + 2}{12\omega_{BD} + 19}\right)^{3/5}\frac{M_\odot}{M_c}. \tag{11.76}
$$

Taking $\omega_{BD} = 600$, the previous limit reads

$$
\omega_{5\ cycles} = 2\pi(1547\ \text{Hz})\frac{M_\odot}{M_c}. \tag{11.77}
$$

A GW excites those vibrational modes of a resonant body having the proper symmetry. In the framework of JBD theory the spheroidal modes with $l = 2$ and $l = 0$ are sensitive to the incoming GW. Thanks to its multimode nature, a single sphere is capable of detecting GWs from all directions and polarizations. We now evaluate the SNR of a resonant-mass detector of spherical shape for its quadrupole mode with $m = 0$ and its monopole mode. In a resonant-mass detector, $S_h(f)$ is a resonant curve and can be characterized by its value at resonance $S_h(f_n)$ and by its half height width [22]. $S_h(f_n)$ can thus be written as

$$
S_h(f_n) = \frac{G}{c^3}\frac{4kT}{\sigma_n Q_n f_n}. \tag{11.78}
$$

Here, σ_n is the cross section associated with the nth resonant mode, T is the thermodynamic temperature of the detector and Q_n is the quality factor of the mode.

The half height width of $S_h(f)$ gives the bandwidth of the resonant mode

$$\Delta f_n = \frac{f_n}{Q_n}\Gamma_n^{-1/2}. \tag{11.79}$$

Here, Γ_n is the ratio of the wideband noise in the nth resonance bandwidth to the narrowband noise.

From the resonant-mass detector viewpoint, the chirp signal can be treated as a transient GW, depositing energy in a timescale short with respect to the detector damping time. We can then consider constant the Fourier transform of the waveform within the band of the detector and write [22]

$$\text{SNR} = \frac{2\pi\,\Delta f_n|H(f_n)|^2}{S_h(f_n)}. \tag{11.80}$$

The cross sections associated with the vibrational modes with $l = 0$ and $l = 2, m = 0$ are respectively [6]

$$\sigma_{(n0)} = H_n\frac{GMv_s^2}{c^3(\omega_{BD} + 2)} \tag{11.81}$$

$$\sigma_{(n2)} = \frac{F_n}{6}\frac{GMv_s^2}{c^3(\omega_{BD} + 2)}. \tag{11.82}$$

All parameters entering the previous equation refer to the detector, M is its mass, v_s the sound velocity and the constants H_n and F_n are given in [6]. The signal-to-noise ratio can be calculated analytically by approximating the waveform with a truncated Taylor expansion around $t = 0$, where $\omega_g(t = 0) = \omega_{nl}$ [20, 23]

$$h_s(t) \approx G\xi_0(t = 0)\sin\left[\omega_{nl}t + \frac{1}{2}\left(\frac{d\omega}{dt}\right)_{t=0}t^2\right]. \tag{11.83}$$

Using quantum limited readout systems, one finally obtains

$$(\text{SNR}_n)_{l=0} = \frac{5 \times 2^{1/3}H_nG^{5/3}}{32(\omega_{BD} + 2)(12\omega_{BD} + 19)\hbar c^3}\frac{M_c^{5/3}Mv_s^2}{r^2\omega_{n0}^{4/3}}\sin^4\gamma \tag{11.84}$$

$$(\text{SNR}_n)_{l=2} = \frac{5 \times 2^{1/3}F_nG^{5/3}}{192(\omega_{BD} + 2)(12\omega_{BD} + 19)\hbar c^3}\frac{M_c^{5/3}Mv_s^2}{r^2\omega_{n0}^{4/3}}\sin^4\gamma \tag{11.85}$$

which are, respectively, the SNR for the modes with $l = 0$ and $l = 2, m = 0$ of a spherical detector.

It has been proposed to realize spherical detectors with 3 m diameter, made of copper alloys, with mass of the order of 100 tons [24]. This proposed detector has resonant frequencies of $\omega_{12} = 2\pi \times 807$ rad s^{-1} and $\omega_{10} = 2\pi \times 1655$ rad s^{-1}. In the case of optimally oriented orbits (inclination angle $\gamma = \pi/2$) and $\omega_{BD} = 600$, the inspiralling of two compact objects of 1.4 solar masses each will then be detected with SNR $= 1$ up to a source distance $r(\omega_{10}) \simeq 30$ kpc and $r(\omega_{12}) \simeq 30$ kpc.

11.4   The hollow sphere

An appealing variant of the massive sphere is a *hollow* sphere [25]. The latter
has the remarkable property that it enables the detector to monitor GW signals in
a significantly *lower frequency range*—down to about 200 Hz—than its massive
counterpart for comparable sphere masses. This can be considered a positive
advantage for a future worldwide network of GW detectors, as the sensitivity
range of such antenna overlaps with that of the large-scale interferometers, now in
a rather advanced state of construction [6,7]. In this section we study the response
of such a detector to the GW energy emitted by a binary system constituted of stars
of masses of the order of the solar mass. A hollow sphere obviously has the same
symmetry of the massive one, so the general structure of its *normal modes* of
vibration is very similar [25] to that of the solid sphere. In particular, the hollow
sphere is very well adapted to sense and monitors the presence of scalar modes in
the incoming GW signal. The extension of the analysis of the previous sections
to a hollow sphere is quite straightforward and in the following we will only give
the main results. Due to the different geometry, the vibrational modes of a hollow
sphere differ from those studied in section 11.2. In the case of a hollow sphere,
we have two boundaries given by the outer and the inner surfaces of the solid
itself. We use the notation a for the inner radius and R for the outer radius. The
boundary conditions are thus expressed by

$$\sigma_{ij}n_j = 0 \quad \text{at } r = R \quad \text{and at } r = a \quad (R \ge a \ge 0), \tag{11.86}$$

(11.3) must now be solved subject to these boundary conditions. The solution
that leads to spheroidal modes is still (11.9) where the radial functions $A_{nl}(r)$ and
$B_{nl}(r)$ have rather complicated expressions:

$$A_{nl}(r) = C_{nl}\left[\frac{1}{q_{nl}^S}\frac{d}{dr}j_l(q_{nl}^S r) - l(l+1)K_{nl}\frac{j_l(k_{nl}^S r)}{k_{nl}^S r}\right.$$
$$\left. + D_{nl}\frac{1}{q_{nl}^S}\frac{d}{dr}y_l(q_{nl}^S r) - l(l+1)\tilde{D}_{nl}\frac{y_l(k_{nl}^S r)}{k_{nl}^S r}\right] \tag{11.87}$$

$$B_{nl}(r) = C_{nl}\left[\frac{j_l(q_{nl}^S r)}{q_{nl}^S r} - K_{nl}\frac{1}{k_{nl}^S r}\frac{d}{dr}\{rj_l(k_{nl}^S r)\}\right.$$
$$\left. + D_{nl}\frac{y_l(q_{nl}^S r)}{q_{nl}^S r} - \tilde{D}_{nl}\frac{1}{k_{nl}^S r}\frac{d}{dr}\{ry_l(k_{nl}^S r)\}\right]. \tag{11.88}$$

Here $k_{nl}^S R$ and $q_{nl}^S R$ are dimensionless *eigenvalues*, and they are the solution
to a rather complicated algebraic equation for the frequencies $\omega = \omega_{nl}$—see [25]
for details. In (11.87) and (11.88) we have set

$$K_{nl} \equiv \frac{C_t q_{nl}^S}{C_l k_{nl}^S}, \quad D_{nl} \equiv \frac{q_{nl}^S}{k_{nl}^S}E, \quad \tilde{D}_{nl} \equiv \frac{C_t F q_{nl}^S}{C_l k_{nl}^S} \tag{11.89}$$

and introduced the normalization constant C_{nl}, which is fixed by the orthogonality properties

$$\int_V (\mathbf{u}^S_{n'l'm'})^* \cdot (\mathbf{u}^S_{nlm}) \varrho_0 \, d^3x = M \delta_{nn'} \delta_{ll'} \delta_{mm'} \tag{11.90}$$

where M is the mass of the hollow sphere:

$$M = \frac{4\pi}{3} \varrho_0 R^3 (1 - \varsigma^3), \quad \varsigma \equiv \frac{a}{R} \le 1. \tag{11.91}$$

Equation (11.90) fixes the value of C_{nl} through the radial integral

$$\int_{\varsigma R}^R [A_{nl}^2(r) + l(l+1)B_{nl}^2(r)] r^2 \, dr = \frac{4\pi}{3} \varrho_0 (1 - \varsigma^3) R^3 \tag{11.92}$$

as can be easily verified by using well known properties of angular momentum operators and spherical harmonics. We shall later specify the values of the different parameters appearing in the above expressions as required in each particular case which will in due course be considered. As seen in [9], a scalar–tensor theory of GWs such as JBD predicts the excitation of the sphere's monopole modes *as well as the* $m = 0$ quadrupole modes. In order to calculate the energy absorbed by the detector according to that theory it is necessary to calculate the energy deposited by the wave in those modes, and this in turn requires that we solve the elasticity equation with the GW driving term included in its right-hand side. The result of such a calculation was presented in full generality in [9], and is directly applicable here because the structure of the oscillation eigenmodes of a hollow sphere is equal to that of the massive sphere—only the explicit form of the wavefunctions needs to be changed. We thus have

$$E_{osc}(\omega_{nl}) = \frac{1}{2} M b_{nl}^2 \sum_{m=-l}^{l} |G^{(lm)}(\omega_{nl})|^2 \tag{11.93}$$

where $G^{(lm)}(\omega_{nl})$ is the Fourier amplitude of the corresponding incoming GW mode, and

$$b_{n0} = -\frac{\varrho_0}{M} \int_a^R A_{n0}(r) r^3 \, dr \tag{11.94}$$

$$b_{n2} = -\frac{\varrho_0}{M} \int_a^R [A_{n2}(r) + 3B_{n2}(r)] r^3 \, dr \tag{11.95}$$

for monopole and quadrupole modes, respectively, and $A_{nl}(r)$ and $B_{nl}(r)$ are given by (11.87). Explicit calculation yields

$$\frac{b_{n0}}{R} = \frac{3}{4\pi} \frac{C_{n0}}{1 - \varsigma^3} [\Lambda(R) - \varsigma^3 \Lambda(a)] \tag{11.96}$$

$$\frac{b_{n2}}{R} = \frac{3}{4\pi} \frac{C_{n2}}{1 - \varsigma^3} [\Sigma(R) - \varsigma^3 \Sigma(a)] \tag{11.97}$$

with

$$\Lambda(z) \equiv \frac{j_2(q_{n0}z)}{q_{n0}R} + D_{n0}\frac{y_2(q_{n0}z)}{q_{n0}R} \tag{11.98}$$

$$\Sigma(z) \equiv \frac{j_2(q_{n2}z)}{q_{n2}R} - 3K_{n2}\frac{j_2(k_{n2}z)}{k_{n2}R} + D_{n2}\frac{y_2(q_{n2}z)}{q_{n2}R} - 3\tilde{D}_{n2}\frac{y_2(k_{n2}z)}{k_{n2}R}. \tag{11.99}$$

The absorption *cross section*, defined as the ratio of the absorbed energy to the incoming flux, can be calculated thanks to an *optical theorem*, as proved, for example, by Weinberg [26]. According to that theorem, the absorption cross section for a signal of frequency ω close to ω_N, say, the frequency of the detector mode excited by the incoming GW, is given by the expression

$$\sigma(\omega) = \frac{10\pi\eta c^2}{\omega^2} \frac{\Gamma^2/4}{(\omega - \omega_N)^2 + \Gamma^2/4} \tag{11.100}$$

where Γ is the *linewidth* of the mode—which can be arbitrarily small, as assumed in the previous section—and η is the dimensionless ratio

$$\eta = \frac{\Gamma_{\text{grav}}}{\Gamma} = \frac{1}{\Gamma}\frac{P_{\text{GW}}}{E_{\text{osc}}} \tag{11.101}$$

where P_{GW} is the energy *re-emitted* by the detector in the form of GWs as a consequence of it being set to oscillate by the incoming signal. In the following we will only consider the case $P_{\text{GW}} = P_{\text{scalar-tensor}}$ with [6, 9]

$$P_{\text{scalar-tensor}} = \frac{2G\omega^6}{5c^5(2\omega_{\text{BD}} + 3)}\left[|Q_{kk}(\omega)|^2 + \frac{1}{3}Q_{ij}^*(\omega)Q_{ij}(\omega)\right] \tag{11.102}$$

where $Q_{ij}(\omega)$ is the quadrupole moment of the hollow sphere:

$$Q_{ij}(\omega) = \int_{\text{Antenna}} x_i x_j \varrho(\boldsymbol{x}, \omega)\, \mathrm{d}^3 x \tag{11.103}$$

and ω_{BD} is Brans–Dicke's parameter.

11.5   Scalar–tensor cross sections

Explicit calculation shows that $P_{\text{scalar-tensor}}$ is made up of two contributions:

$$P_{\text{scalar-tensor}} = P_{00} + P_{20} \tag{11.104}$$

where P_{00} is the scalar, or monopole contribution to the emitted power, while P_{20} comes from the central quadrupole mode which, as discussed in [6] and [9], is excited together with monopole in JBD theory. One must, however, recall that monopole and quadrupole modes of the sphere happen at *different frequencies*, so that cross sections for them only make sense if defined separately. More precisely,

$$\sigma_{n0}(\omega) = \frac{10\pi \eta_{n0} c^2}{\omega^2} \frac{\Gamma_{n0}^2/4}{(\omega - \omega_{n0})^2 + \Gamma_{n0}^2/4} \tag{11.105}$$

$$\sigma_{n2}(\omega) = \frac{10\pi \eta_{n2} c^2}{\omega^2} \frac{\Gamma_{n2}^2/4}{(\omega - \omega_{n2})^2 + \Gamma_{n2}^2/4} \tag{11.106}$$

where η_{n0} and η_{n2} are defined as in (11.101), with all terms referring to the corresponding modes. After some algebra one finds that

$$\sigma_{n0}(\omega) = H_n \frac{GM v_{\mathrm{S}}^2}{(\omega_{\mathrm{BD}} + 2)c^3} \frac{\Gamma_{n0}^2/4}{(\omega - \omega_{n0})^2 + \Gamma_{n0}^2/4} \tag{11.107}$$

$$\sigma_{n2}(\omega) = F_n \frac{GM v_{\mathrm{S}}^2}{(\omega_{\mathrm{BD}} + 2)c^3} \frac{\Gamma_{n2}^2/4}{(\omega - \omega_{n2})^2 + \Gamma_{n2}^2/4}. \tag{11.108}$$

Here, we have defined the dimensionless quantities

$$H_n = \frac{4\pi^2}{9(1 + \sigma_{\mathrm{P}})} (k_{n0} b_{n0})^2 \tag{11.109}$$

$$F_n = \frac{8\pi^2}{15(1 + \sigma_{\mathrm{P}})} (k_{n2} b_{n2})^2 \tag{11.110}$$

where σ_{P} represents the sphere material's Poisson ratio (most often very close to a value of 1/3), and b_{nl} are defined in (11.96); v_{S} is the speed of sound in the material of the sphere.

In tables 11.1 and 11.2 we give a few numerical values of the above cross section coefficients.

As already stressed in [25], one of the main advantages of a hollow sphere is that it enables us to reach good sensitivities at lower frequencies than in a solid sphere. For example, a hollow sphere of the same material and mass as a solid one ($\varsigma = 0$) has eigenfrequencies which are smaller by

$$\omega_{nl}(\varsigma) = \omega_{nl}(\varsigma = 0)(1 - \varsigma^3)^{1/3} \tag{11.111}$$

for any mode indices n and l. We now consider the detectability of JBD GW coming from several interesting sources with a hollow sphere.

The values of the coefficients F_n and H_n, together with the expressions (11.105) for the cross sections of the hollow sphere, can be used to estimate the maximum distances at which a coalescing compact binary system and a gravitational collapse event can be seen with such detector. We consider these in turn.

By taking as a source of GWs a binary system formed by two neutron stars, each of them with a mass of $m_1 = m_2 = 1.4 M_\odot$. The *chirp mass*

Table 11.1. Eigenvalues $k^S_{n0}R$, relative weights D_{n0} and H_n coefficients for a hollow sphere with Poisson ratio $\sigma_P = 1/3$. Values are given for a few different thickness parameters ς.

ς	n	$k^S_{n0}R$	D_{n0}	H_n
0.01	1	5.487 38	−0.000 143 328	0.909 29
	1	12.233 2	−0.001 596 36	0.141 94
	2	18.632 1	−0.005 589 61	0.059 26
	4	24.969 3	−0.001 279	0.032 67
0.10	1	5.454 10	−0.014 218	0.895 30
	1	11.924 1	−0.151 377	0.150 48
	2	17.727 7	−0.479 543	0.049 22
	4	23.541 6	−0.859 885	0.043 11
0.15	1	5.377 09	−0.045 574	0.860 76
	2	11.387 9	−0.434 591	0.176 46
	3	17.105	−0.939 629	0.056 74
	4	23.605	−0.806 574	0.053 96
0.25	1	5.048 42	−0.179 999	0.737 27
	2	10.651 5	−0.960 417	0.305 32
	3	17.819 3	−0.425 087	0.042 75
	4	25.806 3	0.440 100	0.063 47
0.50	1	3.969 14	−0.631 169	0.494 29
	2	13.236 9	0.531 684	0.581 40
	3	25.453 1	0.245 321	0.017 28
	4	37.912 9	0.161 117	0.071 92
0.75	1	3.265 24	−0.901 244	0.430 70
	2	25.346 8	0.188 845	0.662 84
	3	50.371 8	0.093 173	0.003 41
	4	75.469	0.061 981	0.074 80
0.90	1	2.981 41	−0.963 552	0.420 43
	2	62.902 7	0.067 342	0.676 89
	3	125.699	0.033 573	0.000 47
	4	188.519	0.022 334	0.075 38

corresponding to this system is $M_c \equiv (m_1 m_2)^{3/5}(m_1 + m_2)^{-1/5} = 1.22 M_\odot$, and $\nu_{[5\ \text{cycles}]} = 1270$ Hz. Repeating the analysis carried on in section 11.3 we find a formula for the minimum distance at which a measurement can be performed given a certain SNR, for a *quantum limited* detector

$$r(\omega_{n0}) = \left[\frac{5 \times 2^{1/3}}{32} \frac{1}{(\Omega_{BD} + 2)(12\Omega_{BD} + 19)} \frac{G^{5/3} M_c^{5/3}}{c^3} \frac{M v_S^2}{\hbar \omega_{n0}^{4/3} \text{SNR}} H_n \right]^{1/2}$$

$$\tag{11.112}$$

Table 11.2. Eigenvalues $k_{n2}^S R$, relative weights K_{n2}, D_{n2}, $\tilde{D}_{n2}$ and F_n coefficients for a hollow sphere with Poisson ratio $\sigma_P = 1/3$. Values are given for a few different thickness parameters ς.

ς	n	$k_{n2}^S R$	K_{n2}	D_{n2}	$\tilde{D}_{n2}$	F_n
0.10	1	2.638 36	0.855 799	0.000 395	−0.003 142	2.946 02
	2	5.073 58	0.751 837	0.002 351	−0.018 451	1.169 34
	3	10.960 90	0.476 073	0.009 821	−0.071 685	0.022 07
0.15	1	2.611 61	0.796 019	0.001 174	−0.009 288	2.869 13
	2	5.028 15	0.723 984	0.007 028	−0.053 849	1.241 53
	3	8.258 09	−2.010 150	−0.094 986	0.672 786	0.081 13
0.25	1	2.491 22	0.606 536	0.003 210	−0.024 94	2.552 18
	2	4.912 23	0.647 204	0.019 483	−0.138 67	1.550 22
	3	8.242 82	−1.984 426	−0.126 671	0.675 06	0.053 25
	4	10.977 25	0.432 548	−0.012 194	0.022 36	0.035 03
0.50	1	1.943 40	0.300 212	0.003 041	−0.022 68	1.619 78
	2	5.064 53	0.745 258	0.005 133	−0.028 89	2.295 72
	3	10.111 89	1.795 862	−1.697 480	2.982 76	0.197 07
	4	15.919 70	−1.632 550	−1.965 780	−0.309 53	0.171 08
0.75	1	1.449 65	0.225 040	0.001 376	−0.010 17	1.152 91
	2	5.215 99	0.910 998	−0.197 532	0.409 44	1.822 76
	3	13.932 90	0.243 382	0.748 219	−3.201 30	1.089 52
	4	23.763 19	0.550 278	−0.230 203	−0.817 67	0.081 14
0.90	1	1.265 65	0.213 082	0.001 019	−0.007 55	1.038 64
	2	4.977 03	0.939 420	−0.323 067	0.522 79	1.541 06
	3	31.864 29	6.012 680	−0.259 533	4.052 74	1.464 86
	4	61.299 48	0.205 362	−0.673 148	−1.043 69	0.134 70

$$r(\omega_{n2}) = \left[\frac{5 \times 2^{1/3}}{192} \frac{1}{(\Omega_{BD} + 2)(12\Omega_{BD} + 19)} \frac{G^{5/3} M_c^{5/3}}{c^3} \frac{M v_S^2}{\hbar \omega_{n2}^{4/3} \text{SNR}} F_n \right]^{1/2}$$

(11.113)

For a CuAl sphere, the speed of sound is $v_S = 4700$ m s^{-1}. We report in table 11.3 the maximum distances at which a JBD binary can be seen with a 100 ton hollow spherical detector, including the size of the sphere (diameter and thickness factor) for SNR = 1. The Brans–Dicke parameter Ω_{BD} has been given a value of 600. This high value has as a consequence that only relatively nearby binaries can be scrutinized by means of their scalar radiation of GWs. A slight improvement in sensitivity is appreciated as the diameter increases in a fixed mass detector. Vacancies in the tables mean the corresponding frequencies are higher than $v_{[5 \text{ cycles}]}$.

The signal associated with a gravitational collapse can be modelled, within

Table 11.3. Eigenfrequencies, sizes and distances at which coalescing binaries can be seen by monitoring of their emitted JBD GWs. Figures correspond to a 100 ton CuAl hollow sphere.

ς	Φ (m)	ν_{10} (Hz)	ν_{12} (Hz)	$r(\nu_{10})$ (kpc)	$r(\nu_{12})$ (kpc)
0.00	2.94	1655	807	–	29.8
0.25	2.96	1562	771	–	30.3
0.50	3.08	1180	578	55	31.1
0.75	3.5	845	375	64	33
0.90	4.5	600	254	80	40

JBD theory, as a short pulse of amplitude b, whose value can be estimated as [27]

$$b \simeq 10^{-23} \left(\frac{500}{\Omega_{BD}} \right) \left(\frac{M_*}{M_\odot} \right) \left(\frac{10 \, \text{Mpc}}{r} \right) \tag{11.114}$$

where M_* is the collapsing mass.

The minimum value of the Fourier transform of the amplitude of the scalar wave, for a quantum limited detector at unit SNR, is given by

$$|b(\omega_{nl})|_{\min} = \left(\frac{4\hbar}{M v_S^2 \omega_{nl} K_n} \right)^{1/2} \tag{11.115}$$

where $K_n = 2H_n$ for the mode with $l = 0$ and $K_n = F_n/3$ for the mode with $l = 2, m = 0$.

The duration of the impulse, $\tau \approx 1/f_c$, is much shorter than the decay time of the nl mode, so that the relationship between b and $b(\omega_{nl})$ is

$$b \approx |b(\omega_{nl})| f_c \quad \text{at frequency } \omega_{nl} = 2\pi f_c \tag{11.116}$$

so that the minimum scalar wave amplitude detectable is

$$|b|_{\min} \approx \left(\frac{4\hbar \omega_{nl}}{\pi^2 M v_S^2 K_n} \right)^{1/2}. \tag{11.117}$$

Let us now consider a hollow sphere made of molybdenum, for which the speed of sound is as high as $v_S = 5600$ m s^{-1}. For a given detector mass and diameter, equation (11.117) tells us which is the minimum signal detectable with such a detector. For example, a solid sphere of $M = 31$ tons and 1.8 m in diameter, is sensitive down to $b_{\min} = 1.5 \times 10^{-22}$. Equation (11.114) can then be inverted to find which is the maximum distance at which the source can be identified by the scalar waves it emits. Taking a reasonable value of $\Omega_{BD} = 600$, one finds that $r(\nu_{10}) \approx 0.6$ Mpc.

Table 11.4. Eigenfrequencies, sizes and distances at which coalescing binaries can be seen by monitoring of their emitted JBD GWs. Figures correspond to a 3 m external diameter CuAl hollow sphere.

ς	M (ton)	ν_{10} (Hz)	ν_{12} (Hz)	$r(\nu_{10})$ (kpc)	$r(\nu_{12})$ (kpc)
0.00	105	1653	804	—	33
0.25	103.4	1541	760	—	31
0.50	92	1212	593	52	27.6
0.75	60.7	997	442	44.8	23
0.90	28.4	910	386	32	16.3

Table 11.5. Eigenfrequencies, maximum sensitivities and distances at which a gravitational collapse can be seen by monitoring the scalar GWs it emits. Figures correspond to a 31 ton Mb hollow sphere.

ς	ϕ (m)	ν_{10} (Hz)	$|b|_{min}$ (10^{-22})	$r(\nu_{10})$ (Mpc)
0.00	1.80	3338	1.5	0.6
0.25	1.82	3027	1.65	0.5
0.50	1.88	2304	1.79	0.46
0.75	2.16	1650	1.63	0.51
0.90	2.78	1170	1.39	0.6

Table 11.6. Eigenfrequencies, maximum sensitivities and distances at which a gravitational collapse can be seen by monitoring the scalar GWs it emits. Figures correspond to a 1.8 m outer diameter Mb hollow sphere.

ς	M (ton)	ν_{10} (Hz)	$|b|_{min}$ (10^{-22})	$r(\nu_{10})$ (Mpc)
0.00	31.0	3338	1.5	0.6
0.25	30.52	3062	1.71	0.48
0.50	27.12	2407	1.95	0.42
0.75	17.92	1980	2.34	0.36
0.90	8.4	1808	3.31	0.24

Like before, we report, in tables 11.4–11.6, the sensitivities of the detector and consequent maximum distance at which the source appears visible to the device for various values of the thickness parameter ς. In table 11.5 a detector of mass of 31 tons has been assumed for all thicknesses, and in tables 11.4 and 11.6 a constant outer diameter of 3 and 1.8 m has been assumed in all cases.

Acknowledgments

We wish to thank very much the organizers of the school for their kind invitation to present this lecture, and all of my collaborators for sharing their insights on the subject with me.

References

[1] Thorne K S 1987 *Three Hundred Years of Gravitation* ed S W Hawking and W Israel (Cambridge: Cambridge University Press)

[2] Coccia E, Pizzella G and Ronga F 1995 Gravitational wave experiments *Proc. First Edoardo Amaldi Conf. (Frascati, 1994)* (Singapore: World Scientific)

[3] Astone P *et al* 1993 *Phys. Rev.* D **47** 362

[4] Johnson W W 1995 *Proc. First Edoardo Amaldi Conf. (Frascati, 1994)* (Singapore: World Scientific)

[5] Abramovici A *et al* 1992 *Science* **26** 325
Bradaschia C *et al* 1990 *Nucl. Instrum. Methods* A **289** 518

[6] Bianchi M, Coccia E, Colacino C N, Fafone V and Fucito F 1996 *Class. Quantum Grav.* **13** 2865
Bianchi M, Brunetti M, Coccia E, Fucito F and Lobo J A 1998 *Phys. Rev.* D **57** 4525
Brunetti M, Coccia E, Fafone V and Fucito F 1999 *Phys. Rev.* D **59** 044027
Coccia E, Fucito F, Lobo J A and Salvino M 2000 *Phys. Rev.* D **62** 044019

[7] Will C M 1993 *Theory and Experiment in Gravitational Physics* (Cambridge: Cambridge University Press)

[8] Jaerish P 1880 *J.f. Math.* (Crelle) **88**
Lamb H 1882 *Proc. London Mathematical Society* vol 13
Love A E H 1927 *A Treatise on the Mathematical Theory of Elasticity* (London: Cambridge University Press)

[9] Lobo J A 1995 *Phys. Rev.* D **52** 591

[10] Ashby N and Dreitlein J 1975 *Phys. Rev.* D **12** 336

[11] Jackson J D 1975 *Classical Electrodynamics* (New York: Wiley)

[12] Misner C W, Thorne K S and Wheeler J A 1973 *Gravitation* (New York: Freeman)

[13] Wagoner R V and Paik H J 1977 Experimental gravitation *Proc. Int. Symp. held in Pavia 1976* (Roma: Acc. Naz. dei Lincei)

[14] Dhurandhar S V and Tinto M 1989 *Proc. V Marcel Grossmann Meeting* ed D G Blair, M J Buckingham and R Ruffini (Singapore: World Scientific)

[15] Zhou C Z and Michelson P F 1995 *Phys. Rev.* D **51** 2517

[16] Johnson W W and Merkovitz S M 1993 *Phys. Rev. Lett.* **70** 2367

[17] Jordan P 1959 *Z. Phys.* **157** 112
Brans C and Dicke R H 1961 *Phys. Rev.* **124** 925

[18] Miller J, Schneider M, Soffel M and Ruder H 1991 *Astrophys. J. Lett.* **382** L101

[19] Lee D L 1974 *Phys. Rev.* D **10** 2374

[20] Dewey D 1987 *Phys. Rev.* D **36** 1577

[21] Coccia E and Fafone V 1996 *Phys. Lett.* A **213** 16

[22] Astone P, Pallottino G V and Pizzella G 1997 *Class. Quantum Grav.* **14** 2019

[23] Clark J P A and Eardley D M 1977 *Astrophys. J.* **215** 311

[24] Frossati G and Coccia E 1994 *Cryogenics* **34** (ICEC Supplement) 9

Frossati G and Coccia E 1994 *Cryogenics* **34** (ICEC Supplement) 304

[25] Coccia E, Fafone V, Frossati G, Lobo J A and Ortega J A 1998 *Phys. Rev.* D **57** 2051

[26] Weinberg S 1972 *Gravitation and Cosmology* (New York: Wiley)

[27] Shibata M, Nakao K and Nakamura T 1994 *Phys. Rev.* D **50** 7304

Saijo M, Shinkai H and Maeda K 1997 *Phys. Rev.* D **56** 785

PART 3

THE STOCHASTIC GRAVITATIONAL-WAVE BACKGROUND

D Babusci[1], S Foffa[2,3,4], G Losurdo[2,4], M Maggiore[2,3], G Matone[1] and R Sturani[2,3,4]

[1] *INFN, Laboratori Nazionali di Frascati*
[2] *INFN, Sezione di Pisa*
[3] *Dipartimento di Fisica, Universita' di Pisa*
[4] *Scuola Normale Superiore di Pisa*

Chapter 12

Generalities on the stochastic GW background

12.1 Introduction

The stochastic gravitational-wave background (SGWB) is a random noise of gravity waves with no evidence of any sharp specific characters in either the time or frequency domain. The important fact to note is that with the exception of a component associated with the random superposition of many weak signals from binary-star systems, the SGWB could be the result of processes that took place during the early stages of the evolution of the universe.

This potential cosmological origin clearly emerges from the calculation of the time at which the graviton decouples from the evolution of the rest of the universe. For a given particle species the decoupling time t_{dec} is defined as the time when its interaction rate Γ equals the expansion rate of the universe, as measured by the Hubble parameter H. In fact, it can be shown [1] that, under assumptions usually met, the number of interactions that the particle species suffer from t_{dec} onward is less than one. This implies that the spectrum of a particle species produced after the decoupling retains memory of the state of the universe at that time. The only change in the character of these particles is a redshift of the magnitude of their three-momentum due to the expansion of the universe.

On purely dimensional grounds, at a given temperature T the interaction rate for particles that interact only gravitationally is [1]:

$$\Gamma \sim G_N^2 T^5 = \frac{T^5}{M_{Pl}^4}$$

where G is the Newton constant and $M_{Pl} = 1/\sqrt{G} = 1.22 \times 10^{19}$ GeV is the Planck mass (we always use units $\hbar = c = k_B = 1$). Because in the radiation-dominated era (before the time t_{eq} of equal matter and radiation energy density) $H \sim T^2/M_{Pl}$, one has:

$$T_{dec} \sim M_{Pl}.$$

Hence, the gravitons are decoupled below the Planck scale. (At the Planck scale the above estimate of the interaction rate is not valid and nothing can be said without a quantum theory of gravity). Just like the cosmic microwave background (CMB), the gravity-wave background is a randomly polarized relic of the Early universe. The important difference is that the electromagnetic (em) waves decoupled about 4×10^5 years after the big bang, while the gravitational background could come from times as early as the Planck epoch at $t_{\mathrm{Pl}} \simeq 5 \times 10^{-44}$ s. This means that the relic gravitational waves give information on the state of the very early universe and, therefore, on physics at correspondingly high energies, which cannot be accessed experimentally in any other way.

Let us consider the standard Friedmann–Robertson–Walker (FRW) cosmological model, consisting of a radiation-dominated (RD) phase followed by the present matter-dominated (MD) phase, and let us call $a(t)$ the FRW scale factor. The RD phase goes backward in time until some new regime sets in. This could be an inflationary epoch, for example, at the grand unification scale, or the RD phase could go back in time until Planckian energies are reached and quantum gravity sets in ($t \sim t_{\mathrm{Pl}}$). If the correct theory of quantum gravity is provided by string theory, the characteristic mass scale is the string mass which is somewhat smaller than the Planck mass and is presumably in the 10^{17}–10^{18} GeV region, and the corresponding characteristic time is therefore one or two orders of magnitude larger than t_{Pl}. The transition between the RD and MD phases takes place at $t = t_{\mathrm{eq}}$, when the temperature of the universe is of the order of only a few eV, so we are interested in graviton production which takes place well within the RD phase, or possibly at Planckian energies.

A graviton produced with a frequency f_* at a time $t = t_*$[1], within the RD phase has today ($t = t_0$) a red-shifted frequency $f_0 = f_* a_*/a_0$. To compute the ratio a_*/a_0 one uses the fact that during the standard RD and MD phases the universe expands adiabatically: the entropy per unit comoving volume $S = g_S(T) a^3(t) T^3$ is constant, where $g_S(T)$ counts the effective number of species [1]. From this one has [2]

$$f_0 = f_* \left(\frac{g_S(T_0)}{g_S(T_*)} \right)^{1/3} \frac{T_0}{T_*} \simeq 8.0 \times 10^{-14} f_* \left(\frac{100}{g_S(T_*)} \right)^{1/3} \left(\frac{1 \,\mathrm{GeV}}{T_*} \right), \quad (12.1)$$

where we used the fact that $T_0 = 2.728 \pm 0.002$ K [3] and according to the standard model $g_S(T_0) \simeq 3.91$ [1].

The frequency f_* of the graviton produced when the temperature was T_* is determined by the Hubble constant H_*, i.e. the size of the horizon, at the time t_* of production. The horizon size, physically, is the length scale beyond which causal microphysics cannot operate (see chapter 8.4 of [1]), and therefore, for causality reasons, we expect that the production of gravitons or any other particles, at time t_*, with a wavelength longer than H_*^{-1}, will be exponentially suppressed. Therefore, we let $\lambda_* = \epsilon H_*^{-1}$, with $\epsilon \leq 1$. During RD, $H_*^2 = (8\pi/3) G \rho_{\mathrm{rad}}$. The

---

[1] Hereafter, the subscript α denotes the value assumed by a generic quantity at $t = t_\alpha$.

contribution to the energy density from a single species of relativistic particle with g_i internal states (helicity, colour, etc) is $g_i(\pi^2/30)T^4$ for a boson and $(7/8)g_i(\pi^2/30)T^4$ for a fermion. Taking into account that the ith species has in general a temperature $T_i \neq T$ if it already dropped out of equilibrium, we can define a function $g(T)$ from $\rho_{\rm rad} = (\pi^2/30)g(T)T^4$. Then [1]

$$g(T) = \sum_{i={\rm bosons}} g_i \left(\frac{T_i}{T}\right)^4 + \frac{7}{8} \sum_{i={\rm fermions}} g_i \left(\frac{T_i}{T}\right)^4 \qquad (12.2)$$

where the sum runs over relativistic species. This holds if a species is in thermal equilibrium at temperature T_i. If instead it does not have a thermal spectrum (which in general is the case for gravitons) we can still use the above equation, where for this species T_i does not represent a temperature but is defined (for bosons) from $\rho_i = g_i(\pi^2/30)T_i^4$, where ρ_i is the energy density of this species. The quantity $g_S(T)$ used before for the entropy is given by the same expression as $g(T)$, with $(T_i/T)^4$ replaced by $(T_i/T)^3$. We see that both $g(T)$ and $g_S(T)$ give a measure of the effective number of species. For most of the early history of the universe, $g(T) = g_S(T)$, and in the standard model at $T \gtrsim 300\,{\rm GeV}$ they have the common value $g_* = 106.75$, while today $g_0 = 3.36$ [1]. Therefore

$$H_*^2 = \frac{4\pi^3 g_* T_*^4}{45 M_{\rm Pl}^2}, \qquad (12.3)$$

and, using $f_* \equiv H_*/\epsilon$, equation (12.1) can be written as [2]

$$f_0 \simeq 1.65 \times 10^{-7} \frac{1}{\epsilon} \left(\frac{T_*}{1\,{\rm GeV}}\right) \left(\frac{g_*}{100}\right)^{1/6} {\rm Hz}. \qquad (12.4)$$

This simple equation allows us to understand a number of important points concerning the energy scales that can be probed in GW experiments. The simplest estimate of f_* corresponds to taking $\epsilon = 1$ in equation (12.4) [4]. In this case, we would find that a graviton observed today at the frequency $f_0 = 100\,{\rm Hz}$, the scale relevant for VIRGO, was produced when the universe had a temperature $T_* \sim 6 \times 10^8\,{\rm GeV}$. Since in the RD phase one has [1]

$$t = \frac{1}{2H} = \left(\frac{45}{16\pi^3 g(T)}\right)^{1/2} \frac{M_{\rm Pl}}{T^2} \simeq \frac{2.42}{g^{1/2}} \left(\frac{\rm MeV}{T}\right)^2 {\rm s}, \qquad (12.5)$$

this temperature corresponds to a production time $t_* \sim 7 \times 10^{-25}$ s. At this time the graviton had an energy $E_* \sim 3\,{\rm GeV}$.

However, the fact that it makes sense to consider gravitons production only for wavelengths with $\lambda_* \lesssim H_*^{-1}$ does not necessarily mean, in general, that at $t = t_*$ the typical wavelength of GWs produced will be at $\lambda_* \sim H_*^{-1}$ ($\epsilon \sim 1$) even as an order of magnitude estimate. In [5] this point is illustrated with two specific examples, one in which the assumption $\lambda_* \sim H_*^{-1}$ turns out to be basically correct, and one in which it can fail by several orders of magnitudes. Both examples will, in general, illustrate the fact that the argument does not involve only kinematics, but also the dynamics of the production mechanism.

From equation (12.4) we see that the temperatures of the early universe explored detecting today relic GWs at a frequency f_0 are, for constant g^*, smaller by a factor approximately equal to ϵ, compared to those estimates with $\epsilon = 1$. Equivalently, a signal produced at a given temperature T_* could in principle show up today in the VIRGO/LIGO frequency band when a naive estimate with $\epsilon = 1$ suggests that it falls at lower frequencies.

There is, however, another effect, which instead gives hopes of exploring the universe at much *higher* temperatures than naively expected, using GW experiments. In fact, the characteristic frequency that we have discussed is the value of the cut-off frequency in the graviton spectrum. Above this frequency, the spectrum decreases exponentially [6], and no signal can be detected. Below this frequency, however, the form of the spectrum is not fixed by general arguments. Thermal spectra have a low-frequency behaviour $d\rho/df \sim f^2$, but, since the gravitons below the Planck scale interact too weakly to thermalize, there is no *a priori* reason for a $\sim f^2$ dependence. The gravitons will retain the form of the spectrum that they had at the time of production, and this is a very model-dependent feature. However, as shown in [5], from a number of explicit examples and general arguments we learn that spectra flat or almost flat over a large range of frequencies seem to be not at all unconceivable.

This fact has potentially important consequences. It means that, even if a spectrum of gravitons produced during the Planck era has a cut-off at frequencies much larger than the VIRGO/LIGO frequency band, we can still hope to observe in the 10 Hz–1 kHz region the long low-frequency tails of these spectra.

12.2   Definitions

12.2.1 $\Omega_{gw}(f)$ and the optimal SNR

The intensity of a stochastic background of gravitational waves (GWs) can be characterized by the dimensionless quantity

$$\Omega_{gw}(f) = \frac{1}{\rho_c} \frac{d\rho_{gw}}{d\ln f}, \tag{12.6}$$

where ρ_{gw} is the energy density of the stochastic background of gravitational waves, f is the frequency and ρ_c is the present value of the critical energy density for closing the universe. In terms of the present value of the Hubble constant H_0, the critical density is given by

$$\rho_c = \frac{3H_0^2}{8\pi G}. \tag{12.7}$$

The value of H_0 is usually written as $H_0 = h_0 \times 100$ km s^{-1} Mpc^{-1}, where h_0 parametrizes the existing experimental uncertainty. A conservative estimate for this parameter is in the range $0.50 < h_0 < 0.65$ (see [5] and references quoted therein).

It is not very convenient to normalize ρ_{gw} to a quantity, ρ_c, which is uncertain: this uncertainty would appear in all the subsequent formulae, although it has nothing to do with the uncertainties on the GW background. Therefore, we rather characterize the stochastic GW background with the quantity $h_0^2 \Omega_{gw}(f)$, which is independent of h_0. All theoretical computations of a relic GW spectrum are actually computations of $d\rho_{gw}/d \ln f$ and are independent of the uncertainty on H_0. Therefore, the result of these computations is expressed in terms of $h_0^2 \Omega_{gw}$, rather than of Ω_{gw}^2. Under the assumption that the stochastic background of gravitational radiation is isotropic, unpolarized, stationary and Gaussian (see [4,7] for details), it is completely specified by its spectrum $\Omega_{gw}(f)$.

To detect a stochastic GW background the optimal strategy consists in performing a correlation between two (or more) detectors, since, as we will discuss below, the signal will be far too low to exceed the noise level in any existing or planned single detector (with the exception of the space interferometer LISA). The strategy has been discussed in [8–10], and is clearly reviewed in [4,7]. Let us recall the main points of the analysis.

The cross-correlation between the outputs $s_1(t)$ and $s_2(t)$ of two detectors is defined as

$$S = \int_{-T/2}^{T/2} dt \int_{-T/2}^{T/2} dt'\, s_1(t)s_2(t')Q(t, t'), \qquad (12.8)$$

where T is the total integration time (e.g. one year) and Q is a filter function. The output of a single detector characterized by an intrinsic noise $n(t)$ and on which acts a gravitational strain $h(t)$ is of the form $s(t) = n(t) + h(t)$. The gravitational strain can be expressed in terms of the amplitudes $h_{+,\times}$ of the wave in the following way [11]

$$h(t) = F_+ h_+(t) + F_\times h_\times(t), \qquad (12.9)$$

where the detector pattern functions $F_{+,\times}$ are introduced, which depend on the location, orientation and geometry of the detector and the direction of arrival of the GW and its polarization (see section 12.3). These functions have values in the range $0 \le |F_{+,\times}| \le 1$. The noise intrinsic to the detector is assumed stationary, Gaussian and statistically independent on the gravitational strain. As a consequence of the assumed stationariety of both the stochastic GW background and noise, the filter function turns out to be $Q(t, t') = Q(t - t')$. Furthermore, the noises in the two detectors are assumed uncorrelated, i.e. the ensemble average of

---

[2] This simple point has occasionally been missed in the literature, where one can find the statement that, for small values of H_0, Ω_{gw} is larger and therefore easier to detect. Of course, it is larger only because it has been normalized using a smaller quantity.

their Fourier components satisfies

$$\langle \tilde{n}_i^*(f)\tilde{n}_j(f') \rangle = \delta(f - f')\delta_{ij}\tfrac{1}{2}S_n^{(i)}(|f|), \qquad (12.10)$$

where the function $S_n(|f|)$, with dimensions Hz^{-1}, is known as the noise power spectrum[3]. The factor $\tfrac{1}{2}$ is conventionally inserted in the definition so that the total noise power is obtained integrating $S_n(f)$ over the physical range $0 \le f < \infty$, rather than from $-\infty$ to ∞.

As discussed in [4, 7–10], for any given form of the signal, i.e. for any given functional form of $h_0^2\Omega_{\text{gw}}(f)$, it is possible to find explicitly the filter function $Q(t)$ which maximizes the signal-to-noise ratio (SNR). It can be shown [7] that under the above-mentioned assumptions, the Fourier transform of this optimal filter and the corresponding value of the optimal SNR turn out to be, respectively:

$$Q(f) = \lambda \frac{\gamma(|f|)\Omega_{\text{gw}}(|f|)}{|f|^3 S_n^{(1)}(|f|)S_n^{(2)}(|f|)}, \qquad (12.11)$$

$$\text{SNR} = \left[\left(\frac{9H_0^4}{8\pi^4} \right) F^2 T \int_0^\infty df\, \frac{\gamma^2(|f|)\Omega_{\text{gw}}^2(|f|)}{f^6 S_n^{(1)}(|f|)S_n^{(2)}(|f|)} \right]^{1/4}, \qquad (12.12)$$

where F is a normalization factor less than one depending only upon the geometry of the detectors, and λ is a real overall normalization constant that, assuming for the spectrum a power-law $\Omega_{\text{gw}}(f) = \Omega_\beta f^\beta$ (with $\Omega_\beta = \text{constant}$), is fixed by the condition $\langle S \rangle = \Omega_\beta T$. The function $\gamma(f)$ appearing in both formulae is the overlap reduction function introduced in [10], which takes into account the difference in location and orientation of the two detectors. At this stage let us only remark that $\gamma(f)$ is maximum and equal to one in the case of two detectors with the same location and orientation. The detailed expression for F and $\gamma(f)$ will be discussed in section 12.3. Finally, let us note that in equation (12.12) we have taken into account the fact that what has been called S in equation (12.8) is quadratic in the signals and, with usual definitions, it contributes to the SNR squared. This differs from the convention used in [4, 7].

In principle the expression for the SNR, equation (12.12), is all that we need in order to discuss the possibility of detection of a given GW background. However, it is useful, for order of magnitude estimates and for intuitive understanding, to express the SNR in terms of a characteristic amplitude of the stochastic GW-background and of a characteristic noise level, although, as we will see, the latter is a quantity that describes the noise only approximately, in contrast to equation (12.12) which is exact. We will introduce these quantities in the next two subsections.

---

[3] Unfortunately there is not much agreement about notations in the literature. The noise power spectrum, that we denote by $S_n(f)$ following, for example, [10], is called $P(f)$ in [4]. Other authors use the notation $S_h(f)$, which we instead reserve for the power spectrum of the signal. To make things worse, S_n is sometimes defined with or without the factor $\tfrac{1}{2}$ in equation (12.10).

12.2.2 The characteristic amplitude

In a transverse-traceless (TT) gauge, the perturbation to the Minkowski metric of spacetime introduced by the GW background can be written in terms of a plane wave expansion as [4]

$$h_{ij}(t, \vec{r}) = \sum_{A=+,\times} \int_{-\infty}^{\infty} df \int_{S^2} d\hat{\Omega} \int_0^{2\pi} d\psi \, h_A(f, \hat{\Omega}, \psi) e^{i2\pi f(t - \hat{\Omega} \cdot \vec{r})} \varepsilon_{ij}^A(\hat{\Omega}, \psi),$$

(12.13)

where, since h_{ij} is real, the Fourier amplitudes are complex arbitrary functions that satisfy the condition $h_A(-f, \hat{\Omega}, \psi) = h_A^*(f, \hat{\Omega}, \psi)$. The unit vector $\hat{\Omega}$ is along the propagation direction of the wave, and, in terms of the standard polar (θ) and azimuthal (ϕ) angles on the 2-sphere, is:

$$\hat{\Omega} \equiv (\cos\phi \sin\theta, \sin\phi \sin\theta, \cos\theta),$$

(12.14)

with $d\hat{\Omega} = d\cos\theta \, d\phi$. The angle ψ describes the polarization of the wave. By introducing the following pair of orthogonal unit vectors lying in the plane perpendicular to $\hat{\Omega}$

$$\hat{m}(\hat{\Omega}) \equiv (\cos\phi \cos\theta, \sin\phi \cos\theta, -\sin\theta), \quad \hat{n}(\hat{\Omega}) \equiv (\sin\phi, -\cos\phi, 0),$$

(12.15)

ψ is the angle of which is rotated the intrinsic frame of the wave (where, in a TT gauge, $h_{x'y'} = -h_{y'x'}$) respect to the frame $(\hat{m}, \hat{n})$ (see [11], p 367). The polarization tensors can be written as

$$\varepsilon^+(\hat{\Omega}, \psi) = e^+(\hat{\Omega}) \cos 2\psi - e^\times(\hat{\Omega}) \sin 2\psi$$
$$\varepsilon^\times(\hat{\Omega}, \psi) = e^+(\hat{\Omega}) \sin 2\psi + e^\times(\hat{\Omega}) \cos 2\psi,$$

(12.16)

where

$$e^+(\hat{\Omega}) = \hat{m}(\hat{\Omega}) \otimes \hat{m}(\hat{\Omega}) - \hat{n}(\hat{\Omega}) \otimes \hat{n}(\hat{\Omega}), \quad e^\times(\hat{\Omega}) = \hat{m}(\hat{\Omega}) \otimes \hat{n}(\hat{\Omega}) + \hat{n}(\hat{\Omega}) \otimes \hat{m}(\hat{\Omega})$$

(12.17)

with the normalization

$$\text{Tr}\{e^A(\hat{\Omega}) e^{A'}(\hat{\Omega})\} = 2\delta^{AA'}.$$

In the case of a stochastic background, we treat the complex Fourier amplitude h_A as a random variable with zero mean value. If this background is isotropic, unpolarized and stationary, the ensemble average of the product of two Fourier amplitudes satisfies:

$$\langle h_A^*(f, \hat{\Omega}, \psi) h_{A'}(f', \hat{\Omega}', \psi') \rangle = \delta_{AA'} \delta(f - f') \frac{\delta^2(\hat{\Omega}, \hat{\Omega}')}{4\pi} \frac{\delta(\psi - \psi')}{2\pi} \frac{1}{2} S_h(f),$$

(12.18)

where $\delta^2(\hat{\Omega}, \hat{\Omega}') = \delta(\phi - \phi')\delta(\cos\theta - \cos\theta')$. The function $S_h(f)$ defined by the above equation has dimensions Hz^{-1} and satisfies $S_h(f) = S_h(-f)$. With this normalization, $S_h(f)$ is the quantity to be compared with the noise level $S_n(f)$ defined in equation (12.10). Using equations (12.13) and (12.18) we get

$$\langle h_{ij}(t, \vec{r})h^{ij}(t, \vec{r})\rangle = 2\int_{-\infty}^{\infty} df\, S_h(f) = 4\int_{0}^{\infty} df\, S_h(f)$$

$$= 4\int_{f=0}^{f=\infty} d(\ln f)\, f\, S_h(f). \tag{12.19}$$

We now define the characteristic amplitude $h_c(f)$ from

$$\langle h_{ij}(t, \vec{r})h^{ij}(t, \vec{r})\rangle = 2\int_{f=0}^{f=\infty} d(\ln f)\, h_c^2(f). \tag{12.20}$$

Note that $h_c(f)$ is dimensionless, and represents a characteristic value of the amplitude, per unit logarithmic interval of frequency. The origin of the factor of two on the right-hand side of equation (12.20) is carefully explained in [5].

Comparing equations (12.19) and (12.20), we get

$$h_c^2(f) = 2f\, S_h(f). \tag{12.21}$$

We now wish to relate $h_c(f)$ and $h_0^2\Omega_{gw}(f)$. The starting point is the expression for the energy density of gravitational waves, given by the 00-component of the energy-momentum tensor. The energy-momentum tensor of a GW cannot be localized inside a single wavelength (see, e.g., sections 20.4 and 35.7 in [12] for a careful discussion) but it can be defined with a spatial averaging over several wavelengths:

$$\rho_{gw} = \frac{1}{32\pi G}\langle \dot{h}_{ij}\dot{h}^{ij}\rangle. \tag{12.22}$$

For a stochastic background, the spatial average over a few wavelengths is the same as a time average at a given point, which, in Fourier space, is the ensemble average performed using equation (12.18). By inserting equation (12.13) into equation (12.22) and using equation (12.18) one has

$$\rho_{gw} = \frac{4}{32\pi G}\int_{f=0}^{f=\infty} d(\ln f)\, f(2\pi f)^2 S_h(f), \tag{12.23}$$

and, thus

$$\frac{d\rho_{gw}}{d\ln f} = \frac{\pi}{2G}f^3 S_h(f). \tag{12.24}$$

Comparing equations (12.24) and (12.21) we get the important relation

$$\frac{d\rho_{gw}}{d\ln f} = \frac{\pi}{4G}f^2 h_c^2(f), \tag{12.25}$$

or, dividing by the critical density ρ_c,

$$\Omega_{gw}(f) = \frac{2\pi^2}{3H_0^2} f^2 h_c^2(f). \tag{12.26}$$

Inserting the numerical value of H_0, we find ([11], equation (65))

$$h_c(f) \simeq 1.26 \times 10^{-18} \left(\frac{1\,\text{Hz}}{f}\right) \sqrt{h_0^2 \Omega_{gw}(f)}. \tag{12.27}$$

From equations (12.21) and (12.26) one has

$$\Omega_{gw}(f) = \frac{4\pi^2}{3H_0^2} f^3 S_h(f), \tag{12.28}$$

and defining $S_n(f) = (S_n^{(1)}(f) S_n^{(2)}(f))^{1/2}$, equation (12.12) can be written in the following more transparent form:

$$\text{SNR} = \left[2F^2 T \int_0^\infty df\, \gamma^2(f) \frac{S_h^2(f)}{S_n^2(f)}\right]^{1/4}. \tag{12.29}$$

The factor of two in front of the integral can be understood from $\int_{-\infty}^\infty df = 2\int_0^\infty df$.

Finally, we mention another useful formula which expresses $h_0^2 \Omega_{gw}(f)$ in terms of the number of gravitons per cell of the phase space, $n(\vec{x}, \vec{k})$. For an isotropic stochastic background $n(\vec{x}, \vec{k}) = n_f$ depends only on the frequency $f = |\vec{k}|/(2\pi)$ and $\rho_{gw} = \int n_f 2\pi f\, d^3 k/(2\pi)^3 = 8\pi^2 \int_0^\infty d(\ln f) n_f f^4$. Therefore $d\rho_{gw}/d\ln f = 8\pi^2 n_f f^4$, and

$$h_0^2 \Omega_{gw}(f) \simeq 1.8 \left(\frac{n_f}{10^{37}}\right) \left(\frac{f}{1\,\text{kHz}}\right)^4. \tag{12.30}$$

As we will discuss below, to be observable at the VIRGO/LIGO interferometers, we should have $h_0^2 \Omega_{gw} \sim 10^{-6}$ between 1 Hz and 1 kHz, corresponding to n_f of order 10^{31} at 1 kHz and $n_f \sim 10^{43}$ at 1 Hz. A detectable stochastic GW background is therefore exceedingly classical, $n_k \gg 1$.

12.2.3 The characteristic noise level

We have seen in the previous section that there is a very natural definition of the characteristic amplitude of the signal, given by $h_c(f)$, which contains all the information on the physical effects, and is independent of the apparatus. We can, therefore, associate with $h_c(f)$ a corresponding noise amplitude $h_n(f)$, that embodies all the informations on the apparatus, defining $h_c(f)/h_n(f)$ in terms of the optimal SNR.

If, in the integral giving the optimal SNR, equation (12.12) or equation (12.29), we consider only a range of frequencies Δf such that the integrand is approximately constant, we can write

$$\text{SNR} \simeq \left[2F^2 T \Delta f \frac{\gamma^2(f) S_h^2(f)}{S_n^2(f)} \right]^{1/4} = \left[\frac{F^2 T \Delta f \gamma^2(f) h_c^4(f)}{2f^2 S_n^2(f)} \right]^{1/4}. \quad (12.31)$$

The right-hand side of equation (12.31) is proportional to $h_c(f)$, and we can therefore define $h_n(f)$ equating the right-hand side of equation (12.31) to $h_c(f)/h_n(f)$, so that

$$h_n(f) = \frac{1}{(\frac{1}{2}T\Delta f)^{1/4}} \left[\frac{f S_n(f)}{F|\gamma(f)|} \right]^{1/2}. \quad (12.32)$$

From the derivation of equation (12.32) we can understand the limitations implicit in the use of $h_n(f)$. It gives a measure of the noise level only under the approximation that leads from equation (12.29), which is exact, to equation (12.31). This means that Δf must be small compared to the scale on which the integrand in equation (12.29) changes, so that $\gamma(f) S_h(f)/S_n(f)$ must be approximately constant. In a large bandwidth this is non-trivial, and of course depends also on the form of the signal; for instance, if $h_0^2\Omega_{gw}$ is flat, then $S_h(f) \sim 1/f^3$. For accurate estimates of the SNR there is no substitute for a numerical integration of equation (12.12) or equation (12.29), unless the frequency range Δf in which we are interested is sufficiently small. However, for order of magnitude estimates, equation (12.27) for $h_c(f)$ and equation (12.32) for $h_n(f)$ are simpler to use, and they have the advantage of clearly separating the physical effect, which is described by $h_c(f)$, from the properties of the detectors, that enter only in $h_n(f)$.

Equation (12.32) also shows very clearly the advantage of correlating two detectors compared with the use of a single detector. With a single detector, the minimum observable signal, at SNR $= 1$, is given by the condition $S_h(f) \geq S_n(f)$. This means, from equation (12.21), a minimum detectable value for $h_c(f)$ given by

$$h_{\min}^{1d}(f) = (2f S_n(f))^{1/2}, \quad (12.33)$$

where the superscript 1d reminds us that this quantity refers to a single detector. From equation (12.32), by indicating with $\bar{h}_{\min}^{1d}$ the minimum detectable signal for a detector which noise level equals the geometric average of the noise levels of two detectors in coincidence, the minimum detectable signal for this detector pair is:

$$h_{\min}^{2d}(f) = \frac{1}{(2T\Delta f)^{1/4}} \frac{\bar{h}_{\min}^{1d}(f)}{[F|\gamma(f)|]^{1/2}}$$

$$\simeq 1.1 \times 10^{-2} \left(\frac{1\ \text{Hz}}{\Delta f} \right)^{1/4} \left(\frac{1\ \text{year}}{T} \right)^{1/4} \frac{\bar{h}_{\min}^{1d}(f)}{[F|\gamma(f)|]^{1/2}}. \quad (12.34)$$

Of course, the reduction factor in the noise level is larger if the integration time is larger, and if we increase the bandwidth Δf over which a useful coincidence is possible. Note that $h_0^2 \Omega_{gw}$ is quadratic in $h_c(f)$, so that an improvement in sensitivity by two orders of magnitudes in h_c means four orders of magnitude in $h_0^2 \Omega_{gw}$.

12.3   The overlap reduction function

When we consider the correlation between the signals of two GW detectors we have a reduction in sensitivity due to the fact that, in general, these detectors will not be either coincident or coaligned. This effect is quantified by the *overlap reduction function* $\gamma(f)$ appearing in the expression of the SNR (see equation (12.12)). This is a dimensionless function of frequency f which takes into account the relative positions and orientations of the two detectors and for an isotropic and unpolarized background is defined as [10]

$$\gamma(f) = \frac{1}{F} \sum_A \langle e^{i2\pi f \hat{\Omega} \cdot \Delta \vec{r}} F_1^A(\hat{r}_1, \hat{\Omega}, \psi) F_2^A(\hat{r}_2, \hat{\Omega}, \psi) \rangle_{\hat{\Omega}, \psi} \sim \frac{\Gamma(f)}{F} \qquad (12.35)$$

where $\Delta \vec{r} = \vec{r}_1 - \vec{r}_2$ is the separation vector between the two detector sites, F_i^A is the pattern function characterizing the response of the ith detector ($i = 1, 2$) to the $A = +, \times$ polarization, and the following notation

$$\langle \cdots \rangle_{\hat{\Omega}, \psi} = \int_{S^2} \frac{d\hat{\Omega}}{4\pi} \int_0^{2\pi} \frac{d\psi}{2\pi} (\cdots) \qquad (12.36)$$

has been introduced to indicate the average over the propagation direction (θ, ϕ) and the polarization angle ψ of the GW. The normalization factor F is given by:

$$F = \sum_A \langle F_1^A(\hat{r}_1, \hat{\Omega}, \psi) F_2^A(\hat{r}_2, \hat{\Omega}, \psi) \rangle_{\hat{\Omega}, \psi} |_{1 \equiv 2}, \qquad (12.37)$$

where the notation $1 \equiv 2$ is a compact way to indicate that the detectors are coincident and coaligned and, if at least one of them is an interferometer, the angle between its arms is equal to $\pi/2$ (L-shaped geometry). In this situation, by definition, $\gamma(f) = 1$. When the detectors are shifted apart (so there is a phase shift between the signals in the two detectors), or rotated out of coalignment (so the detectors have different sensitivity to the same polarization) it turns out that: $|\gamma(f)| < 1$.

    The pattern functions (or orientation factors) of a GW detector can be written in the following form

$$F^A(\hat{r}, \hat{\Omega}, \psi) = \text{Tr}\{D(\hat{r}) \varepsilon^A(\hat{\Omega}, \psi)\} \qquad (12.38)$$

where the symmetric, traceless tensor $D(\hat{r})$ describes the orientation and geometry of the detector located at $\vec{r}$. In terms of this tensor the gravitational-wave strain sensed by this detector (see equation (12.9)) is given by [10]

$$h(t, \vec{r}) = D^{ij}(\hat{r}) h_{ij}(t, \vec{r}). \tag{12.39}$$

For an interferometer, indicating with $\hat{u}$ and $\hat{v}$ the unit vectors in the directions of its arms, one has:

$$D(\hat{r}) = \tfrac{1}{2}\{\hat{u}(\hat{r}) \otimes \hat{u}(\hat{r}) - \hat{v}(\hat{r}) \otimes \hat{v}(\hat{r})\}. \tag{12.40}$$

For the lowest longitudinal mode of a cylindrical GW antenna with axis in the direction individued by the unit vector $\hat{l}$, one has [13]

$$D(\hat{r}) = \hat{l}(\hat{r}) \otimes \hat{l}(\hat{r}) - \tfrac{1}{3}I, \tag{12.41}$$

where I is the unit matrix. Finally, for the lowest five degenerate quadrupole modes ($m = -2, \ldots, +2$) of a spherical detector, the corresponding tensors are

$$D^{(0)}(\hat{r}) = \frac{1}{2\sqrt{3}}\{e^+(\hat{r}) + 2g^+(\hat{r})\} \sim \frac{1}{2\sqrt{3}}\{2f^+(\hat{r}) - e^+(\hat{r})\}$$
$$D^{(+1)}(\hat{r}) = -\tfrac{1}{2}g^\times(\hat{r}), \quad D^{(-1)}(\hat{r}) = -\tfrac{1}{2}f^\times(\hat{r}) \tag{12.42}$$
$$D^{(+2)}(\hat{r}) = \tfrac{1}{2}e^+(\hat{r}), \quad D^{(-2)}(\hat{r}) = -\tfrac{1}{2}e^\times(\hat{r})$$

where

$$f^+(\hat{r}) = \hat{m}(\hat{r}) \otimes \hat{m}(\hat{r}) - \hat{r} \otimes \hat{r}, \quad f^\times(\hat{r}) = \hat{m}(\hat{r}) \otimes \hat{r} + \hat{r} \otimes \hat{m}(\hat{r})$$
$$g^+(\hat{r}) = \hat{n}(\hat{r}) \otimes \hat{n}(\hat{r}) - \hat{r} \otimes \hat{r}, \quad g^\times(\hat{r}) = \hat{n}(\hat{r}) \otimes \hat{r} + \hat{r} \otimes \hat{n}(\hat{r}),$$

and $e^{+,\times}(\hat{r})$ are the tensors of equation (12.17) written in terms of the unit vectors $\hat{m}(\hat{r})$ and $\hat{n}(\hat{r})$ lying on the plane perpendicular to $\hat{r}$. From these expressions for the tensors D^{ij} and interpreting each of the five modes of a sphere as a single detector, it is possible to show that in the case of coincident detectors one has:

$$\langle F_1^A(\hat{r}, \hat{\Omega}, \psi) F_2^B(\hat{r}, \hat{\Omega}, \psi) \rangle_{\hat{\Omega}, \psi} \sim c_{12}\delta^{AB}, \quad (A, B = +, \times) \tag{12.43}$$

where c_{12} depends only on the geometry and the relative orientations of the two detectors. The corresponding values of F (see equation (12.37)) for the three different geometries considered (interferometer, cylindrical bar, sphere) are summarized in table 12.1.

By introducing the following notation

$$\Delta\vec{r} = d\hat{s}, \quad \eta = 2\pi f d,$$

where $\hat{s}$ is the unit vector along the direction connecting the two detectors and d is the distance between them, it can be shown [10] that the overlap reduction function assumes the following form ($D_k \equiv D(\hat{r}_k)$):

$$\gamma(f) = \rho_0(\eta) D_1^{ij} D_{2ij} + \rho_1(\eta) D_1^{ij} D_{2i}^k s_j s_k + \rho_2(\eta) D_1^{ij} D_2^{kl} s_i s_j s_k s_l \tag{12.44}$$

Table 12.1. The normalization factor F for three different geometries of the detectors: interferometer (ITF), cylindrical bar (BAR) and sphere (SPH). A $\star$ denotes entries that can be obtained from the symmetry of the table.

		ITF	BAR	SPH $m = 0$	SPH $m = \pm1$	SPH $m = \pm2$
ITF		2/5	$\star$	$\star$	$\star$	$\star$
BAR		2/5	8/15	$\star$	$\star$	$\star$
	$m = 0$	0	$2\sqrt{3}/15$	2/5	$\star$	$\star$
SPH	$m = \pm1$	0	0	0	2/5	$\star$
	$m = \pm2$	2/5	2/5	0	0	2/5

where

$$
\begin{bmatrix} \rho_0 \\ \rho_1 \\ \rho_2 \end{bmatrix} (\eta) = \frac{1}{F\eta^2} \begin{bmatrix} 2\eta^2 & -4\eta & 2 \\ -4\eta^2 & 16\eta & -20 \\ \eta^2 & -10\eta & 35 \end{bmatrix} \begin{bmatrix} j_0 \\ j_1 \\ j_2 \end{bmatrix} (\eta), \qquad (12.45)
$$

with $j_k(\eta)$ the standard spherical Bessel functions:

$$
j_0(\eta) = \frac{\sin \eta}{\eta}, \quad j_1(\eta) = \frac{j_0(\eta) - \cos \eta}{\eta}, \quad j_2(\eta) = 3\frac{j_1(\eta)}{\eta} - j_0(\eta).
$$

In the following, this formalism will be applied to treat the case in which the first detector is an interferometer and the other is, in turn, an interferometer, a cylindrical bar and a sphere.

12.3.1 Two interferometers

Following equation (12.40), the detector tensor of the interferometer located at a given point $\vec{r}_k$ can be written as

$$
D_k = \tfrac{1}{4}\{(\cos 2\xi_k - \cos 2\zeta_k)e^+(\hat{r}_k) - (\sin 2\xi_k - \sin 2\zeta_k)e^\times(\hat{r}_k)\} \quad (k = 1, 2)
$$
$$(12.46)$$

where ξ_k and ζ_k are the orientations of the kth interferometer arms measured anticlockwise from the true north. The value of these angles, together with the location of the central (corner) station on the Earth's surface, for different interferometers are reported in table 12.2. By inserting this expression in equation (12.44) it is possible to compute the overlap reduction function for a generic pair of interferometers. The results of this calculation in the case in which one of the interferometers is VIRGO are shown in figure 12.1.

Some important conclusions can be drawn from the inspection of this figure. First of all, owing to the oscillating behaviour with $\eta \propto fd$ of the spherical Bessel

Table 12.2. Site and orientation of Earth-based interferometric gravitational-wave detectors. The latitude is measured in degrees north from the equator, and the longitude in degrees east of Greenwich, England. In the last column is reported the distance with respect to VIRGO.

Project	Corner location		ξ (deg)	ζ (deg)	Distance (km)
	Latitude N (deg)	Longitude E (deg)			
VIRGO	43.63	10.50	71.5	341.5	—
LIGO-LA	30.56	−90.77	108.0	198.0	7905.9
LIGO-WA	46.45	−119.41	36.8	126.8	8158.2
GEO-600	52.25	9.82	25.94	291.61	958.2
TAMA-300	35.68	139.54	90.0	180.0	8864.1

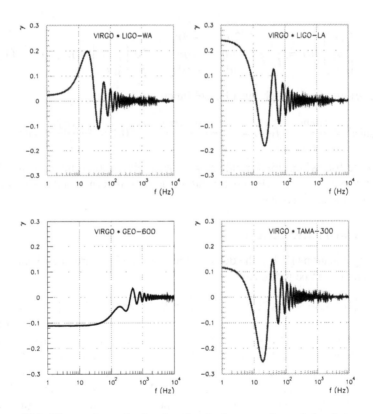

Figure 12.1. The overlap reduction function in the case of correlation between VIRGO and the other major interferometers.

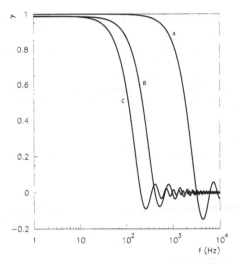

Figure 12.2. The overlap reduction function for the correlation of VIRGO with a coaligned interferometer whose central (corner) station is located at: (A) (43.2° N, 10.9° E), $d = 58.0$ km (Italy); (B) (43.6° N, 4.5° E), $d = 482.7$ km (France); (C) the GEO-600 (see table 12.2).

function intervening in expression (12.45) of ρ_k, there are always frequencies for which $\gamma(f)$ vanishes. Second, in all the considered cases $|\gamma(f)|$ is well below one on the whole frequency range, and practically zero for $f > 100$ Hz. In the case of correlation with the two LIGOs and TAMA-300, $|\gamma(f)|$ becomes its maximum value ($\simeq 0.2$) before the first zero or immediately after it, that is in the frequency region of 10–20 Hz, just where the sensitivity of the interferometric detectors is lower because of the thermal noise. The case is different where the correlation is between VIRGO and GEO-600. In this case, in fact, as a consequence of the relatively small distance between the two detectors, $|\gamma(f)|$, even if low ($\simeq 0.1$), remains constant up to ~ 100 Hz. This implies that from the point of view of the frequency-averaged value of $|\gamma(f)|$, this correlation is more efficient than the others. Let us note that in terms of Ω_{gw} the minimum detectable signal by a pair of detectors is proportional to $|\gamma(f)|^{-1}$ (see equations (12.26) and (12.34)) and, therefore, the low values shown in figure 12.1 lead to a substantial (even one order of magnitude) reduction of the sensitivity of the correlation.

As an example, figure 12.2 shows the overlap reduction function for the correlation of VIRGO with three coaligned interferometers located at about 50 km (probably the minimum distance sufficient to decorrelate local seismic and em noises) (curve A), 480 km (curve B) and 960 km (curve C), respectively, from the VIRGO site. The frequency location of the first zero of $\gamma(f)$ turns out to be

Table 12.3. Site and orientation of the resonant bars nearest to VIRGO. In the last column is reported the distance with respect to VIRGO.

	Location			
Project	Latitude N (deg)	Longitude E (deg)	λ (deg)	Distance (km)
AURIGA	45.35	11.95	39.3	222.84
NAUTILUS	41.80	12.67	39.3	270.12
EXPLORER	46.25	6.25	39.3	443.14

with good approximation inversely proportional to the distance between the two detectors.

12.3.2 Interferometer—bar

The detector tensor of a resonant bar located at a given point $\vec{r}$ turns out to be (see equation (12.41))

$$D(\hat{r}) = \tfrac{1}{2}\{\cos 2\lambda e^{+}(\hat{r}) - \sin 2\lambda e^{\times}(\hat{r}) + b(\hat{r}) - \tfrac{2}{3}I\} \qquad (12.47)$$

where the tensor

$$b(\hat{r}) = \hat{m}(\hat{r}) \otimes \hat{m}(\hat{r}) + \hat{n}(\hat{r}) \otimes \hat{n}(\hat{r})$$

has been introduced, and we have indicated with λ the orientation of the bar axis measured anticlockwise from the true North. The orientation and location of the resonant bars nearest to VIRGO are reported in table 12.3.

The behaviour of the overlap reduction function for the correlation of these detectors with VIRGO is shown in figure 12.3, where, since the resonant detectors can be easily rotated, we considered the case in which the bars are oriented along the direction of the first arm of VIRGO. The frequency dependence of the overlap reduction function is meaningless for narrow-band detectors like bars (and spheres). Therefore, the results shown in figure 12.3 have been calculated by assuming to vary the resonance frequency by varying the length of the bar.

12.3.3 Interferometer—sphere

The same calculation can be repeated by replacing bars with spheres of variable size. The overlap reduction function for the correlation of VIRGO with each of the five modes of a sphere (see equation (12.42)) is shown in figure 12.4. Notice that, since the normalization factors vanish (see table 12.1) for the $m = 0, \pm1$ modes, the quantity reported in figure 12.4 for these modes is the $\Gamma(f)$ function defined in equation (12.35). It is also worthwhile to notice that the result for the $m = 2$ case coincides exactly with that obtained with two interferometers, in complete accordance with the quadrupole nature of the GW excitation.

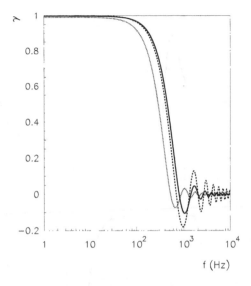

Figure 12.3. The overlap reduction function for the correlation of VIRGO with NAUTILUS (full curve), AURIGA (broken curve), and EXPLORER (dotted curve) in the case in which the angles λ, having the values reported in table 12.3, coincide with the angle ξ of VIRGO.

12.4 Achievable sensitivities to the SGWB

In the following we apply the results of the preceding sections to estimate the expected sensitivities of the correlation among the various detectors to the stochastic background.

12.4.1 Single detectors

To better appreciate the importance of correlating two detectors, it is instructive to consider first the sensitivity that can be obtained using only one detector. In this case a hypothetical signal would manifest itself as an excess noise, and should therefore satisfy $S_h(f) \gtrsim S_n(f)$. By imposing this condition to equation (12.28) and introducing the notation $\tilde{h}_f^2 = S_n(f)$, for the minimum detectable value of $h_0^2 \Omega_{gw}$ one has

$$ h_0^2 \Omega_{gw}^{min}(f) \simeq 1.3 \times 10^{-2} \left(\frac{f}{100 \text{ Hz}} \right)^3 \left(\frac{\tilde{h}_f}{10^{-22} \text{ Hz}^{-1/2}} \right)^2. \qquad (12.48) $$

This function in the case of VIRGO is plotted in figure 12.5, where the following analytical approximation for the noise power spectrum has been used

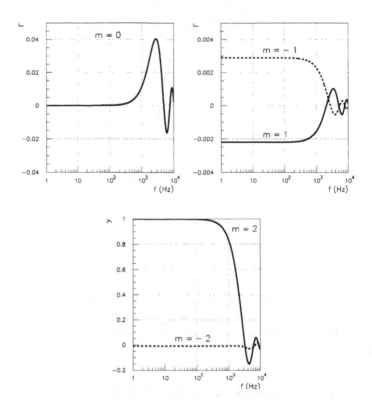

Figure 12.4. The overlap reduction function for the correlation of VIRGO with the five modes of a sphere located at (43.2° N, 10.9° E) ($d = 58.0$ km).

[14]:

$$S_n(f) = S_1 \left(\frac{f_0}{f}\right)^5 + S_2 \left(\frac{f_0}{f}\right) + S_3 \left[1 + \left(\frac{f}{f_0}\right)^2\right] \tag{12.49}$$

where $S_1 = 3.46 \times 10^{-50}$ Hz^{-1}, $S_2 = 6.60 \times 10^{-46}$ Hz^{-1}, $S_3 = 3.24 \times 10^{-46}$ Hz^{-1} and $f_0 = 500$ Hz. Figure 12.5 shows also the minimum detectable value of $h_0^2 \Omega_{\text{gw}}$ for the NAUTILUS detector (target sensitivity $\tilde{h}_f = 8.6 \times 10^{-23}$ Hz$^{-1/2}$ at $f = 907$ Hz) [15], and for a hollow sphere, made of Al 5056, with a mass of 200 ton, an outer radius of 3 m, and an inner radius of 2.1 m (target sensitivity $\tilde{h}_f = 4.3 \times 10^{-24}$ Hz$^{-1/2}$ at 273 Hz; $\tilde{h}_f = 3.4 \times 10^{-24}$ Hz$^{-1/2}$ at 935 Hz) [16].

Unfortunately, all these sensitivity levels are not interesting. As we shall find in section 2, an interesting sensitivity level for $h_0^2 \Omega_{\text{gw}}$ is at least of order 10^{-7}–10^{-6}. To reach such a level with a single interferometer we need, for example, $\tilde{h}_f < (3\text{–}10) \times 10^{-25}$ Hz$^{-1/2}$ at $f = 100$ Hz, or $\tilde{h}_f < (1\text{–}3) \times 10^{-26}$ Hz$^{-1/2}$

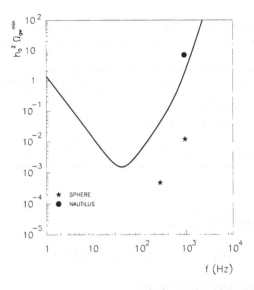

Figure 12.5. The minimum (SNR = 1) detectable value of $h_0^2 \Omega_{gw}$ by a single detector.

at $f = 1$ kHz. These values for $\tilde{h}_f$ are very far from the sensitivity of first generation interferometers, and are in fact even well below the limitation due to quantum noise. A very interesting sensitivity, possibly even of order $h_0^2 \Omega_{gw} \sim 10^{-13}$, could instead be reached with a single detector, with the planned space interferometer LISA [17], at $f \sim 10^{-3}$ Hz.

Note also that, while correlating two detectors the SNR improves with integration time (see equation (12.12)) this is not so with a single detector. So, independently of the low sensitivity, with a single detector it is conceptually impossible to tell whether an excess noise is due to a physical signal or is a noise of the apparatus that has not been properly accounted for. This might not be a great problem if the SNR is very large, but certainly with a single detector we cannot make a reliable detection at SNR of order one, so that the above estimates (which have been obtained setting SNR = 1) are really overestimates.

12.4.2   Two detectors

The statistical treatment of the problem of the extraction of information from the correlation between two detectors has been extensively discussed in [7] in the frame of the so-called 'decision theory' [18]. According to the results of [7], fixed a *false alarm* rate α ($1 - \alpha$ is the fraction of experimental outcomes that the decision rule correctly identifies the absence of a signal) and a minimum *detection rate* δ (the fraction of experimental outcomes that the decision rule correctly identifies the presence of a signal with fixed mean value > 0), the theoretical

Table 12.4. Minimum values of $h_0^2 \Omega_{gw}$ for cross-correlation measurements between VIRGO and the major interferometers for one year of observation, and different values of the pair (δ, α).

Correlation	(0.95, 0.05)	(0.95, 0.10)	(0.90, 0.10)
VIRGO ⋆ LIGO-LA	5.3×10^{-6}	4.7×10^{-6}	4.2×10^{-6}
VIRGO ⋆ LIGO-WA	6.4×10^{-6}	5.7×10^{-6}	5.0×10^{-6}
VIRGO ⋆ GEO-600	7.2×10^{-6}	6.4×10^{-6}	5.6×10^{-6}
VIRGO ⋆ TAMA-300	1.2×10^{-4}	1.1×10^{-4}	9.5×10^{-5}

SNR verifies the following bound

$$\mathrm{SNR}^2 \geq \sqrt{2}[\mathrm{erfc}^{-1}(2\alpha) - \mathrm{erfc}^{-1}(2\delta)]$$

where the complementary error function

$$\mathrm{erfc}(x) \sim \frac{2}{\sqrt{\pi}} \int_x^\infty \mathrm{d}z\, e^{-z^2}$$

has been introduced. Therefore, in the case of a stochastic GW background having a constant frequency spectrum $\Omega_{gw}(f) = \Omega_{gw}$, from equation (12.12) one has:

$$\Omega_{gw} \geq \frac{4\pi^2}{3H_0^2} \frac{1}{F\sqrt{T}} \left[\int_0^\infty \mathrm{d}f\, \frac{\gamma^2(f)}{f^6 S_n^{(1)}(f) S_n^{(2)}(f)} \right]^{-1/2} [\mathrm{erfc}^{-1}(2\alpha) - \mathrm{erfc}^{-1}(2\delta)]. \tag{12.50}$$

Two interferometers

The minimum values of $h_0^2 \Omega_{gw}$ for cross-correlation measurements between different interferometer pairs ($F = 2/5$ in equation (12.50)) in the case $T = 10^7$ s (4 months), $\alpha = 0.05$, and $\delta = 0.95$, are summarized in table II of [7]. In table 12.4 we report the minimum values of $h_0^2 \Omega_{gw}$ for the correlation of VIRGO with the major interferometers for one year of observation and different values of the detection and false alarm rates[4].

We now consider the sensitivity that could be obtained at VIRGO if the planned interferometer were correlated with a second identical interferometer located at a few tens of kilometres from the first, and with the same orientation (the case A at the end of section 12.3.1).

A rough estimate of the sensitivity can be given using the same line of reasoning developed in section 12.2.3. From equation (12.49) we see that we

---

[4] Let us note that α and δ need not sum to one.

can take, for our estimate, $\tilde{h}_f \sim 10^{-22}$ Hz$^{-1/2}$ over a bandwidth $\Delta f \sim 1$ kHz. Moreover, as clearly shown in figure 12.2, the overlap reduction function is approximately equal to one up to $f \sim 1$ kHz. Therefore, from equation (12.31) and (12.27) one has

$$h_0^2 \Omega_{\text{gw}}^{\text{min}}(f) \sim 1.3 \times 10^{-7} \text{ SNR}^2 \left(\frac{1 \text{ year}}{T}\right)^{1/2} \left(\frac{f}{100 \text{ Hz}}\right)^3 \left(\frac{\tilde{h}_f}{10^{-22} \text{ Hz}^{-1/2}}\right)^2.$$

(12.51)

which shows that correlating two VIRGO interferometers for 1 year we can detect a relic spectrum with $h_0^2 \Omega_{\text{gw}}(100 \text{ Hz}) \sim 3 \times 10^{-7}$ at SNR = 1.65, or 10^{-7} at SNR = 1. Compared to the case of a single interferometer with SNR = 1, equation (12.48), we gain five orders of magnitude. As already discussed, to obtain a precise numerical value one must however consider equation (12.12). This involves an integral over all frequencies, (that replaces the somewhat arbitrary choice of Δf made above) and depends on the functional form of $h_0^2 \Omega_{\text{gw}}(f)$. For $h_0^2 \Omega_{\text{gw}}(f)$ independent of the frequency, using the analytical approximation of equation (12.49) for $S_n^{(i)}(f)$ ($i = 1, 2$) and equation (12.44) for $\gamma(f)$, we get[5]

$$h_0^2 \Omega_{\text{gw}}^{\text{min}} \simeq 7 \times 10^{-8} \text{ SNR}^2 \left(\frac{1 \text{ year}}{T}\right)^{1/2}, \quad (h_0^2 \Omega_{\text{gw}}(f) = \text{constant}). \quad (12.52)$$

It is interesting to note that the main contribution to the integral comes from the region $f < 100$ Hz. In fact, neglecting the contribution of the region $f > 100$ Hz, the result for $h_0^2 \Omega_{\text{gw}}^{\text{min}}$ changes only by approximately 4%. Also, the lower part of the accessible frequency range is not crucial. For instance, restricting the numerical integration to the regions 20 Hz $\leq f \leq$ 200 Hz and 30 Hz $\leq f \leq$ 100 Hz the sensitivity on $h_0^2 \Omega_{\text{gw}}$ degrades by 1% and 10%, respectively. This means that the most important source of noise for the measurement of a flat stochastic background is the thermal noise[6]. In particular, the sensitivity is limited by the mirror thermal noise, which dominates in the region 40 Hz $\lesssim f \lesssim$ 200 Hz, while the pendulum thermal noise dominates below

---

[5] The integral has been evaluated numerically in the frequency interval 2 Hz–10 kHz. The analytical fit of equation (12.49) underestimates the noise power spectrum in the region $f < 2$ Hz and, in any case, this frequency region gives no appreciable contribution to the integral. Above 10 kHz the overlap reduction function is negligible (see figure 12.2).

[6] Note also that it is not very meaningful to give more decimal figures in the minimum detectable value of $h_0^2 \Omega_{\text{gw}}^{\text{min}}$. Apart from the various uncertainties which enter the computation of the sensitivity curve, a trivial source of uncertainty is the fact that the computation of the thermal noises are performed using a temperature of 300 K. A 5% variation, corresponding to an equally plausible value of the temperature, gives a 5% difference in $h_0^2 \Omega_{\text{gw}}^{\text{min}}$. Quoting more figures is especially meaningless when the minimum detectable $h_0^2 \Omega_{\text{gw}}$ is estimated using the approximate quantities $h_c(f), h_n(f)$, i.e. approximating the integrand of equation (12.12) with a constant over a bandwidth Δf. For a broadband detector these estimates typically give results which agree with the exact numerical integration of equation (12.12) at best within a factor of two.

Table 12.5. Minimum values of $h_0^2 \Omega_{gw}$ for the correlation VIRGO–VIRGO for one year of observation and different values of the pair (δ, α) (different values of SNR), for the three possible locations of the second VIRGO considered in section 12.3.1.

Correlation	(0.95, 0.05)	(0.95, 0.10)	(0.90, 0.10)
A	2.4×10^{-7}	2.1×10^{-7}	1.8×10^{-7}
B	2.5×10^{-7}	2.2×10^{-7}	1.9×10^{-7}
C	2.8×10^{-7}	2.5×10^{-7}	2.2×10^{-7}

Table 12.6. Minimum values of $h_0^2 \Omega_{gw}(100 \text{ Hz})$ for the correlation VIRGO–VIRGO ($T = 1$ year, SNR = 1), for different values of the exponent β in equation (12.53). The rows refer to the three different locations of the second VIRGO considered in section 12.3.1.

Correlation	$\beta = 0$	$\beta = 1$	$\beta = -1$	$\beta = 3$
A	7.2×10^{-8}	1.1×10^{-7}	3.0×10^{-8}	2.0×10^{-8}
B	7.6×10^{-8}	1.3×10^{-7}	3.1×10^{-8}	9.7×10^{-8}
C	8.5×10^{-8}	1.6×10^{-7}	3.2×10^{-8}	2.5×10^{-7}

approximately 40 Hz. This calculation has been performed also for the other two geographical locations of the second VIRGO interferometer considered in section 12.3.1. Although the range of frequency where $\gamma(f)$ has a sensible value is very different in the three cases (see figure 12.2), it is evident from table 12.5 that, for fixed values of (δ, α), the corresponding sensitivities to a frequency-independent background are practically the same, and approximately one order of magnitude better than the ones achievable with the two LIGO detectors [7].

Let us now consider the case of a frequency-dependent stochastic background, that in the VIRGO frequency band can be parametrized in the following way

$$\Omega_{gw}(f) = \Omega_{gw}(100 \text{ Hz}) \left(\frac{f}{100 \text{ Hz}} \right)^{\beta}. \tag{12.53}$$

The same procedure applied to obtain equation (12.52) gives, in the cases $\beta = \pm 1, 3$, the minimum detectable values for $h_0^2 \Omega_{gw}(100 \text{ Hz})$ reported in table 12.6 ($T = 1$ year, SNR = 1). For the sake of comparison, we also reported the analogous quantities in the case of a frequency-independent background ($\beta = 0$).

Some comments about this table are necessary. First, to compare the sensitivities of a correlation to backgrounds with different frequency behaviour, we have to take into account that in the case $\beta \neq 0$ the spectrum will exhibit a peak within the VIRGO band: at low frequency for $\beta < 0$; at high frequency for

Table 12.7. Minimum values of $h_0^2 \Omega_{\text{gw}}$ for cross-correlation measurements between two bars and between one bar and VIRGO for one year of observation and SNR = 1.

AURIGA ⋆ NAUTILUS	2.0×10^{-4}
AURIGA ⋆ VIRGO	1.6×10^{-4}
NAUTILUS ⋆ VIRGO	2.8×10^{-4}

$\beta > 0$. For example, in the case of the correlation A, according to table 12.6 and from equation (12.53), we find that a spectrum with $\beta = 1$ will be detected, at SNR = 1, if $h_0^2 \Omega_{\text{gw}}(1 \text{ kHz}) = 1.1 \times 10^{-6}$, while a spectrum with $\beta = -1$ will be detected if $h_0^2 \Omega_{\text{gw}}(2 \text{ Hz}) = 1.5 \times 10^{-6}$. So, both for increasing or decreasing spectra, to be detectable in one year at SNR = 1, $h_0^2 \Omega_{\text{gw}}$ must have a peak value, within the VIRGO band, of the order of a few 10^{-6} in the case $\beta = \pm 1$. Under the same conditions, a frequency-dependent spectrum can be detected at the level 7×10^{-8} (see the first column of table 12.6). Clearly, for detecting increasing (decreasing) spectra, the upper (lower) part of the frequency band becomes more important, and this is the reason why the sensitivity degrades compared to flat spectra, since for increasing or decreasing spectra the maximum of the signal is at the edges of the accessible frequency band, where the interferometer sensitivity is worse. Second, for each β the sensitivity on $h_0^2 \Omega_{\text{gw}}(f)$ always degrades going from A to C. Relatively to A, the size of this decay for B and C increases with β, and for $\beta = 3$ the correlation C show a sensitivity one order of magnitude lower than the sensitivity of A. This means that in order to have the same sensitivity, the product $S_n^{(1)} S_n^{(2)}$ in case C has to be two orders of magnitude lower than in case A (see equation (12.50)). The case $\beta = 3$ is particularly important, because this is the frequency dependence of the spectrum predicted by the string cosmology.

Resonant masses and resonant mass-interferometer

Resonant mass detectors includes bars like NAUTILUS, EXPLORER and AURIGA (see, e.g., [15, 21] for reviews). Spherical [16, 22] and truncated icosahedron (TIGA) [23] resonant masses are also being developed or studied. The correlation between two resonant bars and between a bar and an interferometer has been extensively considered in [19, 24–26]. The results obtained in [19] for the minimum values of $h_0^2 \Omega_{\text{gw}}$ in the case of a background having constant frequency spectrum, for one year of integration and SNR = 1, are summarized in table 12.7.

Using resonant optical techniques, it is possible to improve the sensitivity of interferometers at special values of the frequency, at the expense of their broadband sensitivity. Since bars have a narrow band anyway, narrow-banding the interferometer improves the sensitivity of a bar-interferometer correlation by about one order of magnitude [20].

While resonant bars have been taking data for years, spherical detectors are at the moment still at the stage of theoretical studies (although prototypes might be built in the near future), but could reach extremely interesting sensitivities. In particular, the correlation between VIRGO and one sphere with a diameter of 3 m, made of Al 5056 ($M = 38$ ton), gives sensitivities that are, respectively, $h_0^2 \Omega_{\text{gw}} \sim 2 \times 10^{-5}$, if the sphere is located in the AURIGA site, and $h_0^2 \Omega_{\text{gw}} \sim 4 \times 10^{-5}$ if the sphere is at the NAUTILUS site [19]. Instead, the correlation of two spheres of this type but located at the same site could reach a sensitivity $h_0^2 \Omega_{\text{gw}} \sim 4 \times 10^{-7}$ [19]. All these figures improve using a denser material or increasing the sphere diameter, but it might be difficult to build a heavier sphere. Another very promising possibility is given by hollow spheres [16]. The theoretical studies of [16] suggest for the correlation of two 40-ton colocated hollow spheres, made of Al 5056, a sensitivity of $h_0^2 \Omega_{\text{gw}} \sim 6 \times 10^{-8}$ at $f = 218$ Hz (one year of observation and SNR = 1).

12.4.3   More than two detectors

When the outputs of $M > 2$ detectors are available, the information about the magnitude of the stochastic GW background can be extracted in two ways: combining the measurements from each detector pair or directly correlating the outputs of the detectors. Both these techniques have been extensively treated in [7], and here we shall only review the key results obtained in this analysis.

Multiple detector pairs

We indicate with

$$S_1^{(ij)}, S_2^{(ij)}, \ldots, S_{n_{ij}}^{(ij)}, \quad (i, j = 1, \ldots, M)$$

the n_{ij} different measurements, each of length T, of the optimally-filtered cross-correlation signal $S^{(ij)}$ between the ith and jth detectors (see equation (12.8)). Under the following hypothesis:

- $T \gg$ the light travel time between any pair of detectors;
- the optimal filter functions (see equation (12.8)) for each detector pair are normalized in a way that

$$\langle S^{(ij)} \rangle = \frac{1}{n_{ij}} \sum_{k=1}^{n_{ij}} S_k^{(ij)} = \Omega_\beta T, \quad \forall (i, j)$$

for a stochastic background having a power-law spectrum $\Omega_{\text{gw}}(f) = \Omega_\beta f^\beta$;
- the noise is large, i.e. the covariance matrix built from the cross-correlation signals taken during the same time interval T is approximately diagonal

$$C^{(ij)(kl)} = \langle S^{(ij)} S^{(kl)} \rangle - \langle S^{(ij)} \rangle \langle S^{(kl)} \rangle \simeq \delta^{ik} \delta^{kl} (\sigma^{(ij)})^2$$

Table 12.8. Minimum values of $h_0^2 \Omega_{gw}$ for one year of observation, for $(\delta, \alpha) = (0.95, 0.05)$, for the optimal combination of cross-correlation measurements between multiple detector pairs, taken from all possible triples of interferometers containing VIRGO.

VIRGO ⋆ LIGO-WA ⋆ LIGO-LA	2.6×10^{-6}
VIRGO ⋆ LIGO-LA ⋆ GEO-600	4.2×10^{-6}
VIRGO ⋆ LIGO-WA ⋆ GEO-600	4.8×10^{-6}
VIRGO ⋆ LIGO-LA ⋆ TAMA-300	5.4×10^{-6}
VIRGO ⋆ LIGO-WA ⋆ TAMA-300	6.6×10^{-6}
VIRGO ⋆ GEO-600 ⋆ TAMA-300	7.3×10^{-6}

Table 12.9. The same as table 12.8 for the case of quadruples.

VIRGO ⋆ LIGO-WA ⋆ LIGO-LA ⋆ GEO-600	2.4×10^{-6}
VIRGO ⋆ LIGO-WA ⋆ LIGO-LA ⋆ TAMA-300	2.6×10^{-6}
VIRGO ⋆ LIGO-LA ⋆ GEO-600 ⋆ TAMA-300	4.2×10^{-6}
VIRGO ⋆ LIGO-WA ⋆ GEO-600 ⋆ TAMA-300	4.8×10^{-6}

and, therefore, all the measurements can be considered uncorrelated;

the authors of [7] show that the optimal SNR turns out to be[7]

$$\text{SNR}^4 = \sum_{i=1}^{M} \sum_{j>i}^{M} n_{ij} (\text{SNR}^{(ij)})^4 \tag{12.54}$$

where $\text{SNR}^{(ij)}$ is given in equation (12.12). Thus, in the case of a frequency-independent spectrum $\Omega_{gw}(f) = \Omega_{gw}$, the minimum value of Ω_{gw} for given detection (δ) and false alarm (α) rates turns out to be:

$$\left(\frac{1}{\Omega_{gw}(\delta, \alpha)} \right)^2 = \sum_{i=1}^{M} \sum_{j>i}^{M} \left(\frac{1}{\Omega_{gw}^{(ij)}(\delta, \alpha)} \right)^2. \tag{12.55}$$

where $\Omega_{gw}^{(ij)}(\delta, \alpha)$ are the analogous quantities for the ij detector pairs (see equation (12.50)). In the case $(\delta, \alpha) = (0.95, 0.05)$, the minimum values of $h_0^2 \Omega_{gw}$ obtained considering all triples and quadruples of interferometers having in common VIRGO, are reported in tables 12.8 and 12.9, respectively.

[7] Remember that we defined the SNR as the square root of the SNR defined in [7].

Many-detector correlation

The M-detector correlation signal S is the obvious generalization of equation (12.8) for a two-detector:

$$S = \int_{-T/2}^{T/2} dt_1 \ldots \int_{-T/2}^{T/2} dt_M \, s_1(t_1)s_2(t_2)\ldots s_M(t_M)Q(t_1,\ldots,t_M).$$

Let us remark that, as a consequence of the fact that each detector output signal is assumed to be a random Gaussian variable having zero mean value, $\langle S \rangle$ is different from zero only for an even number of detectors ($M = 2N$).

Under the same hypothesis assumed in section 12.2.1 for the treament of the two-detector case, the authors of [7] show that the SNR for the optimally-filtered $2N$-detector correlation is given by

$$\text{SNR}^4 \approx \sum_{\{\ldots\}}(\text{SNR}^{(12)}\text{SNR}^{(34)}\ldots\text{SNR}^{(2N,2N-1)})^4 \qquad (12.56)$$

where the sum run over all the possible permutations of the sequence $\{(ij),(kl),\ldots,(pq)\}$ with $(i < j, k < l,\ldots,p < q)$. In the case of a frequency-independent background, with fixed detection and false alarm rates, the minimum detectable value of Ω_{gw} from data obtained via a $2N$-detector correlation experiment, is given by

$$\left(\frac{1}{\Omega_{\text{gw}}(\delta,\alpha)}\right)^2$$

$$= \left[C(\delta,\alpha)\sum_{\{\ldots\}}\left(\frac{1}{\Omega_{\text{gw}}^{(12)}(\delta,\alpha)\Omega_{\text{gw}}^{(34)}(\delta,\alpha)\ldots\Omega_{\text{gw}}^{(2N-1,2N)}(\delta,\alpha)}\right)^2\right]^{1/N}$$

$$(12.57)$$

with

$$C(\delta,\alpha) = \{\sqrt{2}[\text{erfc}^{-1}(2\alpha) - \text{erfc}^{-1}(2\delta)]\}^{2(N-1)/N}.$$

In table 12.10 are reported the minimum values of $h_0^2\Omega_{\text{gw}}$ obtained from equation (12.57) for the four-interferometer correlations. It is important to note that, as clearly shown from the comparison with tables 12.8 and 12.9, these minimum values are always greater than those for optimal combination of data from multiple detector pairs.

By summarizing the results of this section we can say that the improvement in sensitivity obtained using more detectors is not large compared to the case of a single pair correlation (see table 12.4). This is due to the fact that correlating $2,4,\ldots,2N$ detectors or combining in a optimal way data from multiple detector pairs, one does not change the general dependence that

$$\Omega_{\text{gw}} \sim T_{\text{tot}}^{-1/2}$$

Table 12.10. Minimum values of $h_0^2 \Omega_{gw}$ for one year of observation, for $(\delta, \alpha) = (0.95, 0.05)$, for optimally-filtered four-detector correlations.

VIRGO $\star$ LIGO-WA $\star$ LIGO-LA $\star$ GEO-600	3.7×10^{-6}
VIRGO $\star$ LIGO-WA $\star$ LIGO-LA $\star$ TAMA-300	1.4×10^{-5}
VIRGO $\star$ LIGO-LA $\star$ GEO-600 $\star$ TAMA-300	2.2×10^{-5}
VIRGO $\star$ LIGO-WA $\star$ GEO-600 $\star$ TAMA-300	2.7×10^{-5}

of the minimum detectable value on the total observation time: what changes from one case to another is only the numerical factors multiplying $T_{tot}^{-1/2}$. Although the improvement in sensitivity is limited the availability of the signals from various detectors would be important in ruling out spurious effects.

12.5 Observational bounds

We close this section discussing what is actually known from the observational side about the stochastic background. At present, there are strict limits on this background in only a couple of frequency ranges, but other than that, only one very general constraint. In the following we will only discuss the general one, the nucleosynthesis bound, because it is the only relevant one for the frequency region $1 \text{ Hz} < f < 1 \text{ kHz}$ covered by the ground based detectors. The other bounds are inferred from the timing irregularities in the arrival times of the pulses emitted by some millisecond pulsars and the anisotropies on large angular scales of the CMB. They, respectively, constrain Ω_{gw} in the frequency regions $f \sim 10^{-8}$ Hz and $10^{-18} \text{ Hz} \lesssim f \lesssim 10^{-16}$ Hz, which are far below any frequency band accessible for the present-day ground- or space-based experiments. A complete discussion of these bounds can be found in [4].

Nucleosynthesis successfully predicts the primordial abundances of deuterium, ^{3}He, ^{4}He and ^{7}Li in terms of one cosmological parameter η, the baryon to photon ratio. In the prediction parameters of the underlying particle theory also enter, which are therefore constrained in order not to spoil the agreement. In particular, the prediction is sensitive to the effective number of species at the time of nucleosynthesis, $g_* = g(T \simeq 1 \text{ MeV})$. With some simplifications, the dependence on g_* can be understood as follows. A crucial parameter in the computations of nucleosynthesis is the ratio of the number density of neutrons, n_n, to the number density of protons, n_p. As long as thermal equilibrium is maintained we have (for non-relativistic nucleons, as appropriate at $T \sim 1 \text{ MeV}$) $n_n/n_p = \exp(-Q/T)$ where $Q = m_n - m_p \simeq 1.3$ MeV. Equilibrium is maintained by the process $pe \leftrightarrow n\nu$, with width $\Gamma_{pe \to n\nu}$, as long as this width is greater than H. When the rate drops below the Hubble constant H, the process cannot compete anymore with the expansion of the universe and, apart from occasional weak processes, is dominated by the decay of free neutrons, the ratio n_n/n_p remains frozen at the value $\exp(-Q/T_f)$, where T_f is the value of the temperature at the time of freeze-out. This number, therefore, determines the density of neutrons available for nucleosynthesis, and since practically all neutrons available will eventually form ^{4}He, the final primordial abundance of this nucleus is very sensitive to the freeze-out temperature T_f. If we assume for simplicity $\Gamma_{pe \to n\nu} \simeq G_F^2 T^5$ (which is really appropriate only in the limit $T \gg Q$), where G_F is the Fermi constant, from equation (12.3) turns out that T_f is determined by the condition

$$G_F^2 T_f^5 \simeq \left(\frac{4\pi^3 g_*}{45}\right)^{1/2} \frac{T_f^2}{M_{Pl}}. \qquad (12.58)$$

This shows that $T_f \sim g_*^{1/6}$, at least with the approximation that we used for $\Gamma_{pe \to n\nu}$. A large energy density in relic gravitons gives a large contribution to the total density ρ and therefore to g_*. This results in a larger freeze-out temperature, more available neutrons and then in overproduction of ^{4}He. This is the idea behind the nucleosynthesis bound [28]. More precisely, since the density of ^{4}He increases also with the baryon to photon ratio η, we could compensate an increase in g_* with a decrease in η, and therefore we also need a lower limit on η, which is provided by the comparison with the abundance of deuterium and ^{3}He.

Rather than g_*, N_ν is often used as an 'effective number of neutrino species' and is defined as follows. In the standard model, at $T \sim$ a few MeV, the active degrees of freedom are the photon, $e^\pm$, neutrinos and antineutrinos, and they have the same temperature, $T_i = T$. Then, for N_ν families of light neutrinos, one has

$$g_*(N_\nu) = 2 + \tfrac{7}{8}(4 + 2N_\nu), \qquad (12.59)$$

where the factor of two comes from the two elicity states of the photon, four from $e^\pm$ in the two elicity states, and $2N_\nu$ counts the N_ν neutrinos and the N_ν antineutrinos, each with their single elicity state. According to the standard model, $N_\nu = 3$ and therefore $g_* = 43/4$. Therefore, we can define an 'effective number of neutrino species' N_ν from

$$g_*(N_\nu) \sim \frac{43}{4} + \sum_{i=\text{extra bosons}} g_i \left(\frac{T_i}{T}\right)^4 + \frac{7}{8} \sum_{i=\text{extra fermions}} g_i \left(\frac{T_i}{T}\right)^4. \qquad (12.60)$$

One extra species of light neutrino, at the same temperature as the photons, would contribute one unit to N_ν, but all species, weighted with their energy density, contribute to N_ν, which, of course, in general is not an integer. For $i =$ gravitons, we have $g_i = 2$ and $(T_i/T)^4 = \rho_{gw}/\rho_\gamma$, where $\rho_\gamma = 2(\pi^2/30)T^4$ is the photon energy density. If gravitational waves give the only extra contribution to N_ν, compared to the standard model with $N_\nu = 3$, using equation (12.59), the above equation gives immediately

$$\left(\frac{\rho_{gw}}{\rho_\gamma}\right)_{NS} = \frac{7}{8}(N_\nu - 3), \tag{12.61}$$

where the subscript NS reminds us that this equality holds at the time of nucleosynthesis. If more extra species, not included in the standard model, contribute to $g_*(N_\nu)$, then the equals sign in the above equation is replaced by less than or equal. The same happens if there is a contribution from any other form of energy present at the time of nucleosynthesis and not included in the energy density of radiation, like, for example, primordial black holes.

To obtain a bound on the energy density at the present time, we note that from the time of nucleosynthesis to the present time ρ_{gw} scaled as $1/a^4$, while, as a consequence of the assumed adiabatic expansion of the universe (see section 12.1), $\rho_\gamma \sim T^4 \sim 1/(a^4 g_S^{4/3})$. Therefore, one has

$$\left(\frac{\rho_{gw}}{\rho_\gamma}\right)_0 = \left(\frac{\rho_{gw}}{\rho_\gamma}\right)_{NS} \left(\frac{g_S(T_0)}{g_S(1 \text{ MeV})}\right)^{4/3} = \left(\frac{\rho_{gw}}{\rho_\gamma}\right)_{NS} \left(\frac{3.91}{10.75}\right)^{4/3}. \tag{12.62}$$

Therefore we get the nucleosynthesis bound at the present time,

$$\left(\frac{\rho_{gw}}{\rho_\gamma}\right)_0 \leq 0.227(N_\nu - 3). \tag{12.63}$$

Of course this bound holds only for GWs that were already produced at the time of nucleosynthesis ($T \sim 1$ MeV, $t \sim 1$ s). It does not apply to any stochastic background produced later, like backgrounds of astrophysical origin (see section 13.5). Note that this is a bound on the total energy density in gravitational waves, integrated over all frequencies. Writing $\rho_{gw} = \int d(\ln f) \, d\rho_{gw}/d\ln f$, multiplying both ρ_{gw} and ρ_γ in equation (12.63) by h_0^2/ρ_c, and inserting the numerical value $h_0^2 \rho_\gamma/\rho_c \simeq 2.474 \times 10^{-5}$ [29], we get

$$\int_{f=0}^{f=\infty} d(\ln f) \, h_0^2 \Omega_{gw}(f) \leq 5.6 \times 10^{-6}(N_\nu - 3). \tag{12.64}$$

The bound on N_ν from nucleosynthesis is subject to various systematic errors in the analysis, which have to do mainly with the issues of how much of the observed ^{4}He abundance is of primordial origin, and of the nuclear processing of ^{3}He in stars, and as a consequence over the last five years have been quoted limits on N_ν ranging from 3.04 to around 5. The situation has been recently reviewed in [30]. The conclusion of [30] is that, until current astrophysical uncertainties are clarified, $N_\nu < 4$ is a conservative limit. Using extreme assumptions, a meaningful limit $N_\nu < 5$ still exists, showing the robustness of the argument. Correspondingly, the right-hand side of equation (12.64) is, conservatively, of order 5×10^{-6} and anyway cannot exceed 10^{-5}.

If the integral cannot exceed these values, also its positive definite integrand $h_0^2 \Omega_{gw}(f)$ cannot exceed it over an appreciable interval of frequencies, $\Delta \ln f \sim 1$. One might still have, in principle, a very narrow peak in $h_0^2 \Omega_{gw}(f)$ at some frequency f, with a peak value larger than say 10^{-5}, while still its contribution to the integral could be small enough. However, apart from the fact that such behaviour seems rather implausible, or at least is not suggested by any cosmological mechanism, it would also be probably of little help in the detection at broadband detectors like VIRGO, because even if we gain in the height of the signal we lose because of the reduction of the useful frequency band Δf, see equation (12.12).

These numbers, therefore, give a first idea of what can be considered an interesting detection level for $h_0^2 \Omega_{gw}(f)$, which should be at least a few times 10^{-6}, especially considering that the bound (12.64) refers not only to gravitational waves, but to all possible sources of energy which have not been included, like particles beyond the standard model, primordial black holes, etc.

As a final remark, we note that a very weak bound in the region $f \sim 1$ kHz has been obtained from the analysis of correlation between bar detectors. Preliminary results on a NAUTILUS–EXPLORER correlation, using 12 hours of data, have been reported in [27], and give $h_0^2 \Omega_{gw}(f = 920 \text{ Hz}) \sim 120$.

Chapter 13

Sources of SGWB

Here we review the present knowledge about the potential processes, both of cosmological and astrophysical origin, from which a stochastic GW background might arise. We examine in some detail the mechanisms at work in each case and the features of the corresponding spectrum of GW radiation.

13.1 Topological defects

The concept of the spontaneous symmetry breaking, the idea that there are underlying symmetries of Nature that are not manifest in the structure of the vacuum, play a crucial role in the modern description of the particle interactions. Of particular interest for cosmology is the theoretical expectation that at high temperatures, symmetries that are spontaneously broken today were restored [31]. In the context of the hot big bang cosmology this implies a sequence of phase transitions in the early universe, with critical temperatures related to the corresponding symmetry breaking scales.

To illustrate the phenomenon of the high-temperature symmetry restoration we consider a complex scalar field with a 'Mexican-hat' potential

$$V(\phi) = \tfrac{1}{2}\lambda(\phi^\dagger\phi - \eta^2)^2 \quad (\lambda > 0). \tag{13.1}$$

The Lagrangian is invariant under the group U(1) of the global phase transformations, $\phi \rightarrow e^{i\alpha}\phi$. The minima of the potential are at nonzero values ϕ, and so the symmetry is spontaneously broken and ϕ acquires a vacuum expectation value (VEV)

$$\langle\phi\rangle = \eta e^{i\theta} \tag{13.2}$$

where the phase θ is arbitrary. We thus have a manifold $\mathcal{M}$ of degenerate vacuum states corresponding to different choices of θ, that, in this case, is the circle $|\phi| = \eta$ in the complex ϕ plane.

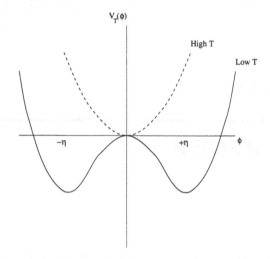

Figure 13.1. The effective potential.

At finite temperature the effective potential for ϕ acquires additional, temperature-dependent terms. In the high-temperature limit, one has:

$$V_T(\phi) = AT^2\phi^\dagger\phi + V(\phi) \tag{13.3}$$

where the dimensionless constant A depend on the self-coupling λ and is assumed to be positive. From equations (13.1) and (13.3) we see that the effective mass of the field ϕ at temperature T is:

$$m^2(T) = 2A(T^2 - T_c^2) \quad \text{where } T_c = \left(\frac{\lambda}{A}\right)^{1/2}\eta.$$

From this expression we see that for $T > T_c$, $m^2(T)$ is positive, the minimum of $V_T(\phi)$ is at $\phi = 0$ and so the symmetry is restored (see figure 13.1). The temperature T_c is the critical temperature of the phase transition from the symmetric ($\langle\phi\rangle = 0$) to the broken-symmetry ($\langle\phi\rangle \neq 0$) phase. Unless λ is very small, $T_c \sim \eta$. In this example the transition is second order: the symmetric phase corresponds to a maximum for V_T at $T < T_c$ and the transition occurs smoothly. More complicated models can lead to first-order transitions, where the symmetric phase remains a local minimum at $T < T_c$ separated by a barrier from the minima at $\phi \neq 0$. In this case the transition occurs through the bubble nucleation.

In a cosmological context, as the universe cools through the critical temperature the Higgs field ϕ develop an expectation value $\langle\phi\rangle$ corresponding to some point in the manifold $\mathcal{M}$. Since all points of this manifold are equivalent, the choice depends on random fluctuations and is different in different regions of space. Therefore, associated to the phase transition there is a length scale ξ representing the maximum distance over which the Higgs field can be correlated.

This correlation length depends upon the details of the phase transition and is temperature dependent. In any case, since correlations cannot be established at speed greater than the speed of light, ξ cannot exceed the causal horizon d_H, the distance travelled by light during the lifetime of the universe. In the standard cosmology $d_H \sim t$ and, thus, one has

$$\xi \le t_c$$

where t_c is the time at which the phase transition is completed. The actual magnitude of ξ at the phase transition and afterwards is determined by complicated dynamical processes and can be much smaller than this causality upper bound.

Since the field must be continous, on the boundaries between different correlation regions ϕ leaves the vacuum manifold $\mathcal{M}$ and assume values corresponding to a high potential energy. For topological reasons, these regions of false vacuum are stable and survive to further evolution of the universe frozen in the form of topological defects.

The cosmological production mechanism described above is known as the Kibble mechanism [32] and is very much akin to the mechanism for production of various defects in solid state and condensed matter systems. Crystal defects, for example, form when the water freezes or when a metal crystallizes. The analogies between defects in particle physics and condensed matter physics are quite deep. Defects form for the same reason: the vacuum manifold is topologically non-trivial. However, the defect dynamics is different. The motion of defects in condensed matter are friction-dominated, whereas the defect in cosmology obey relativistic equations, second order in time derivatives, since they come from a relativistic field theory.

Depending on the topology of the manifold $\mathcal{M}$ the defects can occur in the form of points, lines or surfaces. They are called monopoles, strings and domain walls, respectively [32]. Hybrid defects can be formed in a sequence of phase transitions, for example, the first transition produces monopoles, which get connected by strings at the second phase transition. The main conclusions of the studies about these defects can be summarized as follows.

- Domain walls and monopoles are disastrous for cosmological models and their presence should be avoided.
- Strings cause no harm, but can lead to very interesting cosmological consequences. In particular, they can generate density fluctuations sufficient to explain galaxy formation and can produce a number of distinctive and unique observational effects (anisotropies in the CMB temperature, double images of objects behind them and a stochastic GW background).
- Hybrid defects are transient and eventually decay into relativistic particles. If this happens at a sufficiently early time, the decay products thermalize and we can see no trace of the defects, except perhaps in the form of gravitational waves.

13.1.1 Strings

To illustrate the (cosmic) strings let us come back to the simple model considered at the beginning of the preceding section (see equations (13.1) and (13.2)). Since $\langle \phi \rangle$ is single valued, the total change of the phase θ around any closed path in space must be an integer multiple of 2π. Let us now consider a closed path with $\Delta\theta = 2\pi$. If no singularity is encountered, as the path shrunk to a point, $\Delta\theta$ cannot change continuously from 2π to zero, and, thus, we must encounter at least one point where θ is undefined, i.e. $\langle \phi \rangle = 0$. This means that at least one tube of false vacuum should be caught inside any path with $\Delta\theta \neq 0$. Such tubes of false vacuum, called strings, must either be close or infinite in length, otherwise it would be possible to contract the path to a point without crossing the string.

The simplest strings are produced in the phase transition associated to the spontaneous breaking of a local $U(1)$ symmetry. In this case the Lagrangian contains a gauge field A_μ and a complex Higgs field ϕ which carries $U(1)$ charge g and with self-interaction of the form (13.1):

$$\mathcal{L} = (D_\mu \phi)^\dagger (D^\mu \phi) - \tfrac{1}{4} F_{\mu\nu} F^{\mu\nu} - V(\phi) \tag{13.4}$$

where

$$D_\mu = \partial_\mu - ig A_\mu, \quad F_{\mu\nu} = \partial_\mu A_\nu - \partial_\nu A_\mu.$$

In this case the string solution has a well-defined core outside of which ϕ contains no energy density in spite of non-vanishing gradients $\nabla\phi$: the gauge field A_μ can absorb the gradient, i.e. $D_\mu\phi = 0$ when $\partial_\mu\phi \neq 0$. The radius δ of the string core is determined by the Compton wavelengths of the Higgs and vector bosons. For $m_\phi \ll m_A$, which is usually the case, one has:

$$\delta \sim m_\phi^{-1} = \lambda^{-1/2}\eta^{-1}$$

and the energy of the string per unit length within this width is finite and given by:

$$\mu \sim \lambda\eta^4\delta^2 = \eta^2$$

(independent of the coupling λ). The value of μ (or equivalently η) is the only free parameter of the string. For a phase transition at the grand unification (GUT) energy scale, $\eta = 10^{16}$ GeV and

$$\delta \sim 10^{-30} \text{ cm}, \quad \mu \sim 10^{22} \text{ g cm}^{-1}. \tag{13.5}$$

Strings of cosmological interest have sizes much greater than their width. In this case the internal structure of the string is unimportant and physical quantities of interest, such as the energy-momentum tensor, can be averaged over the cross section. For a long, thin, straight string lying along the z-axis we define

$$\tilde{T}^\nu_\mu = \delta(x)\delta(y) \int dx\, dy\, T^\nu_\mu.$$

It can be shown [33] that, as a consequence of the invariance of the string under Lorentz boost along z and the conservation law $T^\nu_{\mu,\nu}$ one has:

$$\tilde{T}^\nu_\mu = \mu\delta(x)\delta(y)\,\text{diag}(1, 0, 0, 1). \tag{13.6}$$

This expression shows a remarkable property of the strings: the pressure is negative, i.e. is a string tension, and this tension is equal to the mass per unit length μ. One recalls from classical mechanics that small transverse waves in a string with tension T move at speed $(T/\mu)^{1/2}$, so it is apparent that waves move along a string at the velocity of light.

Applied to strings, the Kibble mechanism implies that at the time of phase transition a network of strings with typical length $\xi(t_c)$ will form. According to numerical simulations at formation about one fifth of the initial energy is in small closed loops and the remaining in 'infinite' long strings. The evolution of this network for $t > t_c$ is complicated. The key processes are:

(i) The intercommutation of intersecting string segments, in which the two segments swap partners, rather than passing through one another (see figure below). Based upon numerical simulations it appears that the probability for

this to occur is nearly unity. This process leads to the continual chopping of long strings into smaller segments and/or loops.

(ii) The decay of small loops through the emission of gravitational radiation. The strings oscillate relativistically under their own tension and, thus, a loop of characteristic radius R will radiate gravitational waves at a characteristic frequency $\omega \sim R^{-1}$ due to its time-varying quadrupole moment, $Q \sim \mu R^3$. For an order of magnitude estimate the power radiated in gravitational waves can be calculated using the quadrupole formula (G is Newton's constant)

$$P \sim G\langle\dddot{Q}\dddot{Q}\rangle \sim G\mu^2. \tag{13.7}$$

Because this power is constant, independent of the loop size, the mass-energy of the loop decreases linearly with time. In a characteristic time

$$\tau \sim \frac{R}{G\mu}$$

the loop shrinks to a point and vanishes. From equation (13.5), in the case of a GUT phase transition, one has $G\mu \sim 10^{-6}$.

These two processes combine to create a mechanism by which the infinite string network loses energy (and length as measured in comoving coordinates), preventing the network from dominating the universe. Indeed, based upon computer simulations and analytical arguments, there is strong evidence that the cosmological evolution of the network becomes self-similar, approaching what is called a 'scaling' solution. In the simplest scale-invariant model, the correlation length ξ of the network is proportional to its causality bound:

$$\xi(t) \sim t \sim d_H(t)$$

and, thus, the statistical properties of the network are time independent if all the distances are scaled to the causal horizon.

This scaling property can be used to obtain qualitative relations. For example, the energy density of long strings is given by:

$$\rho_\infty = A\frac{\mu}{d_H^2(t)} \sim A\frac{\mu}{t^2}$$

where A is a dimensionless constant representing the number of long strings present per horizon sized volume. Numerical simulations suggest the value $A = 52$ in the radiation-dominated era, and $A = 31$ in the matter-dominated one. True scale invariance implies that the size of a newly formed loop produced by the network is a fixed fraction of the horizon

$$l(t) = \alpha d_H(t) \sim \alpha t.$$

Although the loops are observed to form with relativistic peculiar velocities v_i (the loop centre of mass is moving with respect to the rest frame of the cosmological fluid), these are rapidly redshifted to zero by the expansion of the universe, leaving a generic loop with only a fraction $f_r = (1 - v_i^2)^{1/2}$ of its initial energy. This redshifting of peculiar velocities does not affect the loop production rate, but it does change the loop size immediately after its formation to:

$$l(t) = f_r \alpha d_H(t). \tag{13.8}$$

Hence, only a fraction f_r of the total loop energy is converted into GWs. By numerical simulations this fraction turns out to be 0.71.

This scale-invariant model is implemented by the assumption that the universe is described by a spatially flat ($\Omega = 1$) FRW cosmology. A full treatment of this model is developed in [34]. In particular, the effect of the string network on the expansion of the universe and the rate of loop formation are calculated.

The spectrum of GWs produced by a network of string loops can be obtained implementing the scale-invariant model described above with a model of the emission of gravitational radiation by string loops. This model has been developed in [34] and is composed of the following three elements:

(i) A loop radiates with power

$$P = \Gamma G \mu^2 \tag{13.9}$$

where Γ is a dimensionless radiation efficiency that does not depend on the loop size, but only on its shape. Recent studies of realistic loop configuration indicate that the distribution of the values of this parameter has mean value $\langle \Gamma \rangle \approx 60$. From equations (13.8) and (13.9), it follows that a loop formed at time t_b at a time $t > t_b$ has length

$$l(t, t_b) = f_r \alpha d_H(t) - \Gamma G \mu (t - t_b).$$

The loop disappears when this length reaches zero, at a time

$$t_d = \left(1 + \frac{f_r \alpha d_H(t_b)}{\Gamma G \mu t_b}\right) t_b = \beta t_b$$

(β is not function of t_b because $d_H(t_b) \sim t_b$).

(ii) The frequency of GWs emitted at time t by a loop formed at time t_b is given by

$$f_n(t, t_b) = \frac{2n}{l(t, t_b)} \quad (n = 1, 2, 3, \ldots). \tag{13.10}$$

(iii) The fraction of the total power emitted in the nth oscillation mode at frequency f_n is given by the coefficient P_n, where

$$\Gamma = \sum_{n=1}^{\infty} P_n. \tag{13.11}$$

Analytic and numerical studies suggest that the radiation efficiency coefficients behave as $P_n \propto n^{-q}$.

This model has several shortcomings. First, the spectral index q has not been well determined. Numerical simulations suggest $q = 4/3$ while from analytic calculations $q = 2$ is obtained. The simulations have limited resolution of the important small-scale features of the long strings and loops and, therefore, the analytic prediction may be more realistic. Second, the effect of the back reaction on the motion of the strings has been ignored. The authors of [35] have argued that this effect result in an effective high-frequency cut off in the oscillation mode number n. Thus, the loop radiates only in a finite range of frequencies and equation (13.11) is replaced by

$$\Gamma = \sum_{n=1}^{n_*} P_n. \tag{13.12}$$

By comparing the back-reaction length scale to the loop size these authors estimate that the cut off should be no larger than $\sim (\Gamma G \mu)^{-1}$.

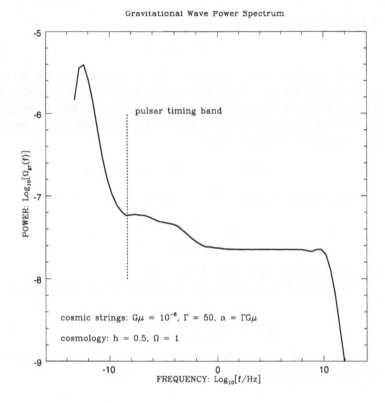

Figure 13.2. The spectrum of gravitational radiation produced by a string network for a given set of dimensionless parameters.

We are now in a position to construct the differential equation describing the rate of change of the energy in loops present in a volume $V(t)$ at time t. The numerical integration of this equation and the method applied to calculate the power spectrum of gravitational radiation are discussed in detail in [34]. The results of this calculation are shown in figure 13.2. In this paper analytic expressions for the latter have also been derived. Even though simplified for convenience, these analytic expressions offer the opportunity to examine the various dependences of the spectrum on string and cosmological parameters.

The spectrum of gravitational radiation produced by a network of strings has two main features:

• A nearly equal gravitational radiation energy density per logarithmic frequency interval in the range $(10^{-8}, 10^{10})$ Hz. This portion ('red noise') of the spectrum corresponds to GWs emitted during the radiation-dominated era and does not show a significant dependence by the spectral index q and the cut-off n_* [35].

- A peak near $f \sim 10^{-12}$ Hz. The shape of this portion depends critically on the model for the emission by a loop. The overall height of the spectrum depends linearly on $G\mu$, while the frequency at which the peaked spectrum merges to the red noise portion depends inversely on α. The important result is that for values $n_* < 10^2$ and $q \geq 4/3$ the spectrum drops off as $1/f$ for any value.

Because the frequency band is accessible to VIRGO, in the following we shall concentrate on the 'red noise' portion of the spectrum, referring to [34, 35] for details about the region of the peak.

An analytic expression for the 'red noise' portion of the GW spectrum is given as follows:

$$\Omega_{gw}(f) = \frac{8\pi}{9} \frac{A\Gamma G^2 \mu^2}{\alpha} (1 - \langle v^2 \rangle) \frac{\beta^{3/2} - 1}{1 + z_{eq}}. \qquad (13.13)$$

Let us remark that the above expression for the spectrum does not account for the reduction in the number of relativistic degrees of freedom that occurs every time the temperature falls through a particle mass threshold. This has the effect of modifying equation (13.13) by a factor

$$N = \left(\frac{g_{*a}}{g_{*b}} \right)^{1/3}$$

where g_{*a} (g_{*b}) is the number of relativistic degrees of freedom at a temperature above (below) the particle mass threshold. Within the standard model $SU(3)_C \otimes SU(2)_L \otimes U(1)_Y$, one has:

$$N = \begin{cases} 1, & f \in (10^{-8}, 10^{-10}\alpha^{-1}) \text{ Hz} \\ \left(\dfrac{3.36}{10.75} \right)^{1/3} = 0.68, & f \in (10^{-10}\alpha^{-1}, 10^{-4}\alpha^{-1}) \text{ Hz} \\ \left(\dfrac{3.36}{106.75} \right)^{1/3} = 0.32, & f \in (10^{-4}\alpha^{-1}, 10^8) \text{ Hz} \end{cases} \qquad (13.14)$$

where we take $\alpha \sim \Gamma G\mu$ in evaluating the above frequency range. Hence, the thermal history of the cosmological fluid reflects on the red noise spectrum by a series of steps down in amplitude with increasing frequency. The detection of such a shift would provide unique insight into the particle content of the early universe (at times much earlier than the electroweak phase transition).

There is a substantial body of astronomical evidence which suggests that the cosmological density parameter Ω_0 is less than one, i.e. the universe is open. The evolution of strings in an open universe will differ from that in the flat case only after the time at which the expansion of the universe becomes curvature-dominated: after this time the linear regime no longer exists. This is important for consideration of GWs created with low frequencies during the matter era, but, apart from a shift in the frequency corresponding to equal matter-radiation has a negligible effect on the red noise spectrum, which is produced in the radiation era [35].

The values of the dimensionless parameters appearing in equation (13.13) are not completely known. Numerical simulations provide a reasonable estimate for A and $\langle v^2 \rangle$ in the radiation-dominated era: $A = 52 \pm 10$, $\langle v^2 \rangle = 0.43 \pm 0.02$. Comparing detailed calculations of large angular scale CMB temperature anisotropies induced by strings [36] with observations, the string mass per unit length has been normalized to

$$G\mu = 1.05^{+0.35}_{-0.20} \times 10^{-6}.$$

This value is below the upper bound obtained from the pulsar timing and nucleosynthesis constraints on the gravitational radiation spectrum [35]. However, the value of α (the size of the loop at formation) is still unknowm. The high-resolution numerical simulations show that $\alpha < 10^{-2}$. This surprisingly small relative size is a result of the small-scale structure on the long strings and, since this is cut off by gravitational back-reaction, we may reasonably expect that $\alpha > \Gamma G\mu \approx 6 \times 10^{-5}$ [33]. This uncertainty in the value of α leads to a wide uncertainty in β that, because this parameter governs the lifetime of the loop, has a large effect on the spectrum. However, it is easy to verify that the spectral density $\Omega_{\mathrm{gw}}(f)$ reach a minimum value when $\alpha \to 0$. Therefore, in the radiation-dominated era ($d_{\mathrm{H}} = 2t$):

$$\Omega_{\mathrm{gw}}(f) \geq \frac{8\pi}{3} A f_r G\mu \frac{1 - \langle v^2 \rangle}{1 + z_{\mathrm{eq}}} N, \quad f \in (10^{-8}, 10^{10}) \text{ Hz.} \qquad (13.15)$$

Since $1 + z_{\mathrm{eq}} = 2.32 \times 10^4 \Omega_0 h_0^2$, in the case $\Omega_0 = 1$ one has:

$$h_0^2 \Omega_{\mathrm{gw}} \geq 5.0 \times 10^{-9} N.$$

In the case of the standard model thermal scenario, the value of N to be inserted in this expression is 0.32 (see equation (13.14)). But, as anticipated in section 12.1, the evolution of N with the temperature is known only up to $T \sim 10^3$, i.e. up to frequencies $f \sim 10^{-3} \alpha^{-1} \sim 10$ Hz. If the particle physics model has more degrees of freedom beyond this temperature ($f \gtrsim 10$ Hz) there could be other steps in the function $N(T)$ associated with other phase transitions. However, the dependence of the spectral density on the number of degrees of freedom is reasonably weak and, thus, we can conservatively estimate

$$h_0^2 \Omega_{\mathrm{gw}} \geq 1.6 \times 10^{-9} \qquad (13.16)$$

in the frequency range explored by VIRGO.

13.1.2 Hybrid defects

If the symmetry breaking responsible for the formation of the strings occur as a part of a much more complicated breaking scheme, it is likely that hybrid systems composed by topological defects of different dimensionality may form. In the case when the hybrid system does not annihilate immediately this may lead to a stochastic background in a similar way to that for the strings [38]. The interesting feature of these objects is that they evade the constraints from the CMB and pulsar timing allowing for larger values of $G\mu$ and hence larger contributions to the SGWB in the detectable range of frequencies.

Walls bounded by strings

Domain walls are formed when a discrete symmetry is broken. The simplest model of this sort is that of real scalar field with a potential

$$V(\phi) = \tfrac{1}{2}\lambda(\phi^2 - \eta_w^2)^2.$$

The reflection symmetry group Z_2 of the Lagrangian (invariance under $\phi \to -\phi$) is spontaneously broken when ϕ takes on the VEV $\langle\phi\rangle = \pm\eta$, and so the manifold $\mathcal{M}$ consists of only two points. As we go from a region with $\langle\phi\rangle = \eta$ to a region with $\langle\phi\rangle = -\eta$, we should necessarily pass through $\langle\phi\rangle = 0$ and, thus, the two regions must be separated by a wall of false vacuum. Therefore, the simplest sequence of phase transitions that results in walls bounded by strings is

$$G \to H \otimes Z_2 \to H$$

where at first transition ($T \sim \eta_s$) strings form and at the second ($T \sim \eta_w$) each string gets attached to a domain wall.

Before the formation of walls the evolution of strings is as in the standard scenario described above. After a period of overdamped motion, the strings start moving relativistically at time t_s and approach a scaling regime where the characteristic scale of the network is comparable to the horizon. After the time t_w at which the domain walls form the evolution of the network depends on the ratio between the string tension $\mu \sim \eta_s^2$ and wall surface tension $\sigma \sim \eta_w^3$. The walls become dynamically important at $t \sim \mu/\sigma$, when they pull the strings towards one another, and the network breaks into pieces of wall bounded by string. Alternatively, if $t_w > \mu/\sigma$ the breakup of the network occurs immediately after the wall formation. By indicating with $t_* = \max\{\mu/\sigma, t_w\}$, the typical size of the pieces is expected to be $\sim\alpha t_*$.

Gravitational waves emitted by oscillating loop of strings during the period $t \in (t_s, t_*)$ form a SGWB with a nearly flat spectrum extending over the frequency range $f \in [2/\alpha(t_* t_{eq})^{1/2}, 2/\alpha(t_s t_{eq})^{1/2}]$. For this frequency range to overlap with the one covered by VIRGO the strings must decay before the decoupling, and, thus, the only constraint on this background comes from big bang nucleosynthesis. It can be shown [38] that:

$$\Omega_{gw}(f) \sim 6 \times 10^{-3} G \mu h_0^{-2} \quad \text{with } G\mu \lesssim \frac{8 \times 10^{-4}}{\ln(t_*/t_s)}. \tag{13.17}$$

Although t_s and t_* are model-dependent (t_s through the assumed cosmological scenario; t_* through μ and σ), they do not have to be separated by many orders of magnitude. Following [38] if we assume $\ln(t_*/t_s) \lesssim 1$ equation (13.17) gives:

$$h_0^2 \Omega_{gw} \lesssim 5 \times 10^{-6}. \tag{13.18}$$

At $t > t_*$ these string loops spanned by domain walls will collapse into gravitational radiation (and other decay products) in about a Hubble time. This process takes place over a relatively short frequency range and may lead to a sharp peak in the spectrum. However, since the exact nature of this contribution display a strong dependence on model phenomenology, it is not possible at present to predict whether or where such a peak should occur [37].

Strings connected by monopoles

Monopoles are point defects which form when the manifold $\mathcal{M}$ of equivalent vacua contains unshrinkable surfaces. A simple model that illustrates the monopole solution is that of an $SO(3)$ gauge theory spontaneously broken down to $U(1)$ by a Higgs triplet ϕ^a with a potential

$$V(\phi) = \tfrac{1}{2}\lambda(\phi^a\phi_a - \eta_m^2)^2, \quad (a = 1, 2, 3).$$

The magnitude of $\langle\phi^a\rangle$ is fixed by the minimization of the potential to $|\phi| = \eta_m$, but its direction in group space is not and the manifold $\mathcal{M}$ is a sphere in this space. If we choose $\phi(\vec{r}) = (0, 0, \eta_m)$ the $SO(3)$ symmetry is broken down to $U(1)$ because ϕ is invariant under rotations about the $\hat{e}_3$-axis. Another, less trivial, choice is represented by

$$\phi^i(\vec{r}) = \eta_m\hat{r}_i$$

where $\hat{r}_i$ is a unit vector in coordinate space. For this configuration of the field there must be a point in space where $\phi = 0$ and the energy density is non-vanishing. This point of false vacuum is a monopole.

Therefore, the prototypical sequence of symmetry breakings resulting in monopoles connected by strings is

$$G \rightarrow H \otimes U(1) \rightarrow H$$

where monopoles formed at first transition ($T \sim \eta_m$) and strings get connected by monopole/antimonopole ($M\bar{M}$) pairs at the second ($T \sim \eta_s$). If the monopole-forming transition occurs after any period of inflation which can have taken place, then the average separation of the monopoles is always smaller than the

Hubble radius and when the $M\bar{M}$ pairs get connected by strings the hybrid system collapse in less than one Hubble time by dissipating energy into friction with the cosmological fluid.

For our purposes, the most interesting scenario is when the monopoles are formed during the inflation but are not completely inflated away. The strings are formed later with a length scale ξ that is much smaller than the average monopole separation d. In the course of the evolution ξ grows like t and eventually becomes comparable to d, so that at some time t_m we are left with $M\bar{M}$ pairs connected by strings. If the strings are formed during inflation, soon after the monopoles, they do not go through a period of relativistic evolution and no gravitational radiation is produced prior to t_m. In contrast, if the strings are formed in the post-inflationary epoch, they can have a period of relativistic evolution and a nearly flat stochastic background identical to that for walls bounded by strings is produced before t_m.

At $t > t_m$, the $M\bar{M}$ pairs oscillate and gradually convert their energy in gravitational radiation. This process has been studied in detail in [38] for the simplest case of a straight string connecting the monopoles. It was found that the spectral density of the radiation emitted by this simple configuration verify the bound

$$h_0^2 \Omega_{\text{gw}} \lesssim 2 \times 10^{-8} \tag{13.19}$$

in the frequency range of VIRGO.

13.2 Inflation

It is well known that many of the shortcomings of the standard cosmological model (such as isotropy of the CMBR, structure formation, flatness and monopole problems) can be successfully faced in the framework of the so-called inflationary models (see [39,40] for a review).

The basic idea shared by almost all the various models [41–44] of inflation is that in early times the universe was dominated by the vacuum energy of some scalar field, which provided an exponential growth of the scale factor of the universe; then, as a result of a phase transition (maybe associated with a spontaneous symmetry breaking), the scalar field was captured by the true minimum of its potential, made some oscillations and finally settled down in it. The energy previously stored in the false vacuum was converted into the decay products of the scalar field, produced mainly during the oscillatory stage: this process is responsible for the reheating of the universe, thus providing the usual 'hot' initial conditions for the beginning of the radiation-dominated era.

Despite the fact that a complete and satisfying model of inflation is not yet at hand, the 'inflationary paradigm', i.e. the idea of a primordial stage of accelerated expansion, is widely accepted and considered as a necessary ingredient of every cosmological model. Moreover, an inflationary stage provides a very interesting mechanism of amplification of perturbations (see [47]) that, as well as being important for what concerns the problem of the structures formation, generates also a stochastic background of gravitational waves. The basic concept is quite simple: zero point quantum fluctuations of every sufficiently light field can be amplified when the size of the perturbation becomes larger than the event horizon. This is the classical interpretation of the well-known fact that a time-dependent gravitational field produces particles even if the initial state contains no quanta (see [45] for details).

13.2.1 Classical picture

Since the final number of quanta created is typically very large, it comes out that a classical analysis of the phenomenon is appropriate and gives the right answer. So, let us show how this amplification works by studying the classical equation of motion for gravitational perturbations; we will return to the quantum description later.

The starting point is to separate the metric into a 'background'[1] and a 'propagating' part:

$$g_{\mu\nu} \sim R^2(\eta)(-d\eta^2 + d\vec{x}^2) + h_{\mu\nu},$$

where R is the scale factor of the universe, depending only on the conformal time η (in terms of the usual cosmic time $d\eta = dt/R(\eta)$). The equation of motion for $h_{\mu\nu}$ in the FRW background is obtained by taking the first-order variation of the Einstein equations. One can define

$$h^{\mu}_{\nu} \sim \frac{1}{R(\eta)} \varepsilon^{\mu}_{\nu}(\vec{k}) \psi(\eta) e^{i\vec{k}\cdot\vec{x}}, \tag{13.20}$$

where $\varepsilon^{\mu}_{\nu}(\vec{k})$ is the polarization tensor. It comes out that, with a suitable gauge choice, ψ obeys the following equation:

$$\psi''(\eta) + \left(k^2 - \frac{R''(\eta)}{R(\eta)}\right)\psi \sim 0, \tag{13.21}$$

where '$''$' denotes the derivative with respect to the conformal time. This is a Schrödinger-like equation with the potential given by $V = R''/R$. For $k \gg V(\eta)$ we have for ψ the obvious plane wave solution $\psi \sim e^{-ik\eta}$, so that $|h^{\mu}_{\nu}| \sim 1/R$; if, instead, $k \ll V(\eta)$, we get the two following solutions:

$$\psi_1 \sim R(\eta), \quad \psi_2 \sim R(\eta) \int \frac{d\eta}{R^2(\eta)}.$$

As will be seen later, in the case of our interest ψ_1 is the dominant solution, so, in this case, $|h^{\mu}_{\nu}| \sim 1$. This means that a solution characterized by a long wavelength ($k \ll V(\eta)$) is amplified with respect to a short wavelength one by a factor $R(\eta)$;

---

[1] For the sake of simplicity we consider a spatially flat universe.

as a first approximation we can assume that this amplification takes place as long as $k < V(\eta)$. The last inequality, apart from irrelevant numerical factors due to our approximation, turns out to have the same analytical form of the condition:

$$\frac{k}{R} < \frac{R'}{R^2} \equiv H(\eta),$$

that relates the physical momentum of the perturbation to the Hubble parameter H, whose inverse corresponds to the horizon size: this is just the condition anticipated above, i.e. that a perturbation is amplified when its wavelength becomes larger than the event horizon.

13.2.2 Calculation of the spectrum

To calculate the spectrum of perturbations we need to find particular solutions of the Schrödinger-like equation previously introduced. Let us, then, specify our model: the form of the potential V is determined by the evolution of $R(\eta)$; for our purposes it is sufficient to consider a three-stage model in which, as shown in figure 13.3, a de Sitter phase (characterized by a constant Hubble parameter H_{ds}) is followed by a radiation-dominated (RD) and, then, by a matter-dominated (MD) era:

$$R(\eta) \sim \begin{cases} -\dfrac{1}{H_{ds}\eta}, & \eta < \eta_1 < 0 \quad \text{de Sitter} \\[2mm] \dfrac{\eta - 2\eta_1}{H_{ds}\eta_1^2}, & \eta_1 < \eta < \eta_2 \quad \text{radiation} \\[2mm] \dfrac{(\eta + \eta_2 - 4\eta_1)^2}{4H_{ds}\eta_1^2(\eta_2 - 2\eta_1)}, & 0 < \eta_2 < \eta < \eta_0 \quad \text{matter.} \end{cases} \qquad (13.22)$$

The form of this expression is chosen in order to make R and R' continuous at the transition points η_1 and η_2 (while we have indicated present conformal time by η_0). Now we have to find a solution of equation (13.21) that reduces to a pure positive frequency mode at early times, and see what is the coefficient of the corresponding negative frequency mode at late times. Hence, our solution has the following form:

$$\psi(\eta) \sim \begin{cases} \psi_i^+, & \eta < \eta_1 \\ \alpha_k \psi_r^+ + \beta_k \psi_r^-, & \eta_1 < \eta < \eta_2 \\ \gamma_k \psi_m^+ + \delta_k \psi_m^-, & \eta_2 < \eta \end{cases} \qquad (13.23)$$

with

$$\psi_i^+ = e^{-ik\eta}\left(1 - \frac{i}{k\eta}\right)$$

$$\psi_r^{\pm} = e^{\mp ik\eta} \qquad (13.24)$$

$$\psi_m^{\pm} = e^{\mp ik\eta}\left(1 - \frac{i}{2k(\eta + \eta_2 - 4\eta_1)}\right).$$

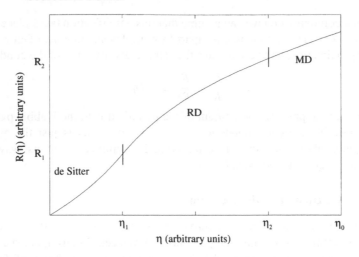

Figure 13.3. The scale factor as a function of the conformal time. The universe undergoes a transition from a de Sitter- to a radiation-dominated phase at the time η_1, and from a radiation to a matter dominated phase at the time η_2. The present epoch corresponds to η_0.

At first sight, the first and the third modes written in equation (13.24), seem not to be pure plane waves solutions. This comes from the fact that we are in curved spacetime: it can be easily verified that in the limit $k \to \infty$, i.e. when the wavelength is so short that the particle does not 'feel' the curvature of spacetime, all the modes in equation (13.24) reduce to the standard plane-wave form. The fact that $\psi_r^{\pm}$ is already in the standard form is due to the conformal invariance of the radiation-dominated spacetime.

The coefficients α, β, γ and δ in equation (13.23) are calculated by requiring the overall solution $\psi(\eta)$ to be continuous with its first derivative at transition times $\eta_{1,2}$. Equation (13.21) says that this problem is similar to the well-known quantum mechanics problem of tunnelling through a potential barrier $V(\eta)$. Therefore, α, β, γ and δ can be considered as the transmission/reflection coefficients and one can show that the number of created gravitons with comoving frequency k is

$$N_k \sim |\delta_k|^2 = \left(\frac{3H_{\mathrm{ds}}^3 R(\eta_1)^4}{8R(\eta_2)} \right)^2 \frac{1}{k^6}. \tag{13.25}$$

This expression should be taken with care, because the simple model described by equation (13.22) does not take into account two important physical effects that force equation (13.25) to hold only in a limited range of frequencies. First of all, the perturbations whose physical wavelength is greater than the present Hubble length ($R(\eta_0)/k > 1/H_0$) do not contribute to the energy density; so $k_0 = R_0'/R_0$, corresponding to a physical frequency $f_0 \sim 10^{-19}$ Hz, provides the lower bound to the extension of the spectrum. To find the upper bound one must note that in our model the transition between the different regimes takes place instantaneously; in a more realistic situation such transitions have a typical duration time Δt, and consequently there is a cut-off physical frequency of order $1/\Delta t$. In our case the typical time of the transitions is of the order of the Hubble parameters at the transition points $(H(\eta_1), H(\eta_2))$. This means that for $k > k_2 = R_2'/R_2$, i.e. for $f \geq 10^{-16}$ Hz, equation (13.25) is not the correct expression for N_k; above this frequency the radiation-matter transition does not affect the spectrum, but the inflation-radiation one is still important. Hence, the number of gravitons created is given by the corresponding equation, with δ replaced by β

$$N_k = |\beta_k|^2 = \left(\frac{H_{ds}^2 R(\eta_1)^2}{2k^2}\right)^2. \tag{13.26}$$

In turn, equation (13.26) has a high frequency cut-off given by $k_1 = R_1'/R_1 = -1/\eta_1$; η_1, and consequently the corresponding value of the physical cut-off frequency f_1, is an almost free parameter of the model and depends on the reheating temperature. A typical value for f_1 is the one reported in [46] ($f_1 \sim 10^{10}$ Hz), but other (also much lower) values are possible, depending on the model; in general one has $f_1 > 1$ kHz. Beyond f_1 there is no amplification mechanism at all and the spectrum goes rapidly to zero.

Once obtained the result for N_k, it is straightforward to obtain its contribution to ρ_{gw}, and consequently to Ω_{gw}

$$d\rho_{gw}(f) = 2 \cdot 2\pi f \cdot N_k(f) \cdot 4\pi f^2 \, df,$$

where the factor of two is due to polarization, and $f = k/2\pi R_0$ is the physical frequency. The final result for Ω_{gw} in terms of the physical frequency is then

$$\Omega_{gw}(f) = \frac{8}{3\pi H_0^2 M_{Pl}^2}\begin{cases} \left(\frac{3H_{ds}^3}{16\pi}\right)^2 \left(\frac{R_1}{R_2}\right)^2 \left(\frac{R_1}{R_0}\right)^6 \frac{1}{f^2}, & k_0 < 2\pi f < k_2 \\ \frac{H_{ds}^4}{4}\left(\frac{R_1}{R_0}\right)^4, & k_2 < 2\pi f < k_1 \\ 0, & 2\pi f > k_1 \end{cases} \tag{13.27}$$

where M_{Pl} is the Planck mass and H_0 is the present value of the Hubble parameter.

Because of the flatness of $\Omega_{gw}(f)$ the strongest constraint turns out to be the one related to the observed anisotropy in CMBR. This constrains the spectrum as follows

$$h_0^2 \Omega_{gw} \leq 7 \times 10^{-11} \left(\frac{H_0}{f}\right)^2 \quad \text{for } f \in (10^{-19}, 10^{-16}) \text{ Hz}, \tag{13.28}$$

which in turn implies

$$H_{ds} \leq 10^{39} \text{ Hz} \simeq 5 \times 10^{14} \text{ Gev} \simeq 5 \times 10^{-5} M_{Pl}.$$

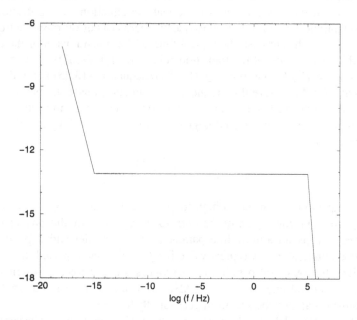

Figure 13.4. $h_0^2 \Omega_{gw}$ against the physical frequency f (logarithmic scales). Here $f_1 = 100$ kHz.

Given the shape of the spectrum (see figure 13.4) one obtains in the frequency region of interest for VIRGO the following bound:

$$h_0^2 \Omega_{gw} < 8 \times 10^{-14}, \tag{13.29}$$

many orders of magnitude below the sensitivity limit of the planned detectors.

Finally we remark that the result reported in equation (13.27) is valid if the Hubble parameter during inflation is strictly constant; the calculation can be done also for other, more realistic models, but the final result is not significantly different from the one reported above. For instance, in the so-called 'slow-roll' inflation the Hubble parameter is slightly decreasing during the inflationary stage and makes the spectrum slightly tilted, instead of constant, in the radiation-dominated region. Anyway the tilt is so small that the value of Ω_{gw} at interesting frequencies is different from the one of equation (13.29) only for an order of magnitude (in addition this correction has the 'wrong' sign, i.e. it lowers the bound!).

13.3   String cosmology

The basic mechanism of generation of relic gravitational waves in cosmology has been discussed in the preceding section about inflation where the main concepts have been exposed. The crucial point of the outcome was the flatness of the spectrum which, combined with the COBE bound at very low frequency, gives a very strong constraint on the spectrum even at high frequency. To satisfy the COBE bound and still have a chance of being observable at VIRGO or LIGO the spectrum must grow with frequency. A spectrum of this kind has been found in a cosmological model suggested by *string theory* [48].

In string theory the fundamental objects are one-dimensional extended entities, i.e. strings. Their fundamental excitation of given energy and angular momentum are particles of given mass and spin. String theory has one only fundamental (dimensionful) constant: the *string tension* T which can be traded for a fundamental length $\lambda_s \equiv \sqrt{\hbar c / T}$. The mass scale of the excitation of the string is therefore $\sqrt{T}$ which, as string theory includes gravity, should be near the Planck mass[2]. The gauge couplings are not constant (neither at classical level) but depend on the expectation value of a scalar field, the *dilaton*. For example, Newton's gravitational constant G is given by

$$G \sim \frac{\lambda_s^2}{8\pi} e^{\phi},$$

while all the gauge couplings are proportional to $e^{\phi/2}$. In the regime where all couplings and derivatives are small string physics can be suitably described by a *field theory* action which describes the dynamics of the light (massless) fields of the string (the graviton $g_{\mu\nu}$ and the dilaton ϕ), which is

$$S = -\frac{1}{2\lambda_s^2} \int d^4x \sqrt{-g} [e^{-\phi}(\mathcal{R} + \partial_\mu \phi \partial^\mu \phi) + \text{higher derivatives}]$$
$$+ [\text{higher order in } e^\phi]. \tag{13.30}$$

where $g = \det \|g_{\mu\nu}\|$, and $\mathcal{R}$ is the Ricci scalar. Higher derivative terms are relevant whenever spacetime derivatives become of the order of one (in λ_s units), whereas the higher orders in e^ϕ acquire importance when $e^\phi \sim 1$, thus signalling the beginning of a full stringy-quantum regime. The first terms without corrections reproduce Einsteinian gravity for constant dilaton.

To analyse a situation of cosmological interest it suffices to study spatially homogeneous fields, i.e. $\phi = \phi(t)$ and the metric is chosen so that the line element ds^2 can be written as

$$ds^2 = dt^2 - R^2(t)\, d\vec{x}^2, \tag{13.31}$$

---

[2] This is not a compelling argument as models have been presented in which $\sqrt{T}$ can be consistently put some orders of magnitude below the Planck mass (see, e.g., [49]), but we will not treat these models.

which is a suitable ansatz for the metric describing an evolving homogeneous isotropic universe. The cosmological field equations derived from the low-energy part of (13.30) have the symmetry (in four dimensions)

$$R(t) \rightarrow \frac{1}{R(-t)}, \quad \phi(t) \rightarrow \phi(-t) - 6\ln[R(-t)]$$

which relates ordinary FRW cosmology characterized by $H \equiv \dot{R}/R > 0$, $\dot{H} < 0$, constant ϕ (where a dot means a derivative with respect to t) at $t > 0$, with an inflationary one with $H > 0$, $\dot{H} > 0$ and $\dot{\phi} > 0$ at $t < 0$. This symmetry may suggest that the universe started its evolution from the state of *perturbative vacuum*, i.e. empty, cold, flat and decoupled with increasing Hubble parameter and eventually emerged in the standard cosmology at $t > 0$ with decreasing H and 'frozen' dilaton. The dual cosmology is called *pre-big bang* (PBB) phase and it does not need a beginning time as $H \rightarrow 0$ and $\phi \rightarrow -\infty$ for $t \rightarrow -\infty$. However, the low-energy equations of motion do *not* smoothly interpolate between these two phases. Instead they lead to singularities, in the past for the FRW phase (as ordinary cosmology) and in the future for the PBB phase; in both cases at $t = 0$ where big bang should be placed. However, the approximations on which the validity of equation (13.30) relies, break down whenever $H \sim O(\lambda_s - 1)$ or $\phi \sim 0$. The corrections indicated in equation (13.30) may prevent reaching the singularity, allowing a smooth transition to standard hot big bang and FRW cosmology. Actually there are indications (see, e.g., [50–54]) that a regularization mechanism can be really provided by taking into account of quantum and stringy effects on the evolution of the system. The big bang should then be identified with the epoch of maximum but not infinite curvature which should be followed by a post-big bang evolution, by which we mean a standard evolution with $\Omega = 1$, as it has been achieved by the long period of PBB inflation. The above described scenario is displayed in figure 13.5.

Nevertheless in the strong coupling regime physics is really new and the perturbative approach proposed by equation (13.30) is no more valid. One should resort to a new picture, with an effective action written in terms of the new light modes appropriate to the strong coupling regime [55]. Anyway, admitting that a regularization is still possible makes this picture noteworthy from both the theoretical and the phenomenological point of view. From the theoretical side it addresses issues left open by standard inflationary models such as the initial singularity and the theoretical origin of the field driving inflation; from the phenomenological side in the next subsection the production of gravitational waves (studied in [56, 57]) will be analysed.

13.3.1   The model

The solution of the equations of motion derived from equation (13.30) which presents the initial conditions typical of PBB phase are (for time-dependent only

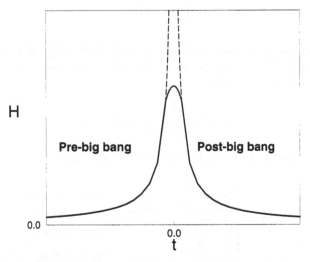

Figure 13.5. Qualitative behaviour of H in the suggested cosmological model. Broken curves indicate the lowest order solution singular at $t = 0$. At around $t = 0$ a regular evolution interpolating between the two phases is displayed.

fields):

$$R(\eta) = -\frac{1}{H_s \eta_s} \left(\frac{\eta - (1 - \alpha)\eta_s}{\alpha \eta_s} \right)^{-\alpha}, \tag{13.32}$$

and

$$\phi(\eta) = \phi_s - \gamma \ln \frac{\eta - (1 - \alpha)\eta_s}{\eta_s},$$

where η is the conformal time (see section 13.2) ranging between $-\infty$ and $\eta_s < 0$, $\alpha = (\sqrt{3} - 1)/2$, and $\gamma = \sqrt{3}$.

At a value $\eta = \eta_s$ the curvature becomes of order of the string scale and the lowest order effective action does not give any more a good description of physics: we enter a 'full stringy' regime. One expects that higher order corrections to the effective action tame the growth of the curvature, and both H and $d\phi/dt$ stay approximatively constant. In terms of conformal time this means

$$R(\eta) = -\frac{1}{H_s \eta}, \quad \phi(\eta) = \phi_s - 2\beta \ln \frac{\eta}{\eta_s}, \tag{13.33}$$

where β is a free parameter which measures the growth of the dilaton in this phase, which lasts for $\eta_s < \eta < \eta_1 < 0$. It is conceivable that at this stage other effects become important triggering a *graceful exit* to a standard radiation dominated phase characterized by

$$R(\eta) = \frac{1}{H_s\eta_1^2}(\eta - 2\eta_1), \quad \phi = \phi_0. \tag{13.34}$$

The time-dependent component of a metric perturbation h^{ν}_{μ} can be expressed as in equation (13.20), where ψ obeys the Schrödinger-like equation (13.21) with potential given, in this case, by:

$$V(\eta) = \frac{e^{\phi/2}}{R}\frac{d^2}{d\eta^2}(Re^{-\phi/2}).$$

Taking into account equations (13.32) and (13.34) one finds

$$V(\eta) = \begin{cases} \dfrac{1}{4}\dfrac{4\nu^2 - 1}{[\eta - (1 - \alpha)\eta_s)]^2}, & -\infty < \eta < \eta_s \\[2ex] \dfrac{1}{4}\dfrac{4\mu^2 - 1}{\eta^2}, & \eta_s < \eta < \eta_1 \\[2ex] 0, & \eta_1 < \eta < \eta_r \end{cases} \tag{13.35}$$

where $2\mu = |2\beta - 3|$, $2\nu = |2\alpha - \gamma + 1|$. We consider the solutions of equation (13.21) which reduce for $k \to \infty$ to usual positive (ψ^+) and negative (ψ^-) frequency mode. Let ψ^+_{PBB} be the positive frequency mode corresponding to the PBB phase. Matching the solutions (and their first derivatives) between the PBB and the 'full stringy' (FS) phase one imposes:

$$\psi^+_{PBB} = \alpha_k\psi^+_{FS} + \beta_k\psi^-_{FS}$$
$$(\psi^+_{PBB})' = \alpha_k(\psi^+_{FS})' + \beta_k(\psi^-_{FS})'$$

and subsequently between the 'full stringy' and the radiation dominated (RD) phase

$$\alpha_k\psi^+_{FS} + \beta_k\psi^-_{FS} = \gamma_k\psi^+_{RD} + \delta_k\psi^-_{RD},$$
$$\alpha_k(\psi^+_{FS})' + \beta_k(\psi^-_{FS})' = \gamma_k(\psi^+_{RD})' + \delta_k(\psi^-_{RD})'.$$

As previously stated (see section 13.2), the mean occupation number N_k per unit phase-space cell of gravitons with wavevector k in the radiation dominated phase is given by[3]

$$N_k \sim |\delta_k|^2.$$

The analytical expression can be computed explicitly [57], and it turns out to be a rather complicated formula whose main features can be understood as follows. The parameters of the spectrum are: μ, which is completely free and is determined

---

[3] We neglect the effect on the spectrum of the radiation to matter dominated phase transition, as it only affects the very low-frequency part of the spectrum ($f \lesssim 10^{-16}$ Hz), where string cosmology predicts a negligible value of Ω_{gw}.

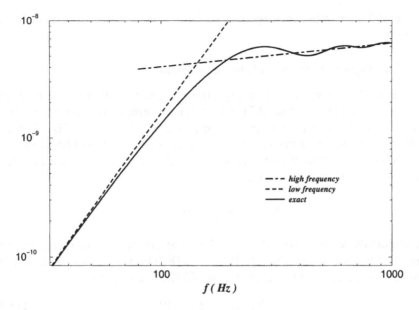

Figure 13.6. $h_0^2 \Omega_{gw}(f)$ as a function of f per $f_s = 100$ Hz, $H_s = 0.15 M_{Pl}$, $f_1 = 4.3 \times 10^7$ kHz, $\mu = 1.4$ compared with the low- and high-frequency limits (from [57]).

by the growth of the dilaton during the string phase; H_s, the constant value of the Hubble parameter during the string phase, which we expect to be of the order of $\lambda_s - 1$, η_s and η_1. It is convenient to trade η_s and η_1 with the associated frequencies $f_s = 1/(2\pi R_0 |\eta_s|)$, being R_0 the present value of the scale factor of the universe, and f_1 defined analogously; f_1 can be estimated to be $f_1 \sim 10$ GHz and it corresponds to the maximum amplified frequency, whereas the only constraint on f_s is $f_s < f_1$. As shown in figure 13.6, the spectrum presents an f^3 raise for $f < f_s$ (and $f_s \ll f_1$) and a series of oscillations around the line of slope $3 - 2\mu$ for $f_s < f < f_1$, while for frequency higher than f_1 the spectrum is exponentially suppressed, as it corresponds to the maximum of the potential in equation (13.35). In the most favourable case ($\mu = 1.5$ and f_s less than the smaller frequency in the VIRGO frequency range), the maximum value of the spectrum is reached as long as $f > f_s$ [57]

$$h_0^2 \Omega_{gw}^{max} \simeq 3.0 \times 10^{-7} \left(\frac{H_s}{0.15 M_{Pl}} \right) \left(\frac{t_1}{\lambda_s} \right)^2,$$

being t_1 the time of transition to the post-big bang phase.

This is just *an* example of possible cosmological dynamics driven by string theory, as it stands it cannot be identified as *the* prediction of string cosmology. Nevertheless the low-frequency behaviour, i.e. the power-law raise, is a quite general feature which belongs to any PBB-dictated spectrum of gravitational waves.

13.3.2 Observational bounds to the spectrum

Because of the power raise of the spectrum for low frequency the COBE bound is easily evaded (see figure 13.7) and the same statement holds for the constraint derived from pulsar timing observation as long as $f_s < 10^{-7}$ Hz. The most stringent bound is then the nucleosynthesis one given in section 12.5. In the most favourable case (a flat spectrum $\Omega_{gw}(f) = \Omega_{gw}^{max}$ for $f_s < f < f_1$, which means $\mu = 1.5$) that bound translates into

$$h_0^2 \Omega_{gw}^{max} \ln \frac{f_1}{f_s} < 6.3 \times 10^{-6}.$$

In order to have experimentally interesting values, the spectrum must have already reached the maximum value Ω_{gw}^{max} in the VIRGO frequency range. Under this assumption, a favourable choice of f_s is 100 Hz, that leads to

$$h_0^2 \Omega_{gw}^{max} < 3.2 \times 10^{-7}. \tag{13.36}$$

The effect of the observational bound on the parameters of the model is displayed in figure 13.8, where the independent parameters are chosen to be $\eta_s/\eta_1 = f_1/f_s$ and β (see equation (13.33)).

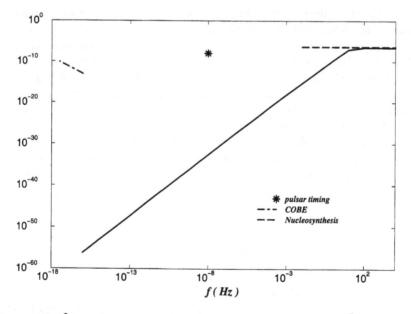

Figure 13.7. $h_0^2 \Omega_{gw}(f)$ as a function of f per $f_s = 10$ Hz, $f_1 = 4.3 \times 10^7$ kHz, $\mu = 1.5$ compared with observational bounds (from [57]).

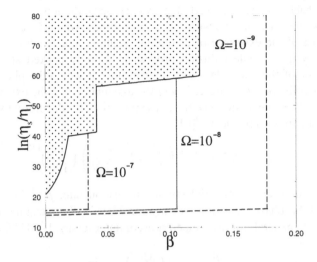

Figure 13.8. The forbidden region in the parameter space is the shaded area. Along the chain line $\Omega \equiv h_0^2 \Omega_{\text{gw}}(1 \text{ kHz}) = 10^{-7}$, along the dotted line $\Omega = 10^{-8}$ and along the dashed line $\Omega = 10^{-9}$ (from [58]).

The maximum value obtained for Ω_{gw} is still below the experimental planned sensibility of VIRGO and LIGO, but hopefully within the planned sensitivity of the advanced projects.

13.4 First-order phase transitions

First-order phase transitions are thought to have occurred in the early stage of the expansion of the universe, each time its temperature dropped sufficiently below the critical temperature T_c of a transition. Candidates for such a phase transition include GUT-symmetry breaking ($T_c = 10^{15 \pm 1}$ GeV), electroweak-symmetry breaking ($T_c \sim 200$ GeV), and phase transitions yet to be discussed.

In a first-order phase transition, the universe starts in a metastable (high-energy-density) false-vacuum state. If the energy barrier separating this state to the (low-energy-density) true-vacuum state is sufficiently large, significant supercooling occurs and the transition between these two states proceeds via nucleation [59] of bubbles of true vacuum within the false vacuum phase [60]. Once nucleated, each bubble larger than a critical size begins to expand, being driven by the energy difference (latent heat) between its true-vacuum interior and the false-vacuum exterior. This energy difference is converted into kinetic energy of the bubble wall, which become thinner and more energetic as the bubble expands, and rapidly approaches a velocity near the speed of light. As a consequence of the high velocities and energy densities involved, when, within a Hubble expansion time, these bubbles collide a large fraction of the energy that was in the bubble walls is converted in gravitational radiation.

Unlike the case of a network of cosmic strings, where $\Omega_{gw}(f)$ is flat as a consequence of the existence of a 'scaling' solution, the spectrum of the GWs produced by this bubble collision process is strongly peaked at a frequency characteristic of the particular cosmological time at which the phase transition and bubble collisions took place. Under the assumption that the expansion of the universe has been adiabatic since the phase transition, for the present value of this characteristic frequency one has [61]:

$$f_{\max} \approx 5.2 \times 10^{-8} \left(\frac{\beta}{H_*} \right) \left(\frac{T_*}{1\,\text{GeV}} \right) \left(\frac{g_*}{100} \right)^{1/6} \qquad (13.37)$$

where, assuming an exponential bubble nucleation rate, β^{-1} is, roughly, the duration of the phase transition (see section 12.1 for the meaning of the other quantities). In general, for the temperatures of interest (say, 1–10^{16} GeV), it turns out to be [62]:

$$\frac{\beta}{H_*} \sim 4 \ln \left(\frac{M_{\text{Pl}}}{T_*} \right) \sim 100, \qquad (13.38)$$

and, because g_* is also of order 100 in typical GUT models, from equation (13.37) we find that the most 'promising' cosmological phase transition, from the point of view of the VIRGO sensitivity band, would be one that occurred at a temperature comprised between 10^7 and 10^8 GeV (for which we have no compelling candidate).

The amplitude of the spectrum depends mostly upon the difference in the free energy between the inside and the outside of the bubble, driving the expansion of the bubble. By indicating with v the propagation velocity of the bubble walls, and with α the ratio of vacuum energy to the thermal energy in the symmetric phase (the high-temperature phase before the transition), the amplitude at the frequency $f_{\max}$ is approximately [61]:

$$h_0^2 \Omega_{gw}(f_{\max}) \approx 1.1 \times 10^{-6} \kappa^2 \left(\frac{H_*}{\beta} \right)^2 \left(\frac{\alpha}{1+\alpha} \right)^2 \left(\frac{v^3}{0.24+v^3} \right) \left(\frac{100}{g_*} \right)^{1/3},$$
$$(13.39)$$

where k is an increasing function of α quantifying the fraction of the available vacuum energy that goes into kinetic (rather than thermal) energy of the fluid. The parameter α characterizes the strength of the phase transition, and the limits $\alpha \to 0$, $\alpha \to \infty$ correspond to very weak and very strong first-order phase transitions, respectively. For α ranging between these two extremes, typically, $0.01 < \kappa < 1$. For strongly first-order phase transitions it turns out to be also $v \to 1$, and, thus, for a transition at $T \sim 2 \times 10^7$ GeV, according to equations (13.37) and (13.39), one finds:

$$h_0^2 \Omega_{gw}(f \sim 100\,\text{Hz}) \approx 10^{-10} \qquad (13.40)$$

which is three orders of magnitude too small to be observed by any of the ground-based interferometers presently under construction.

13.5   Astrophysical sources

To the stochastic background of gravitational radiation that should pervade our universe, the GWs associated with unresolved astrophysical sources also contribute. Since the generation of these waves dates back to more recent epochs, when galaxies and stars started to form and evolve, their contribution to Ω_{gw} is not subject to the nucleosynthesis bound. Therefore, our first concern is whether this component of the stochastic background can give a contribution to $h_0^2 \Omega_{gw}(f)$ larger than the bound (12.64), or anyway larger than the expected relic signal, thereby masking the background of cosmological origin.

A first observation is that there is a maximum frequency at which astrophysical sources can radiate. This comes from the fact that a source of mass M, even if very compact, will be at least as large as its gravitational radius $2GM$, the bound being saturated by black holes. Even if its surface were rotating at the speed of light, its rotation period would be at least $4\pi GM$, and the source cannot emit waves with a period much shorter than that. Therefore, we have a maximum frequency [63],

$$f \lesssim \frac{1}{4\pi GM} \sim 10^4 \frac{M_\odot}{M} \text{ Hz.} \tag{13.41}$$

To emit near this maximum frequency an object must presumably have a mass of the order of the Chandrasekhar limit $\sim 1.2 M_\odot$, which gives a maximum frequency of the order of 10 kHz [63], and this limit can be saturated only by very compact objects (see [64] for a recent review of GWs emitted in the gravitational collapse to black holes, with typical frequencies $f \lesssim 5$ kHz). The same numbers, apart from factors of order one, can be obtained using the fact that for a self-gravitating Newtonian system with density ρ, radius R, there is a natural dynamical frequency [65]

$$f_{dyn} = \frac{1}{2\pi} (\pi G\rho)^{1/2} = \left(\frac{3GM}{16\pi^2 R^3} \right)^{1/2}. \tag{13.42}$$

With $R \geq 2GM$ we recover the same order of magnitude estimate apart from a factor $(3/8)^{1/2} \simeq 0.6$. This is already an encouraging result, because it shows that the natural frequency domains of cosmological and astrophysical sources can be very different. We have seen in section 12.1 that the natural frequency scale for Planckian physics is the GHz, while no astrophysical objects can emit above, say, (6–10) kHz. Therefore, a stochastic background detected above these frequencies would be unambiguously of cosmological origin.

However, ground-based interferometers have their maximum sensitivity around 100 Hz, where astrophysical sources hopefully produce interesting radiation (since these sources were the original motivation for the construction of interferometers). The radiation from a single source is not a problem, since it is easily distinguished from a stochastic background. The problem arises if there are many unresolved sources. (Of course this is a problem from the point of view of the cosmological background, but the observation of the astrophysical background would be very interesting in itself; techniques for the detection of this background with a single interferometer using the fact that it is not isotropic and exploiting the sideral modulation of the signal have been discussed in [66].)

The stochastic background from rotating neutron stars has been discussed in [67] and references therein. The main uncertainty comes from the estimate of the typical ellipticity ϵ of the neutron star, which measures its deviation from sphericity. An upper bound on ϵ can be obtained assuming that the observed slowing down of the period of known pulsars is entirely due to the emission of gravitational radiation. This is almost certainly a gross overestimate, since most of the spin down is probably due to electromagnetic losses, at least for Crab-like pulsars. With realistic estimates for ϵ, [67] gives a value of $h_0^2 \Omega_{gw}(100 \text{ Hz}) \sim 10^{-15}$. This is very far from the sensitivity of even the advanced experiments. An absolute upper bound can be obtained assuming that the spin down is due only to gravitational losses, and this gives $h_0^2 \Omega_{gw}(100 \text{ Hz}) \sim 10^{-7}$, but again this value is probably a gross overestimate. Very recently the gravitational background has also been estimated [68] produced by a cosmological population of hot, young and rapidly rotating neutron stars. Within a reasonable range of values of the main parameters which characterize the energy spectrum of a single source, the authors of [68] find that $h_0^2 \Omega_{gw}(f)$ show a long plateau extending from ≈ 300 Hz up to ≈ 1.7 kHz, with an amplitude of $\approx (2.2–3.3) \times 10^{-8}$.

In [69] the stochastic background has been considered emitted by a cosmological population of core-collapse supernovae for a range of progenitor masses leading to a black hole. The expected frequencies in this case are of the order of kHz or lower, depending on the redshift when these objects are produced. Using the observational data on the star formation rate, it turns out that the duty cycle, i.e. the ratio between the duration of a typical burst and the typical time interval between successive bursts is low, of order 0.01. Therefore, this background is not stochastic, but rather like a 'pop noise', and can be distinguished from a really stochastic background. The value of $h_0^2 \Omega_{gw}$ for the background from supernovae have been computed in [69] assuming axially symmetric collapse, and assuming that all sources have the same value of $a = J/(GM^2)$, where J is the angular momentum. The results depend on the value of a, and on h_0. Assuming $h_0 = 0.5$ and typical values of a, one finds that $\Omega_{gw}(f)$ has a maximum amplitude ranging between $10^{-11}–10^{-10}$ in the frequency interval (1.5–2.5) kHz.

These results suggest that astrophysical backgrounds might not be a problem for the detection of a relic background at VIRGO/LIGO frequencies. The situation is different in the LISA frequency band [63, 65, 67, 70, 71]. LISA can reach a sensitivity of order $h_0^2 \Omega_{gw} \sim$ a few $\times 10^{-13}$ at $f \sim 10^{-3}$ Hz (see figure 1.3

of [71]). However, for frequencies below a few mHz, one expects a stochastic background due to a large number of galactic white dwarf binaries. The estimate of this background depends on the rate of white dwarf mergers, which is uncertain. With rates of order 4×10^{-3} per year (which should be a secure upper limit [67]), the background can be as large as $h_0^2 \Omega_{gw} \sim 10^{-8}$ at $f = 10^{-3}$ Hz. This number is actually quite uncertain, and in [71], it is used another plausible rate, which gives for instance $h_0^2 \Omega_{gw} \sim 10^{-11} - 10^{-10}$ at $f = 10^{-3}$ Hz. Above a frequency of the order of a few times 10^{-2} Hz, most signals from galactic binaries can be resolved individually and no continuous background of galactic origin is presently known at the level of sensitivity of LISA.

It should be observed that, even if an astrophysical background is present, and masks a relic background, not all hopes are lost. If we understand well enough the astrophysical background, we can subtract it, and the relic background would still be observable if it is much larger than the uncertainty that we have on the astrophysical background. In fact, LISA should be able to subtract the background due to white dwarf binaries, since there is a large number of binaries close enough to be individually resolvable [71]. This should allow us to predict with some accuracy the space density of white dwarf binaries in other parts of the Galaxy, and therefore to compute the stochastic background that they produce. Furthermore, any background of galactic origin is likely to be concentrated near the galactic plane, and this is another handle for its identification and subtraction. The situation is more uncertain for the contribution of extragalactic binaries, which again can be relevant at LISA frequencies. The uncertainty in the merging rate is such that it cannot be predicted reliably, but it is believed to be lower than the galactic background [67]. In this case the only handle for the subtraction would be the form of the spectrum. In fact, even if the strength is quite uncertain, the form of the spectrum may be quite well known [71].

References

[1] Kolb E and Turner M 1990 *The Early Universe* (New York: Addison-Wesley)
[2] Kosowsky A, Kamionkowski M and Turner M 1994 *Phys. Rev.* D **49** 2837
[3] Fixsen D J *et al* 1996 *Astrophys. J.* **473** 576
[4] Allen B 1996 Relativistic gravitation and gravitational radiation *Proc. Les Houches School on Astrophysical Sources of Gravitational Waves* ed J Marck and J Lasota (Cambridge: Cambridge University Press)
     Allen B 1996 *Preprint* gr-qc/9604033
[5] Maggiore M 1998 High-energy physics with gravitational-wave experiments gr-qc/9803028
[6] Birrel N and Davies P C W 1982 *Quantum Fields in Curved Space* (Cambridge: Cambridge University Press)
[7] Allen B and Romano J 1999 *Phys. Rev.* D **59** 102001
[8] Michelson P 1987 *Mon. Not. R. Astron. Soc.* **227** 933
[9] Christensen N 1992 *Phys. Rev.* D **46** 5250
[10] Flanagan E 1993 *Phys. Rev.* D **48** 2389

[11] Thorne K S 1987 *300 Years of Gravitation* ed S Hawking and W Israel (Cambridge: Cambridge University Press) p 330

[12] Misner C, Thorne K and Wheeler J 1973 *Gravitation* (New York: Freeman)

[13] Zhou C Z and Michelson P 1995 *Phys. Rev.* D **51** 2517

[14] Cuoco E, Curci G and Beccaria M 1997 Adaptive identification of VIRGO-like noise spectrum gr-qc/9709041

[15] Astone P *et al* 1997 *Astropart. Phys.* **7** 231

[16] Coccia E, Fafone V, Frossati G, Lobo J and Ortega J 1998 *Phys. Rev.* D **57** 2051

[17] Bender P *et al* 1995 LISA *Pre-Phase A Report* unpublished

[18] Poor H V 1994 *An Introduction to Signal Detection and Estimation* (Berlin: Springer)

[19] Vitale S, Cerdonio M, Coccia E and Ortolan A 1997 *Phys. Rev.* D **55** 1741

[20] Compton K and Schutz B 1997 Bar-interferometer observing *Gravitational Waves; Sources and Detectors* ed I Ciufolini and F Fidecaro (Singapore: World Scientific) p 173

[21] Coccia E 1997 *Proc. 14th Int. Conf. on General Relativity and Gravitation* ed M Francaviglia *et al* (Singapore: World Scientific) p 103
Prodi G A *et al* 1997 *Gravitational Waves; Sources and Detectors* ed I Ciufolini and F Fidecaro (Singapore: World Scientific) p 166
Cerdonio M *et al* 1997 *Class. Quantum Grav.* **14** 1491

[22] Coccia E, Pizzella G and Ronga F 1994 *Proc. 1st Edoardo Amaldi Conf. (Frascati)* (Singapore: World Scientific)
Bianchi M, Coccia E, Colacino C, Fafone V and Fucito F 1996 *Class. Quantum Grav.* **13** 2865
Coccia E 1997 *Gravitational Waves; Sources and Detectors* ed I Ciufolini and F Fidecaro (Singapore: World Scientific) p 201

[23] Johnson W and Merkowitz S 1993 *Phys. Rev. Lett.* **70** 2367

[24] Astone P, Lobo J and Schutz B 1994 *Class. Quantum Grav.* **11** 2093

[25] Astone P, Pallottino G and Pizzella G 1997 *Class. Quantum Grav.* **14** 2019

[26] Astone P, Frasca S, Papa M A and Ricci F 1997 Spectral detection strategy of stochastic gravitational wave search in VIRGO *VIRGO Note* VIR-NOT-ROM-1390-106, unpublished

[27] Astone P 1999 *Proc. 2nd Edoardo Amaldi Conf. on Gravitational Waves* ed E Coccia *et al* (Singapore: World Scientific)

[28] Schwartzmann V F 1969 *JETP Lett.* **9** 184

[29] Particle Data Group 1996 Review of particle properties *Phys. Rev.* D **54** 66

[30] Copi C J, Schramm D N and Turner M S 1997 *Phys. Rev.* D **55** 3389

[31] Kirzhnits D A 1972 *JETP Lett.* **15** 745

[32] Kibble T W B 1976 *J. Phys. A: Math. Gen.* **9** 1387

[33] Vilenkin A and Shellard S 1994 *Cosmic Strings and other Topological Defects* (Cambridge: Cambridge University Press)

[34] Caldwell R R and Allen B 1992 *Phys. Rev.* D **45** 3447

[35] Caldwell R R *et al* 1996 *Phys. Rev.* D **54** 7146

[36] Allen B *et al* 1996 *Phys. Rev. Lett.* **77** 3061

[37] Battye R A *et al* 1997 Gravitational waves from cosmic strings *Topological Defects in Cosmology* ed F Melchiorrei and M Signore (Singapore: World Scientific) pp 11–31 (astro-ph/9706013)

[38] Martin X and Vilenkin A 1996 *Phys. Rev. Lett.* **77** 2879
Martin X and Vilenkin A 1997 *Phys. Rev.* D **55** 6054

[39] Liddle A R 2000 An introduction to cosmological inflation: high energy physics and cosmology *Proc. ICTP Summer School* ed A Masiero and A Smirnov (Singapore: World Scientific) astro-ph/9901124

[40] Turner M S 1997 Ten things everyone should know about inflation *Generation of Cosmological Large Scale Structure* ed P Galeotti and D Schramm (Dordrecht: Kluwer) pp 153–92 (astro-ph/9704062)

[41] Guth A H 1981 *Phys. Rev.* D **23** 347

[42] Linde A 1982 *Phys. Lett.* B **108** 389

[43] Albrecht A and Steinhardt P 1982 *Phys. Rev. Lett.* **48** 1220

[44] Linde A 1983 *Phys. Lett.* B **129** 177

[45] Birrel N D and Davies P C W 1982 *Quantum Field Theory in Curved Space* (Cambridge: Cambridge University Press)

[46] Allen B 1988 *Phys. Rev.* D **37** 2078

[47] Allen B 1996 *Proc. Les Houches School on Astrophysical Sources of Gravitational Waves* ed J Marck and J Lasota (Cambridge: Cambridge University Press)
Allen B 1996 *Preprint* gr-qc/9604033

[48] Veneziano G 1991 *Phys. Lett.* B **265** 287
Gasperini M and Veneziano G 1993 *Astropart. Phys.* **1** 317
Gasperini M and Veneziano G 1993 *Mod. Phys. Lett.* A **8** 3701 (an up-to-date collection of references on string cosmology can be found at http://www.to.infn.it/~gasperin/

[49] Antoniadis I, Arkani-Hamed N, Dimopoulos S and Dvali G 1998 *Phys. Lett.* B **429** 263

[50] Gasperini M, Maggiore M and Veneziano G 1997 *Nucl. Phys.* B **494** 315

[51] Maggiore M 1998 *Nucl. Phys.* B **525** 413

[52] Brustein R and Madden R 1997 *Phys. Lett.* B **410** 110
Brustein R and Madden R 1998 *Phys. Rev.* D **57** 712

[53] Foffa S, Maggiore M and Sturani R 1999 *Phys. Rev.* D **59** 043507

[54] Foffa S, Maggiore M and Sturani R 1999 *Nucl. Phys.* B **552** 395

[55] Maggiore M and Riotto A 1999 *Nucl. Phys.* B **548** 427

[56] Brustein R, Gasperini M, Giovannini M and Veneziano G 1995 *Phys. Lett.* B **361** 45

[57] Buonanno A, Maggiore M and Ungarelli C 1997 *Phys. Rev.* D **55** 3330

[58] Maggiore M and Sturani R 1997 *Phys. Lett.* B **415** 335

[59] Coleman S 1977 *Phys. Rev.* D **15** 2929
Callan C G and Coleman S 1977 *Phys. Rev.* D **16** 1762

[60] Turner M S *et al* 1992 *Phys. Rev.* D **46** 2384
See also Hogan C J 1983 *Phys. Lett.* B **133** 172

[61] Kamionkowsky M *et al* 1994 *Phys. Rev.* D **49** 2837

[62] Kosowsky A *et al* 1992 *Phys. Rev. Lett.* **69** 2026

[63] Thorne K S 1995 *Proc. Snowmass 1994 Summer Study on Particle and Nuclear Astrophysics and Cosmology* ed E Kolb and R Peccei (Singapore: World Scientific) p 398

[64] Ferrari V and Palomba C 1998 Is the gravitational collapse to black holes and interesting source for VIRGO? *VIRGO Note* VIR-NOT-ROM-1390-109, unpublished

[65] Schutz B F 1997 Low-frequency sources of gravitational waves: a tutorial *Proc. 1997 Alpbach Summer School on Fundamental Physics in Space* ed A Wilson (ESA)

[66] Giazotto A, Bonazzola S and Gourgoulhon E 1997 *Phys. Rev.* D **55** 2014

[67] Postnov K 1997 Astrophysical sources of stochastic gravitational radiation in the universe astro-ph/9706053
[68] Ferrari V, Matarrese S and Schneider R 1999 *Mon. Not. R. Astron. Soc.* **303** 258
[69] Ferrari V, Matarrese S and Schneider R 1999 *Mon. Not. R. Astron. Soc.* **303** 247
[70] Bender P and Hils D 1997 *Class. Quantum Grav.* **14** 1439
[71] Bender P *et al* 1995 LISA *Pre-Phase A Report* unpublished

PART 4

THEORETICAL DEVELOPMENTS

Hermann Nicolai, Alessandro Nagar, Donato Bini,
Fernando De Felice, Maurizio Gasperini and
Luc Blanchet

Chapter 14

Infinite-dimensional symmetries in gravity

Hermann Nicolai[1] *and Alessandro Nagar*[2]
[1] *Max-Planck-Institut für Gravitationsphysik
(Albert-Einstein-Institut), Mühlenberg 1, D-14476 Golm,
Germany*
[2] *Dipartimento di Fisica, Università degli Studi di Parma*

14.1   Einstein theory

14.1.1   Introduction

In these lectures we review the symmetry properties of Einstein's theory when it is reduced from four to two dimensions. We explain how, in this reduction, the theory acquires an infinite-dimensional symmetry group, the Geroch group, whose associated Lie algebra is the affine extension of $SL(2, \mathbb{R})$. The action of the Geroch group, which is nonlinear and non-local, can be linearized, thereby permitting the explicit construction of many solutions of Einstein's equations with two commuting Killing vectors ∂_2 and ∂_3. A non-trivial example of this method for a colliding plane wave metric is given.

The lectures review some well-known material at a pedagogical level. Therefore, rather than including references in the text, we have chosen to collect some basic references at the end, which readers are invited to use as a guide to the vast literature on the subject of exact solutions, on the integrability of Einstein's equations in the reduction to two dimensions, and finally on the generalization of these structures to other theories, including supergravity.

14.1.2   Mathematical conventions

Our main interest is in studying the structural properties of Einstein's theory and its generalizations. We will first formulate it in D dimensions, with coordinates

245

$x^M = (x^0, \ldots, x^{D-1})$. The metric can be expressed in terms of the vielbein as

$$g_{MN} = E_M^A E_N^B \eta_{AB} \tag{14.1}$$

with the flat metric $\eta_{AB} \equiv (+, -, \ldots, -)$. For the following it will be important that the vielbein can be viewed as an element of a coset space according to

$$E_M^A \in GL(D, \mathbf{R})/SO(1, D-1). \tag{14.2}$$

The metric must be covariantly conserved

$$D_N(\Gamma)g_{MP} = 0 \tag{14.3}$$

where Γ is the Christoffel symbol of the metric g_{MN}. We next introduce a spin connection one-form, with coefficients $\omega_{MA}{}^B$. The vielbein postulate, that is the covariant constancy of the vielbein, which agrees exactly with Cartan's structure equation for the torsion two-form, is

$$D_M(\omega, \Gamma)E_N{}^A = 0. \tag{14.4}$$

Writing out this equation, we have

$$\partial_{[M} E_{N]}{}^A + \omega_M{}^A{}_B E_N{}^B = \Gamma_{[MN]}{}^P E_P{}^A. \tag{14.5}$$

We assume there is no torsion, so the Christoffel symbols are symmetric in spacetime indices, hence

$$\partial_{[M} E_{N]}{}^A + \omega_{[M}{}^A{}_{N]} = 0. \tag{14.6}$$

The coefficients of the anholonomy are

$$\Omega_{AB}{}^C = 2E_{[A}{}^M E_{B]}{}^N \partial_M E_N{}^C. \tag{14.7}$$

Using the torsion-free condition for the spin connection and permuting the indices of the coefficients of the anholonomy we obtain the following equations

$$\begin{aligned} \Omega_{ABC} + \omega_{ACB} - \omega_{BCA} &= 0 \\ -\Omega_{BCA} - \omega_{BAC} + \omega_{CAB} &= 0 \\ \Omega_{CAB} + \omega_{CBA} - \omega_{ABC} &= 0. \end{aligned} \tag{14.8}$$

Employing then the property of the spin-connection $\omega_{ABC} = -\omega_{ACB}$, due to the fact that the generators of the algebra of the D-Lorentz Group are totally antisymmetric matrices, we have the expression of the spin connection as a function of Ω_{ABC}

$$\omega_{ABC} = \tfrac{1}{2}(\Omega_{ABC} - \Omega_{BCA} + \Omega_{CAB}). \tag{14.9}$$

The Riemann tensor (the curvature two-form) can be defined by

$$[D_M(\omega), D_N(\omega)]V^A = R_{MN}{}^A{}_B V^B. \tag{14.10}$$

The explicit expression in terms of the spin connection is

$$R_{MNA}{}^B = 2\partial_{[M}\omega_{N]A}{}^B + 2\omega_{[MA}{}^C\omega_{N]C}{}^B. \tag{14.11}$$

From the Riemann tensor the Ricci tensor and the curvature scalar are obtained in the usual way: $R_{MN} := R_{MPN}{}^P$ and $R := g^{MN}R_{MN}$. The metric determinant is

$$E = \det E_M{}^A = \sqrt{-g}. \tag{14.12}$$

14.1.3 The Einstein–Hilbert action

Now we have all the elements to define Einstein theory. The Einstein–Hilbert action is

$$S = \int d^4x\, \mathcal{L} \tag{14.13}$$

and the Lagrangian $\mathcal{L}$ can be expressed in function of the spin connection

$$\begin{aligned}
\mathcal{L} &= -\tfrac{1}{4}ER = -\tfrac{1}{4}E E_A{}^M E_B{}^N R_{MN}{}^{AB} \\
&= -\tfrac{1}{2}E E_A{}^M E_B{}^N \partial_M \omega_N{}^{AB} - \tfrac{1}{4}E\omega_A{}^{AC}\omega_{BC}{}^B + \tfrac{1}{4}\omega_{BAC}\omega^{ACB}.
\end{aligned} \tag{14.14}$$

Substituting now the expression $\omega = \omega(\Omega)$, integrating by parts and dropping total derivatives, we arrive at

$$-\tfrac{1}{4}ER = \tfrac{1}{16}E(\Omega_{ABC}\Omega^{ABC} - 2\Omega^{ABC}\Omega_{BCA} - 4\Omega_{AC}{}^C\Omega^A{}_D{}^D). \tag{14.15}$$

This is the expression best suited for dimensional reduction of Einstein's theory. In the remainder, we will now set $D = 4$, i.e. work in four spacetime dimensions.

14.1.4 Dimensional reduction $D = 4 \rightarrow D = 3$

'Dimensional reduction' is equivalent to searching for solutions of Einstein's equations with one Killing vector, which we take to be $\xi^M \partial_M \equiv \partial_3$. For this purpose, we proceed from the 'Kaluza–Klein ansatz' for the vierbein.

$$E_M{}^A = \begin{pmatrix} \Delta^{-1/2}e_m{}^a & \Delta^{1/2}B_m \\ 0 & \Delta^{1/2} \end{pmatrix}, \quad E_A{}^M = \begin{pmatrix} \Delta^{1/2}e_a{}^m & -e_a{}^n B_n \Delta^{1/2} \\ 0 & \Delta^{-1/2} \end{pmatrix}. \tag{14.16}$$

The matrix $e_m{}^a$ is the three-bein; B_m is called the Kaluza–Klein vector and Δ the Kaluza–Klein scalar. The ansatz fixes a part of the $SO(1, 3)$ Lorentz symmetry. The residual symmetry group preserving the gauge condition $E_3{}^a = 0$ is the gauge group $SO(1, 2)$.

After some algebra, we find

$$\Omega_{abc} = \Delta^{\frac{1}{2}}(\Omega_{abc}^{(3)} - e_{[a}{}^m\eta_{b]c}\Delta^{-1}\partial_m\Delta) \qquad (14.17)$$

$$\Omega_{ab}{}^3 = \Delta^{3/2}e_a{}^m e_b{}^n B_{mn} \qquad (14.18)$$

$$\Omega_{3b}{}^3 = -\tfrac{1}{2}e_b{}^m\Delta^{-1/2}\partial_m\Delta \qquad (14.19)$$

$$\Omega_{3b}{}^c = 0. \qquad (14.20)$$

Substituting the ansatz for the vierbein in the field action and making use of the above decomposition of the Einstein action, after some calculations we arrive at the following result

$$-\tfrac{1}{4}E R(E) = -\tfrac{1}{4}e R^{(3)}(e) - \tfrac{1}{16}e\Delta^2 B^{mn}B_{mn} + \tfrac{1}{8}e g^{mn}\Delta^{-2}\partial_m\Delta\partial_n\Delta \qquad (14.21)$$

where $B_{mn} = \partial_m B_n - \partial_n B_m$.

Duality transformation

The very special feature of three dimensions is that the Kaluza–Klein vector field can be dualized to a scalar. This is achieved by adding to the Einstein–Hilbert Lagrangian the expression

$$\mathcal{L}' = \tfrac{1}{8}e\tilde{\epsilon}^{mnp}B_{mn}\partial_p B \qquad (14.22)$$

where B is a Lagrange multiplier and $\tilde{\epsilon}^{mnp}$ the Levi-Civita totally antisymmetric symbol. The dualization makes the Lagrangian depend only on B. So, adding $\mathcal{L}'$ to $\mathcal{L}$ and varying B_n leads to

$$e\Delta^2 B^{mn} = \epsilon^{mnp}\partial_p B \qquad (14.23)$$

modulo a numerical constant. Here we have set $\epsilon^{mnp} = e\tilde{\epsilon}^{mnp}$. When we substitute this expression in the three-dimensional reduced Einstein–Hilbert Lagrangian, we get a new one with two scalar fields

$$\mathcal{L} = -\tfrac{1}{4}e R^{(3)}(e) + \tfrac{1}{8}e g^{mn}\Delta^{-2}(\partial_m\Delta\partial_n\Delta + \partial_m B\partial_n B). \qquad (14.24)$$

This is consistent with the equation of motion $\partial_m(e\Delta^2 B^{mn}) = 0$. In fact, the term we add to the Lagrangian, which is now three dimensional, can be dropped by an integration by parts and the use of the three-dimensional Bianchi identities for the tensor B^{mn}.

14.1.5 Dimensional reduction $D = 3 \rightarrow D = 2$

Then we perform a dimensional reduction from three to two, i.e. we have two Killing commuting vectors (∂_3 and ∂_2) and there is no dependence on x^2 at all.

$$x^m = (x^\mu, x^2). \qquad (14.25)$$

Repeating the same steps as before, the two-bein now takes the form

$$e_m{}^a = \begin{pmatrix} e_\mu{}^\alpha & \rho A_\mu \\ 0 & \rho \end{pmatrix}, \qquad e_a{}^m = \begin{pmatrix} e_\alpha{}^\mu & -e_\alpha^\sigma A_\sigma \\ 0 & \rho^{-1} \end{pmatrix}. \tag{14.26}$$

Detailed calculation shows that

$$-\tfrac{1}{4}e^{(3)}R^{(3)} = -\tfrac{1}{4}\rho e R^{(2)} - \tfrac{1}{16}\rho^3 e A_{\mu\nu}A^{\mu\nu} \tag{14.27}$$

with

$$e R^{(2)} = -2\partial_\mu(ee^{\alpha\mu}\Omega_{\alpha\beta}{}^\beta) \tag{14.28}$$

where e is the determinant of the two-bein. At this point we can write the equations of motion for the theory. The equation of motion for the Kaluza–Klein vector is given by

$$\partial_\mu(\rho^3 e A^{\mu\nu}) = 0. \tag{14.29}$$

In two dimensions, a Maxwell field does not propagate, as there are no transverse degrees of freedom. Neglecting topological effects (i.e. non-vanishing holonomies) we can, therefore, set $A_\mu = 0$.

For the remaining equations of motion, we can fix the gauge, and then calculate them in a particular gauge, called the conformal gauge. The term $e R^{(2)}(e)$ is Weyl-invariant. To see why this is so, let us consider the term

$$-\tfrac{1}{4}\rho R^{(2)} = \tfrac{1}{2}\rho\partial_\nu(ee_\alpha{}^\nu\Omega^{\alpha\gamma}{}_\gamma). \tag{14.30}$$

An integration by parts gives

$$-\tfrac{1}{4}\rho R^{(2)} \doteq -\tfrac{1}{2}e\Omega^{\alpha\gamma}{}_\gamma e_\alpha{}^\mu \partial_\mu\rho. \tag{14.31}$$

Then, using the definition of the anholonomy, we get

$$= -\tfrac{1}{2}e(e_\alpha{}^\nu e_\gamma{}^\tau \partial_\nu e_\tau{}^\gamma - e_\gamma{}^\nu e_\alpha{}^\tau \partial_\nu e_\tau{}^\gamma)e^{\alpha\mu}\partial_\mu\rho \tag{14.32}$$

$$= -\tfrac{1}{2}eg^{\mu\nu}e_\gamma{}^\tau \partial_\nu e_\tau{}^\gamma \partial_\mu\rho - \tfrac{1}{2}e\partial_\nu e_\alpha{}^\nu e^{\alpha\mu}\partial_\mu\rho \tag{14.33}$$

where another integration by parts and the definition of the two-bein have been used.

Now we can set the gauge, i.e. the 2D diffeomorphisms, by a condition on the two-bein. So we write

$$e_\mu{}^\alpha = \lambda\tilde{e}_\mu{}^\alpha \tag{14.34}$$

with $\det\tilde{e}_\mu{}^\alpha = 1$ and $\lambda = \lambda(x)$; hence, we are not considering the whole group $GL(2, \mathbf{R})$ but only its restriction to unimodular matrices $SL(2, \mathbf{R})$.

As we said before, we can set a particular gauge, the conformal gauge, by imposing the following condition on the two-bein. It is given by

$$\tilde{e}_\mu{}^\alpha = \delta_\mu^\alpha. \tag{14.35}$$

Then, after an integration by parts, we get

$$-\tfrac{1}{4}\rho R^{(2)} \doteq - \tilde{g}^{\mu\nu}\lambda^{-1}\partial_\nu\lambda\partial_\mu\rho + \tfrac{1}{2}\tilde{e}_\alpha{}^\nu\partial_\nu(\tilde{e}^{\alpha\mu}\partial_\mu\rho). \qquad (14.36)$$

In this gauge it is obviously

$$\tilde{g}^{\mu\nu} = \tilde{e}_\alpha{}^\mu\tilde{e}^{\nu\alpha} = \lambda^2 g^{\mu\nu}. \qquad (14.37)$$

So we have three fields: the dilaton ρ, the λ and the unimodular two-bein $\tilde{e}_\mu{}^\alpha$. We can calculate the equations of motion varying the Lagrangian with respect to all these fields. Varying it with respect to λ we get

$$\partial_\mu(\tilde{g}^{\mu\nu}\partial_\nu\rho) \equiv \Box\rho = 0 \qquad (14.38)$$

because in conformal gauge $\tilde{g}^{\mu\nu} = \eta^{\mu\nu}$. The solution of this equation is

$$\rho(x) = \rho_+(x^+) + \rho_-(x^-) \qquad (14.39)$$

with $x^\pm = x^0 \pm x^1$.

The dilaton can be dualized: in two dimensions, the dual of a scalar field is again a scalar field. We will refer to the dual of the dilaton field as the 'axion'; it is defined by

$$\partial_\mu\rho + \epsilon_{\mu\nu}\partial^\nu\tilde{\rho} = 0 \qquad (14.40)$$

where $\tilde{\rho}$ is just the axion. In the conformal gauge this field is

$$\tilde{\rho}(x) = \rho_+(x^+) - \rho_-(x^-). \qquad (14.41)$$

The equation obtained by varying ρ is

$$\partial_\mu(\tilde{g}^{\mu\nu}\lambda^{-1}\partial_\nu\lambda) = \text{matter contribution.} \qquad (14.42)$$

Note that with matter contribution we refer to the fields Δ and B coming out from dimensional reduction. The terminology *matter part* will be clear in the following section, where we will be able to identify this fields with the fields of a bosonic nonlinear σ-model Lagrangian.

Before writing the complete Lagrangian of the two-dimensional reduced gravity we must still consider the equation that is obtained from (14.36) by variation with respect to the unimodular two-bein. The corresponding equations must be interpreted as constraint equations (in standard conformal field theory, they would just correspond to the Virasoro constraints). We have

$$-\delta\tilde{e}_\alpha{}^\mu\tilde{e}^{\alpha\nu}\lambda^{-1}\partial_{(\mu}\lambda\partial_{\nu)}\rho + \tfrac{1}{2}\delta\tilde{e}_\alpha{}^\mu\partial_\mu(\tilde{e}^{\alpha\nu}\partial_\nu\rho) - \tfrac{1}{2}\partial_\mu\tilde{e}_\alpha{}^\mu\delta\tilde{e}^{\alpha\nu}\partial_\nu\rho + \text{matter} = 0. \qquad (14.43)$$

In conformal gauge, $\tilde{e}_\alpha{}^\mu = \delta_\alpha^\mu$, this expression becomes

$$-\tfrac{1}{2}\delta\tilde{g}^{\mu\nu}\lambda^{-1}\partial_\mu\lambda\partial_\nu\rho + \tfrac{1}{4}\delta\tilde{g}^{\mu\nu}\partial_\mu\partial_\nu\rho = \text{matter} \qquad (14.44)$$

where $\delta\tilde{g}^{\mu\nu} = 2\delta\tilde{e}_\alpha{}^\mu \tilde{e}^{\alpha\nu}$ has been used. The metric is diagonal in this gauge, so the equations for the conformal factors become

$$\text{(traceless part of)}\{-\tfrac{1}{2}\lambda^{-1}\partial_\mu\lambda\partial_\nu\rho + \tfrac{1}{4}\partial_\mu\partial_\nu\rho\} = \text{matter}. \qquad (14.45)$$

We have not explicitly written the matter sector yet. This will be done in the next section.

The pure gravity action written in two dimensions in the conformal gauge reads

$$-\tfrac{1}{4}e^{(3)}R^{(3)} = -\tfrac{1}{2}\lambda^{-1}\partial_\mu\lambda\partial^\mu\rho \qquad (14.46)$$

where the equation $\Box\rho = 0$ has been used to make the second term of (14.36) vanish. The whole Lagrangian is then

$$\mathcal{L}_E = -\tfrac{1}{2}\lambda^{-1}\partial_\mu\lambda\partial^\mu\rho + \tfrac{1}{8}\rho\Delta^{-2}(\partial_\mu\Delta\partial^\mu\Delta + \partial_\mu B\partial^\mu B) \qquad (14.47)$$

where the second term with the Kaluza–Klein scalar and the dual of the Kaluza–Klein vector has been obtained considering only the two-dimensional part of the action. The subscript E stands for Ehlers, who did this analysis for the first time in the 1950s.

Here we have treated one possible way of performing dimensional reduction: we have seen it consists of many steps. One first reduces from four to three; then, dualizes the vector field and performs the reduction to two dimensions. However, this is not the whole story: actually, it is possible also to get the two-dimensional Lagrangian directly from the three-dimensional one, without dualization.

The procedure for doing the calculation is as follows: first we express the Kaluza–Klein vector in the form

$$B_m = (B_\mu, B_2 \equiv \tilde{B}) \qquad (14.48)$$

and then we perform directly the dimensional reduction in conformal gauge by using the previous choice of the three-bein in triangular form. Proceeding in this way, we meet two electromagnetic fields in two dimensions, A_μ and B_μ, which can be set to zero because they do not propagate (we have already used this argument before) and there is no cosmological constant.

Let us note now the various steps of the calculation. The new writing for the Kaluza–Klein vector and the considerations on two-dimensional electromagnetism imply

$$B_\mu = 0 \rightarrow B_{\mu\nu} = 0. \qquad (14.49)$$

So the only non-vanishing terms of the three-dimensional Lagrangian before the reduction are

$$\mathcal{L} = -\tfrac{1}{4}e^{(3)}R^{(3)}(e) - \tfrac{1}{8}e^{(3)}\Delta^2 B_{\mu 2}B^{\mu 2} + \tfrac{1}{8}e^{(3)}g^{mn}\Delta^{-2}\partial_m\Delta\partial_n\Delta. \qquad (14.50)$$

Then we reduce the dimensions as before (we set $A_\mu = 0$) and choose the conformal gauge. Keeping in mind that

$$g^{\mu\nu} = \lambda^2\eta^{\mu\nu}, \quad g^{22} = -\rho^2 \qquad (14.51)$$

we get a new 2D Lagrangian

$$\mathcal{L}_{\text{MM}} = -\tfrac{1}{2}\lambda^{-1}\partial_\mu\lambda\partial^\mu\rho + \tfrac{1}{8}\rho\Delta^{-2}\partial_\mu\Delta\partial^\mu\Delta + \tfrac{1}{8}\rho^{-1}\Delta^2\partial_\mu\tilde{B}\partial^\mu\tilde{B} \qquad (14.52)$$

for the fields $\tilde{B}$ and Δ.

The subscript MM stands for Matzner–Misner, who performed this analysis at the end of the 1960s. It is worth noting that the link between the fields B and $\tilde{B}$ is given by three-dimensional duality. To see why it is so, let us consider the duality relation

$$e\Delta^2 B^{mn} = \epsilon^{mnp}\partial_p B \qquad (14.53)$$

and then reduce the dimensionality using the properties of the Kaluza–Klein vector B_m. The duality relation becomes then

$$\rho^{-1}\Delta^2\partial_\mu\tilde{B} = \epsilon_{\mu\nu}\partial^\nu B. \qquad (14.54)$$

The deep relation between these two distinct reduced Lagrangians will be explained in the next section, treating nonlinear σ-models. In the language of the nonlinear sigma models, these two reduced actions correspond to two distinct $SL(2, \boldsymbol{R})/SO(2)$ models.

14.2   Nonlinear σ-models

In this section, we introduce nonlinear σ-models and discover that the reduced gravity is a certain nonlinear σ-model, connected to a certain symmetry group. The expression of the Lagrangian of the model depends on this symmetry group.

Let us start from a non-compact Lie group G and consider the maximal compact Lie subgroup H of G. The Lie algebra decomposition is

$$\boldsymbol{G} = \boldsymbol{H} \oplus \boldsymbol{K} \qquad (14.55)$$

with the following commutation rules

$$[\boldsymbol{H}, \boldsymbol{H}] \subset \boldsymbol{H}, \quad [\boldsymbol{K}, \boldsymbol{K}] \subset \boldsymbol{H}, \quad [\boldsymbol{H}, \boldsymbol{K}] \subset \boldsymbol{K}. \qquad (14.56)$$

This decomposition is invariant under the symmetric space automorphism

$$\tau(\boldsymbol{H}) = \boldsymbol{H}, \quad \tau(\boldsymbol{K}) = -\boldsymbol{K} \qquad (14.57)$$

which can alternatively be formulated in terms of Lie group elements g directly through

$$\tau(g) = \eta^{-1}(g^{\text{T}})^{-1}\eta \qquad (14.58)$$

where the matrix η depends on the group G (e.g. $\eta = 1$ for $G = SL(n, \boldsymbol{R})$).

Example: $G = SL(2, \mathbf{R})$, $H = SO(2)$

The generators of the group are

$$Y^1 = \begin{pmatrix} 1 & 0 \\ 0 & -1 \end{pmatrix}, \quad Y^2 = \begin{pmatrix} 0 & 1 \\ 1 & 0 \end{pmatrix}, \quad Y^3 = \begin{pmatrix} 0 & 1 \\ -1 & 0 \end{pmatrix}. \tag{14.59}$$

We have

$$\mathrm{Tr}(Y^1)^2 = \mathrm{Tr}(Y^2)^2 = -\mathrm{Tr}(Y^3)^2 = 2 \tag{14.60}$$

so Y^1 and Y^2 are the non-compact generators, while Y^3 generates the $SO(2)$ subgroup.

The group can be decomposed on its generators, as

$$H = RY^3, \quad K = RY^1 \oplus RY^2. \tag{14.61}$$

Let us introduce now an element of the group $v(x) \in G$ with the property

$$v(x) \to v'(x) = g^{-1}v(x)h(x) \tag{14.62}$$

where g is a rigid G transformation and $h(x)$ a local H transformation. This type of transformation is needed for the necessity of preserving gauge choice. In fact, you can fix the gauge choosing a particular element of the group v. Then, when you act on v by an arbitrary g, that gauge choice will be lost. To restore the gauge you have to introduce the local transformation $h(x)$ so that the rotation g can be compensated. It follows that h does not depend only on the coordinates x, but also on the vector v and the rotation g.

Therefore, equation (14.62) is called a nonlinear realization of symmetries, because h depends nonlinearly on v.

This is important for the following calculation, because we can fix a gauge, called triangular gauge, such that

$$v(x) = \exp \varphi(x), \quad \varphi(x) \in K \to v \in G/H. \tag{14.63}$$

The next step is the construction of a Lagrangian with the required symmetry. To this aim, let us consider the Lie algebra valued expression

$$v^{-1}\partial_m v = Q_m + P_m, \quad Q_m \in H, \ P_m \in K \tag{14.64}$$

or equivalently

$$v^{-1}D_m v = v^{-1}(\partial_m v - vQ_m) = P_m \tag{14.65}$$

which defines the H-covariant derivative D_m. It is straightforward to verify that Q_m transforms like a gauge field with respect to the local group H, namely $Q'_m = h^{-1}Q_m h + h^{-1}\partial_m h$ and that $P'_m = h^{-1}P_m h$. The formula (14.62) implies the integrability relations

$$\partial_m Q_m - \partial_n Q_m + [Q_m, Q_n] = -[P_m, P_n]$$
$$D_m P_n - D_n P_m = 0.$$

The Lagrangian is given by

$$\mathcal{L} = \tfrac{1}{4} e g^{mn} \operatorname{Tr} P_m P_n \qquad (14.66)$$

and then the field equations for P_m read

$$D_m(\sqrt{g}\, g^{mn} P_n) = 0. \qquad (14.67)$$

In the following section we will show how it is possible to reproduce the Lagrangians obtained by dimensional reduction from this general construction of nonlinear σ-models.

14.2.1   Ehlers Lagrangian as a nonlinear σ-model

To link these arguments to the previous discussion, let us consider the groups

$$G = SL(2, \mathbf{R}), \quad H = SO(2). \qquad (14.68)$$

The quotient space has only two degrees of freedom. We enforce the triangular gauge choosing for v the following expression

$$v = \begin{pmatrix} \Delta^{1/2} & B\Delta^{-1/2} \\ 0 & \Delta^{-1/2} \end{pmatrix} \qquad (14.69)$$

and then

$$v^{-1}\partial_m v = \begin{pmatrix} \tfrac{1}{2}\Delta^{-1}\partial_m \Delta & \Delta^{-1}\partial_m B \\ 0 & -\tfrac{1}{2}\Delta^{-1}\partial_m \Delta \end{pmatrix}$$

$$= P_m^1 Y^1 + P_m^2 Y^2 + Q_m Y^3 \qquad (14.70)$$

where the coefficients of the generators of the algebra are given by

$$P_m^1 = \tfrac{1}{2}\Delta^{-1}\partial_m \Delta, \quad P_m^2 = Q_m = \tfrac{1}{2}\Delta^{-1}\partial_m B. \qquad (14.71)$$

The evaluation of the Lagrangian is straightforward, and we get

$$\mathcal{L} = \tfrac{1}{4} e g^{mn} \operatorname{Tr} P_m P_n = \tfrac{1}{8} e g^{mn} \Delta^{-2}(\partial_m \Delta \partial_n \Delta + \partial_m B \partial_n B). \qquad (14.72)$$

This result matches exactly with the matter part of the Einstein–Hilbert Lagrangian found in the previous section. We have found that this expression can be directly reduced to two dimensions, and then, coupled to gravity, it becomes simply the Ehlers Lagrangian $\mathcal{L}_\mathrm{E}$ seen before.

The Ehlers Lagrangian after dimensional reduction is

$$\mathcal{L}_\mathrm{E} = \text{gravity} + \tfrac{1}{4}\rho e^{(2)} g^{\mu\nu} \operatorname{Tr}(P_\mu P_\nu)$$

$$= -\tfrac{1}{2}\lambda^{-1}\partial_\mu \lambda \partial^\mu \rho + \tfrac{1}{8}\rho e^{(2)}\Delta^{-2} g^{\mu\nu}(\partial_\mu \Delta \partial_\nu \Delta + \partial_\mu B \partial_\nu B). \quad (14.73)$$

We saw in the previous section that, by another type of dimensional reduction, we got a different reduced Lagrangian, the Matzner–Misner one.

This one can be constructed as a nonlinear σ-model too: we need only a different gauge choice, as we will see in the next section; before this, let us look at the equations of motion derived from the Ehlers Lagrangian.

14.2.2 The Ernst equation

The equations of motion for the fields Δ and B from the Lagrangian $\mathcal{L}_E$ are

$$\Delta \partial_\mu (\rho \partial^\mu \Delta) = \rho (\partial_\mu \Delta \partial^\mu \Delta - \partial_\mu B \partial^\mu B) \tag{14.74}$$

$$\Delta \partial_\mu (\rho \partial^\mu B) = 2\rho \partial^\mu \Delta \partial_\mu B. \tag{14.75}$$

Defining a complex function $\mathcal{E} = \Delta + iB$ called the Ernst potential, these equations can be combined into a single one, called the 'Ernst equation':

$$\Delta \partial_\mu (\rho \partial^\mu \mathcal{E}) = \rho \partial^\mu \mathcal{E} \partial_\mu \mathcal{E}. \tag{14.76}$$

This equation figures prominently in studies of exact solutions of Einstein's equations.

Here we have got the Ernst equation from the fields equations for Δ and B. Actually equation (14.67) *is* the Ernst equation, in the sense that it reduces to it choosing the Ehlers triangular form for v in the conformal gauge.

14.2.3 The Matzner–Misner Lagrangian as a nonlinear σ-model

Recalling the shape of the Matzner–Misner Lagrangian as written before in the conformal gauge

$$\mathcal{L}_{MM} = -\tfrac{1}{2}\lambda^{-1}\partial_\mu \lambda \partial^\mu \rho + \tfrac{1}{8}\rho \Delta^{-2}\partial_\mu \Delta \partial^\mu \Delta + \tfrac{1}{8}\rho^{-1}\Delta^2 \partial_\mu B_2 \partial^\mu B_2. \tag{14.77}$$

This can be thought of as a nonlinear σ-model, too. We suppose our Lagrangian to be composed by a term of pure gravity, but reduced to two dimensions, and a term coming from a two-dimensional nonlinear σ-model. We are in conformal gauge, namely $e_\mu{}^\alpha = \lambda \delta_\mu{}^\alpha$ and $\tilde{g}_{\mu\nu} = \eta_{\mu\nu}$. The Lagrangian is

$$\mathcal{L} = -\tfrac{1}{2}\tilde{\lambda}^{-1}\partial^\mu \tilde{\lambda} \partial_\mu \rho + \tfrac{1}{4}\rho \eta^{\mu\nu} \operatorname{Tr} \tilde{P}_\mu \tilde{P}_\nu. \tag{14.78}$$

We have to choose a proper gauge, namely an expression for v, such that the two Lagrangians match together.

We refer now to the generators of $SL(2, \mathbf{R})$ introduced at the beginning of this section. Let us choose for $\tilde{v}$ the following triangular form

$$\tilde{v} = \begin{pmatrix} (\rho/\Delta)^{1/2} & B_2(\Delta/\rho)^{1/2} \\ 0 & (\Delta/\rho)^{1/2} \end{pmatrix}, \quad \tilde{v}^{-1} = \begin{pmatrix} (\Delta/\rho)^{1/2} & -B_2(\Delta/\rho)^{1/2} \\ 0 & (\rho/\Delta)^{1/2} \end{pmatrix}.$$

Evaluating now the matrix product $\tilde{v}^{-1}\partial_\mu \tilde{v}$ and decomposing it on the algebra generators. Following the standard procedure seen before, the Lagrangian is built using only the non-compact elements of this decomposition. After calculation, we have

$$\tilde{v}^{-1}\partial_\mu \tilde{v} = \tilde{P}_\mu^1 + \tilde{P}_\mu^2 + \tilde{Q}_\mu = \frac{1}{2}(\rho^{-1}\partial_\mu \rho - \Delta^{-1}\partial_\mu \Delta) \begin{pmatrix} 1 & 0 \\ 0 & -1 \end{pmatrix}$$

$$+ \frac{1}{2}\left(\frac{\Delta}{\rho}\right)\partial_\mu B_2 \begin{pmatrix} 0 & 1 \\ 1 & 0 \end{pmatrix} + \frac{1}{2}\left(\frac{\Delta}{\rho}\right)\partial_\mu B_2 \begin{pmatrix} 0 & 1 \\ -1 & 0 \end{pmatrix}. \tag{14.79}$$

Then, the trace is

$$\tfrac{1}{4}\,\mathrm{Tr}\,\tilde{P}_\mu \tilde{P}^\mu = \tfrac{1}{4}(\tilde{P}_\mu^1 \tilde{P}^{\mu 1} + \tilde{P}_\mu^2 \tilde{P}^{\mu 2}) = \tfrac{1}{8}\rho^{-1}\partial_\mu\rho(\rho^{-1}\partial^\mu\rho - 2\Delta^{-1}\partial^\mu\Delta)$$

$$+ \frac{1}{8}\Delta^{-2}\partial_\mu\Delta\partial^\mu\Delta + \frac{1}{8}\left(\frac{\Delta}{\rho}\right)^2 \partial_\mu B_2 \partial^\mu B_2. \tag{14.80}$$

Now, the two Lagrangians coincide if $\tilde{\lambda}$ satisfies the condition

$$-\tfrac{1}{2}\tilde{\lambda}^{-1}\partial^\mu\tilde{\lambda}\partial_\mu\rho + \tfrac{1}{8}\rho^{-2}\partial_\mu\rho(\rho^{-1}\partial_\mu\rho - 2\Delta^{-1}\partial^\mu\Delta) = -\tfrac{1}{2}\lambda^{-1}\partial^\mu\lambda\partial_\mu\rho \tag{14.81}$$

namely if

$$\tilde{\lambda} \equiv \lambda\rho^{1/4}\Delta^{-1/2}. \tag{14.82}$$

Therefore, the two-dimensional reduced gravity in conformal gauge is given by a part of pure two-dimensional gravity, characterized by the conformal factor λ and the dilaton ρ, and a matter part given by the bosonic fields Δ and B, or $\tilde{B}$: this one has the structure of a nonlinear G/H sigma model.

Following the first section of this paper, the complete Lagrangian reduced to two dimensions in conformal gauge, for any G/H σ-model is

$$\mathcal{L} = -\tfrac{1}{2}\lambda^{-1}\partial^\mu\lambda\partial_\mu\rho + \tfrac{1}{4}\rho\,\mathrm{Tr}(P_\mu P^\mu) \tag{14.83}$$

and we can recover, as before, the field equation for the conformal factor λ, this time with the general σ-model matter part. It is given by *the traceless part of*

$$\lambda^{-1}\partial_\mu\lambda\partial_\nu\rho = \tfrac{1}{2}\,\mathrm{Tr}(P_\mu P_\nu) + \tfrac{1}{2}\partial_\mu\partial_\nu\rho. \tag{14.84}$$

This will be useful in the foregoing sections when recovering the colliding plane wave solutions of Einstein's theory.

The Kramer–Neugebauer transformation

Note now that the two models, that of Ehlers and that of Matzner–Misner, are related by the Kramer–Neugebauer transformation, defined by

$$\Delta \leftrightarrow \frac{\rho}{\Delta}, \quad B \leftrightarrow B_2$$

It is worth remembering that the fields B and B_2 are related by duality too, namely

$$\epsilon_{\mu\nu}\partial^\nu B = \frac{\Delta^2}{\rho}\partial_\mu B_2. \tag{14.85}$$

To sum up: in this section we have seen that the dimensional reduction of Einstein theory from $D = 4$ to $D = 2$ can be done in two ways, leading to two different $SL(2, \boldsymbol{R})/SO(2)$ σ-models.

We discover two different isometry groups, that of Ehlers and that of Matzner–Misner

$$SL(2, \boldsymbol{R})_\mathrm{E}, \quad SL(2, \boldsymbol{R})_\mathrm{MM}. \tag{14.86}$$

Combining these two groups, one gets the (infinite-dimensional) Geroch group.

14.3 Symmetries of nonlinear σ-models

We have seen that for preserving the gauge choice, in particular the triangular gauge, the symmetry

$$v \to v' = g^{-1}v, \quad g \in G \tag{14.87}$$

must be realized in a nonlinear way, namely

$$v \to v'(x) = g^{-1}v(x)h(x), \quad g \in G, \ h \in H. \tag{14.88}$$

Now consider the infinitesimal form of (14.88). The infinitesimal variation of v is

$$\delta v(x) = -\delta g^{-1}v(x) + v(x)\delta h(x) \tag{14.89}$$

applying now this linearized transformation to the two σ-models seen before.

Considering in particular the Chevalley–Serre generators for the $SL(2, \mathbb{R})$ Lie algebra

$$e \equiv T^+ = \begin{pmatrix} 0 & 1 \\ 0 & 0 \end{pmatrix}, \quad f \equiv T^- = \begin{pmatrix} 0 & 0 \\ 1 & 0 \end{pmatrix}, \quad h \equiv T^3 = \begin{pmatrix} 1 & 0 \\ 0 & -1 \end{pmatrix} \tag{14.90}$$

endowed with the following commutation rules

$$[h, e] = 2e, \quad [h, f] = -2f, \quad [e, f] = h \tag{14.91}$$

one can check that this nonlinear transformation has been introduced to preserve the gauge. Let us now analyse the action of the Ehlers and Matzner–Misner groups in turn.

14.3.1 Nonlinear realization of $SL(2, R)_E$

We use the Chevalley–Serre generators for the algebra. Considering the triangular gauge

$$v = \begin{pmatrix} \Delta^{1/2} & B\Delta^{-1/2} \\ 0 & \Delta^{-1/2} \end{pmatrix} \tag{14.92}$$

we now linearize the transformation (14.87). The variation of v is only due to the algebra element a:

$$\delta v = v' - v = -av. \tag{14.93}$$

Then, given the triangular form of v, it follows also

$$\delta v = \begin{pmatrix} \frac{1}{2}\Delta^{-1/2}\delta\Delta & -\frac{1}{2}\Delta^{-3/2}B\delta\Delta + \Delta^{-1/2}\delta B \\ 0 & -\frac{1}{2}\Delta^{-3/2}\delta\Delta \end{pmatrix}. \tag{14.94}$$

In the following we will refer to the variation $\delta\Delta$ by, for example, the generator e with the compact notation $e(\Delta)$ or $e(B)$ for B. Now, we realize the transformation using the Chevalley–Serre algebra generators, e, h and f. For e we have

$$e_1 : -\begin{pmatrix} 0 & 1 \\ 0 & 0 \end{pmatrix}\begin{pmatrix} \Delta^{1/2} & B\Delta^{-1/2} \\ 0 & \Delta^{-1/2} \end{pmatrix} = \begin{pmatrix} 0 & -\Delta^{-1/2} \\ 0 & 0 \end{pmatrix} \tag{14.95}$$

where the subscript 1 refers to the Ehlers group. From (14.94), one deduces

$$e_1(\Delta) = 0, \quad e_1(B) = -1 \tag{14.96}$$

and the triangular gauge is preserved. The calculation is analogous for the generator h

$$h_1 : -\begin{pmatrix} 1 & 0 \\ 0 & -1 \end{pmatrix}\begin{pmatrix} \Delta^{1/2} & B\Delta^{-1/2} \\ 0 & \Delta^{-1/2} \end{pmatrix} = \begin{pmatrix} -\Delta^{1/2} & -B\Delta^{-1/2} \\ 0 & \Delta^{-1/2} \end{pmatrix} \tag{14.97}$$

with $h_1(\Delta) = -2\Delta$ and $h_1(B) = -2B$. The triangular gauge is still preserved. This is not so for the third generator, f. Repeating the above steps we find

$$f_1 : -\begin{pmatrix} 0 & 0 \\ 1 & 0 \end{pmatrix}\begin{pmatrix} \Delta^{1/2} & B\Delta^{-1/2} \\ 0 & \Delta^{-1/2} \end{pmatrix} = \begin{pmatrix} 0 & 0 \\ -\Delta^{1/2} & -B\Delta^{-1/2} \end{pmatrix} \tag{14.98}$$

namely the triangular gauge is not preserved. Therefore, we have to introduce a compensating term, i.e. we need the transformation rule (14.89). We introduce a local H transformation parametrized by a function ω, which is determined in such a way as to preserve the gauge. Remember that the H generator is Y^3:

$$f_1 : -f_1 v + v(-\omega Y^3) = \begin{pmatrix} 0 & 0 \\ -\Delta^{1/2} & -B\Delta^{-1/2} \end{pmatrix} + \omega \begin{pmatrix} B\Delta^{-1/2} & -\Delta^{1/2} \\ \Delta^{-1/2} & 0 \end{pmatrix}. \tag{14.99}$$

The triangular gauge is defined by the condition

$$-\sqrt{\Delta} + \frac{\omega}{\sqrt{\Delta}} = 0 \rightarrow \omega = \Delta \tag{14.100}$$

and so the transformation reads

$$f_1 : \delta v = \begin{pmatrix} B\Delta^{1/2} & -\Delta^{3/2} \\ 0 & -B\Delta^{-1/2} \end{pmatrix}. \tag{14.101}$$

Hence the variations of the fields Δ and B are

$$f_1(\Delta) = 2\Delta B, \quad f_1(B) = B^2 - \Delta^2 \tag{14.102}$$

clearly not linear in the fields.

Note that the $SL(2, \boldsymbol{R})$ transformations leave the fields ρ and λ unchanged, i.e.

$$\delta\lambda = 0, \quad \delta\rho = 0.$$

14.3.2   Nonlinear realization of $\boldsymbol{S}L(2, R)_{\mathrm{MM}}$

On the other side, identical calculations can be done to evaluate the action of $SL(2, \boldsymbol{R})_{\mathrm{MM}}$ on the fields (Δ, B_2). Also in this case the symmetry is realized in a nonlinear way. We have (with the suffix 0 for Matzner–Misner)

$$e_0(\Delta) = 0, \quad e_0(B_2) = -1 \tag{14.103}$$

$$h_0(\Delta) = 2\Delta, \quad h_0(B_2) = -2B_2 \tag{14.104}$$

$$f_0(\Delta) = -2\Delta B_2, \quad f_0(B_2) = B_2^2 - \left(\frac{\rho}{\Delta}\right)^2. \tag{14.105}$$

Again, the generator f_0 acts nonlinearly.

14.4   The Geroch group

The aim of this section is to combine the two groups, $SL(2, \mathbb{R})_E$ with fields (Δ, B) and $SL(2, \mathbf{R})_{MM}$, with (Δ, B_2), into a unified group, the infinite-dimensional Geroch group. The associated Lie algebra is an affine Kac–Moody algebra.

We return first to duality relation

$$\rho^{-1}\Delta^2 \partial_\mu B_2 = \epsilon_{\mu\nu} \partial^\nu B \tag{14.106}$$

which is invariant under the Kramer–Neugebauer transformation. We need this equation because we now have to evaluate the action of $SL(2)_E$ on B_2 and of $SL(2)_{MM}$ on B.

14.4.1   Action of $SL(2, R)_E$ on $\tilde{\lambda}$, B_2

Keeping in mind that $\delta\rho = 0$, we have

$$B \to B + \delta B \Rightarrow \epsilon_{\mu\nu} \partial^\nu(\delta B) = \delta(\Delta^2 \rho^{-1} \partial_\mu B_2) \tag{14.107}$$

after the functional differentiation and the usage of duality

$$\partial_\mu(\delta B_2) = \rho \epsilon_{\mu\nu}(\partial^\nu \delta B - 2\Delta^{-3}\delta\Delta). \tag{14.108}$$

Consequently, from the change of B calculated before, we have the variation of B_2 due to the $SL(2)_E$ generators.

$$e_1 : 0 = \partial_\mu(\delta B_2) \Rightarrow e_1(B_2) = c_1 (= \text{constant}) \tag{14.109}$$

$$h_1 : \partial_\mu(\delta B_2) = 2\rho\Delta^{-2}\epsilon_{\mu\nu}\partial^\nu B \Rightarrow h_1(B_2) = 2B_2 \tag{14.110}$$

$$f_1 : \epsilon^{\mu\nu}\partial_\nu(\delta B_2) = 2\rho(\Delta^{-2}B\partial^\nu B + \Delta^{-1}\partial^\nu \Delta) \Rightarrow f_1(B_2) = 2\phi_1. \tag{14.111}$$

Here a dual potential ϕ_1 has been introduced, which is defined such that

$$\rho^{-1}\epsilon_{\mu\nu}\partial^\nu \phi_1 = \Delta^{-2}(B\partial_\mu B + \Delta\partial_\mu \Delta). \tag{14.112}$$

Careful inspection of these relations now shows the following. The contributions due to e_1 and h_1 are linear in the fields and local; the difference is in the transformation generated by f_1, which is clearly nonlinear and non-local, because one has to perform an integration to calculate explicitly the dual potential.

We then evaluate also the action on $\tilde{\lambda}$. From the definition of (14.82), and observing that $\delta\lambda = 0$, it follows that

$$\tilde{\lambda}^{-1}\delta\tilde{\lambda} = -\tfrac{1}{2}\Delta^{-1}\delta\Delta \qquad (14.113)$$

14.4.2   Action of $SL(2, R)_{\text{MM}}$ on λ, B

Exactly the same analysis has to be done for the other group, with generators (e_0, h_0, f_0)

$$e_0 :\Rightarrow e_0(B) = c_0 \qquad (14.114)$$
$$h_0 :\Rightarrow h_0(B) = 2B \qquad (14.115)$$
$$f_0 :\Rightarrow f_0(B) = 2\phi_0 \qquad (14.116)$$

with

$$\rho\epsilon_{\mu\nu}\partial^\nu\phi_0 = -\Delta^2 B_2\partial_\mu B_2 + \rho\Delta\partial_\mu\left(\frac{\rho}{\Delta}\right). \qquad (14.117)$$

14.4.3   The affine Kac–Moody $SL(2, R)$ algebra

The transformations we have just derived are to be identified with an affine $SL(2, R)$ Kac–Moody algebra. The latter is characterized by the Cartan matrix

$$A_{ij} = \begin{pmatrix} 2 & -2 \\ -2 & 2 \end{pmatrix} \qquad (14.118)$$

and the standard Chevalley–Serre presentation defining the algebra which can be read off from the Cartan matrix:

$$[h_i, h_j] = 0$$
$$[h_i, e_j] = A_{ij}e_j$$
$$[h_i, f_j] = -A_{ij}f_j$$
$$[e_i, f_j] = \delta_{ij}h_j$$
$$[e_i[e_i[e_i, e_j]]] = 0$$
$$[f_i[f_i[f_i, f_j]]] = 0.$$

Here $i = j = 0, 1$; note that there is no summation on repeated indices and that the first relation defines the Cartan subalgebra. To see the relation with the $SL(2, R)$ transformations dealt with before, we make the identifications

$$e_1 = T_0^+, \quad f_1 = T_0^-, \quad h_1 = T_0^3 \qquad (14.119)$$
$$e_0 = T_1^-, \quad f_0 = T_{-1}^+, \quad h_0 = c - T_0^3. \qquad (14.120)$$

Using the known commutation rules of the Chevalley–Serre generators of $SL(2, R)$ it is possible to directly check the algebra.

For example, it is straightforward to see that

$$[h_1, e_0] = [T_0^3, T_1^-] = -2T_1^- = -2e_0 \qquad (14.121)$$

or that

$$[e_1, e_0] = [T_0^+, T_1^-] = 2T_1^3 \qquad (14.122)$$

$$[e_1, [e_1, e_0]] = -4T_1^+ \Rightarrow [e_1[e_1[e_1, e_0]]] = 0$$

and so on for the other commutators.

The full current algebra is now built by taking multiple commutators in all possible ways. The Lie algebra element $c = h_0 + h_1$ is the central charge. It has a trivial action on the fields Δ, B, B_2, $\tilde{\lambda}$, λ.

14.5   The linear system

The aim of this section is to linearize and localize the action of the Geroch group seen in the previous section.

Let us start from the Lagrangian for arbitrary G/H in three dimensions and then reduce to two

$$\mathcal{L} = -\tfrac{1}{4}\rho e R(e) + \tfrac{1}{4}\rho e g^{mn} \operatorname{Tr} P_m P_n. \qquad (14.123)$$

We pick now the conformal gauge for the three-bein, as before

$$e_m^a = \begin{pmatrix} \lambda \delta_\mu^\alpha & 0 \\ 0 & \rho \end{pmatrix} \qquad (14.124)$$

where we have dropped the two-dimensional Kaluza–Klein vector because it carries no physical degrees of freedom any more. It is well known that the choice $e_\mu^\alpha = \lambda e_\mu^\alpha$ is preserved under conformal diffeomorphisms

$$\delta x^+ = \xi_-(x^+), \quad \delta x^- = \xi_+(x^-) \qquad (14.125)$$

with the light cone coordinates $x^\pm \equiv \frac{1}{\sqrt{2}}(x^0 \pm x^1)$. This residual coordinate freedom can be gauged away, for example, by employing the dilaton and the axion fields. One can fix the residual conformal diffeomorphisms by identifying the field ρ or $\tilde{\rho}$ with one of the coordinates.

14.5.1   Solving Einstein's equations

Let us now focus our attention on the way of solving Einstein's equation. First note that by substituting the gauge (14.124) into the scalar equation (14.67), we arrive at

$$\rho^{-1} D^\mu (\rho P_\mu) = 0. \qquad (14.126)$$

The dependence of this equation on ρ is all that remains of three-dimensional gravity. Equation (14.126) reduces to the Ernst equation for $G = SL(2, \mathbb{R})$, but we will return to this later. For the moment, note only that this equation works for the σ-model degrees of freedom, namely on Δ.

The remaining equations, which follow from higher dimensions, are the equations for the dilaton ρ and for the conformal factor λ. ρ is a free field in two dimensions which can be solved for in terms of two arbitrary functions (left-movers and right-movers)

$$\Box \rho = 0 \Rightarrow \rho(x) = \rho_+(x^+) + \rho_-(x^-). \qquad (14.127)$$

The equations of motion for the conformal factor in light-cone coordinates

$$\rho^{-1}\partial_\pm \rho \lambda^{-1}\partial_\pm \lambda = \tfrac{1}{2}\operatorname{Tr}(P_\pm P_\pm) + \tfrac{1}{2}\rho^{-1}\partial_\pm^2 \rho \qquad (14.128)$$

can be written as

$$\partial_\pm \rho \partial_\pm \hat{\sigma} = \tfrac{1}{2}\rho \operatorname{Tr} P_\pm P_\pm \qquad (14.129)$$

where the second term on the right-hand side of (14.128) has been reabsorbed into the Liouville scalar $\hat{\sigma} = \lambda(\partial_+\rho)^{-\frac{1}{2}}(\partial_-\rho)^{-\frac{1}{2}}$. Note that this equation determines λ only up to a constant factor. Observe also that this equation has no analogue in flat space theories, and this, together with the presence of ρ, makes a great difference. For instance, we cannot simply put $\rho = $ constant, for this would imply the vanishing of the right-hand side of (14.129), which by the positivity of the Killing metric on the subalgebra $\boldsymbol{K}$ would imply $P_\pm = 0$ and leave us only with the trivial solution $v = $ constant (modulo H gauge transformations).

Now specializing to general relativity, i.e. $G/H = SL(2, \mathbb{R})/SO(2)$ coset space. As anticipated before, we start from the equation of motion (14.126), employ the triangular gauge in the Ehlers form, so to have explicit expressions for P_μ and Q_ν. After a little algebra we have again the Ernst equation

$$\Delta \partial_\mu (\rho \partial^\mu \mathcal{E}) = \rho \partial_\mu \mathcal{E} \partial^\mu \mathcal{E} \qquad (14.130)$$

in terms of the complex potential $\mathcal{E} = \Delta + iB$. Solving Einstein's equations is now simply a matter of choosing the appropriate $\rho(x)$, finding a solution of the nonlinear partial differential equation (14.130) and finally determining the conformal factor λ by integration of (14.129). For the colliding plane wave solutions, that will be recovered in the next sections, one distinguishes waves with collinear polarization, where $B = 0$ and waves with non-collinear polarization. For collinearly polarized waves, the nonlinear Ernst equation can be reduced to a *linear* partial differential equation through the replacement $\Delta = \exp \psi$.

So, for collinearly polarized waves, with $B = 0$, the four-bein is

$$E_M{}^A = \begin{pmatrix} \lambda \Delta^{-1/2} & 0 & 0 & 0 \\ 0 & \lambda \Delta^{-1/2} & 0 & 0 \\ 0 & 0 & \rho \Delta^{-1/2} & 0 \\ 0 & 0 & 0 & \Delta^{1/2} \end{pmatrix} \qquad (14.131)$$

and then the four dimensional line element is

$$ds^2 = 2\Delta^{-1}\lambda^2\,dx^+\,dx^- - \Delta^{-1}\rho^2(dx^2)^2 - \Delta(dx^3)^2 \qquad (14.132)$$

where Δ, ρ and λ depend only on x^+ and x^-.

14.5.2 The linear system

The integrability of the nonlinear equation of motion (14.126) is reflected in the existence of a linear system. This means that there is a set of *linear* differential equations, whose compatibility conditions yield just the nonlinear equations that one tries to solve.

To formulate the linear system one must introduce a so-called spectral parameter t as an extra variable and replace $v(x)$ by a matrix $\hat{v}(x)$ which also depends on t.

$$v(x^0, x^1) \rightarrow \hat{v}(x^0, x^1; t). \qquad (14.133)$$

We postulate

$$\hat{v}^{-1}\partial_\mu\hat{v} = Q_\mu + \frac{1+t^2}{1-t^2}P_\mu + \frac{2t}{1-t^2}\epsilon_{\mu\nu}P^\nu. \qquad (14.134)$$

This is a generalization of $v^{-1}\partial_\mu v = Q_\mu + P_\mu$, which is obtained from (14.134) in the case $t = 0$. (14.134) is equivalent to

$$\hat{v}\partial_\pm\hat{v} = Q_\pm + \frac{1\mp t}{1\pm t}P_\pm. \qquad (14.135)$$

Here we have an integrability condition, written as

$$\partial_+(\hat{v}^{-1}\partial_-\hat{v}) - \partial_-(\hat{v}^{-1}\partial_+\hat{v}) + [\hat{v}^{-1}\partial_+\hat{v}, \hat{v}^{-1}\partial_-\hat{v}] = 0 \qquad (14.136)$$

which using (14.135) can be directly checked by calculation, making use of the integrability condition seen before and of the equation of motion for ρ. We define explicitly

$$\mathcal{A} = \partial_+(\hat{v}^{-1}\partial_-\hat{v}) - \partial_-(\hat{v}^{-1}\partial_+\hat{v}) \qquad (14.137)$$
$$\mathcal{B} = [\hat{v}^{-1}\partial_+\hat{v}, \hat{v}^{-1}\partial_-\hat{v}]. \qquad (14.138)$$

Employing (14.135) these relations become

$$\mathcal{A} = \partial_+Q_- - \partial_-Q_+ + \frac{1+t}{1-t}\partial_+P_- - \frac{1-t}{1+t}\partial_-P_+$$
$$+ \partial_+\left(\frac{1+t}{1-t}\right)P_- - \partial_-\left(\frac{1-t}{1+t}\right)P_+ \qquad (14.139)$$

$$\mathcal{B} = [Q_+, Q_-] + [P_+, P_-] + \frac{1+t}{1-t}[Q_+, P_-] - \frac{1-t}{1+t}[Q_-, P_+]. \quad (14.140)$$

The sum now reads

$$\mathcal{A} + \mathcal{B} = \frac{1+t}{1-t} D_+ P_- - \frac{1-t}{1+t} D_- P_+$$
$$+ \frac{2t}{(1-t)^2} t^{-1} \partial_+ t P_- + \frac{2t}{(1+t)^2} t^{-1} \partial_- t P_+ \qquad (14.141)$$

where the integrability relation seen in section 14.2 has been used. Now let us postulate

$$t^{-1} \partial_\pm t = \frac{1 \mp t}{1 \pm t} \rho^{-1} \partial_\pm \rho \qquad (14.142)$$

so it follows

$$\mathcal{A} + \mathcal{B} = \frac{1+t^2}{1-t^2} (D_+ P_- - D_- P_+) + \frac{2t}{1-t^2} (D_+ P_- + D_- P_+)$$
$$+ \frac{2t}{1-t^2} (\rho^{-1} \partial_+ \rho P_- + \rho^{-1} \partial_- \rho P_+). \qquad (14.143)$$

Now the first term is null for the integrability relation $D_+ P_- = D_- P_+$ and the second for the equation (14.126). Therefore, the integrability condition is checked.

Let us now focus on equation (14.142): it is integrable once one has a solution of $\Box \rho = 0$. This can be explicitly verified, as it follows. First, let us multiply (14.142) by $(1 - t^2)$; after a little algebra this equation reduces to

$$\partial_\pm \left[\rho \left(t + \frac{1}{t} \right) - 2\tilde{\rho} \right] = 0 \qquad (14.144)$$

where the axion $\tilde{\rho}$ has been introduced. So one must have

$$\frac{1}{2} \rho \left(t + \frac{1}{t} \right) - \tilde{\rho} = w \qquad (14.145)$$

where w is an integration constant. When we substitute in this relation the explicit expression of the dilaton and the axion as functions of incoming and outgoing fields, we get

$$t(x; w) = \frac{\sqrt{w + \rho_+(x^+)} - \sqrt{w - \rho_-(x^-)}}{\sqrt{w + \rho_+(x^+)} + \sqrt{w - \rho_-(x^-)}}. \qquad (14.146)$$

For fixed x, the function $t(x; w)$ lives on a two-sheeted Riemann surface over the complex w-plane, with an x-dependent cut extending from $\rho_-(x^-)$ to $\rho_+(x^+)$. The integration constant w can be regarded as an alternative spectral parameter.

The inverse of the spectral parameter is also important

$$y \equiv \frac{1}{w} = \frac{2t}{\rho(1+t^2) - 2t\tilde{\rho}} = \begin{cases} \dfrac{2t}{\rho} + \cdots, & t \sim 0 \\[2ex] \dfrac{2}{\rho t} + \cdots, & t \sim \infty \end{cases} \qquad (14.147)$$

where we consider the expansion around zero and infinity. What is the significance of the replacement (14.133)? A spectral parameter is required if one wants to enlarge the finite Lie group to its affine extension, and the appearance of t in (14.133) fits nicely with this expectation. There is now an infinite hierarchy of fields, as one can see by expanding $\hat{v}$ in t. For convenience let us pick a generalized triangular gauge, defined by the requirement that $\hat{v}$ should be regular at $t = 0$, or

$$\hat{v}(x; t) = \exp \sum_{n=0}^{\infty} t^n \varphi_n(x). \tag{14.148}$$

Another important feature of the linear system is the invariance under a generalization of the symmetric space automorphism. Let us define it for $\eta = 1$

$$\tau^{\infty} \hat{v}(t) = (\hat{v}^{\mathrm{T}})^{-1} \left(\frac{1}{t}\right). \tag{14.149}$$

In terms of the Lie algebra, the action of τ^{∞} reads

$$Q_{\mu} \rightarrow Q_{\mu}, \quad P_{\mu} \rightarrow -P_{\mu}. \tag{14.150}$$

It is straightforward to verify that

$$\tau^{\infty}(\hat{v}^{-1} \partial_{\mu} \hat{v}) = \hat{v}^{-1} \partial_{\mu} \hat{v}. \tag{14.151}$$

We can say that it is $\hat{v} \partial_{\mu} \hat{v} \in H^{\infty}$, which is the subalgebra of the Geroch group G_t^{∞} which is t^{∞}-invariant, as happens for finite-dimensional symmetric spaces. It is worth noticing that this property does not hold for $v^{-1} \partial_{\mu} v$ if we replace τ^{∞} with the transformation τ defined in section two.

14.5.3 Derivation of the colliding plane metric by factorization

At this point we can convince ourselves that the results obtained so far can be used to construct exact solutions of Einstein's equations. Of central importance for this task is the monodromy matrix, which is defined as follows

$$\mathcal{M} = \hat{v}(x; t) \hat{v}^{\mathrm{T}} \left(x; \frac{1}{t}\right). \tag{14.152}$$

A short calculation reveals that

$$\partial_{\mu} \mathcal{M} = \hat{v}(\hat{v}^{-1} \partial_{\mu} \hat{v} - \tau^{\infty}(\hat{v}^{-1} \partial_{\mu} \hat{v})) \tau^{\infty} \hat{v}^{-1} = 0 \tag{14.153}$$

where the relation (14.151) was used. Consequently, $\mathcal{M}$ can only depend on w. The solutions generating procedure now consists in choosing a matrix $\mathcal{M}(w)$ and finding a factorization as in (14.152).

The simplest non-trivial example, that will be considered here, permits us to recover the Ferrari–Ibanez colliding plane wave metric. Let us consider for this aim the monodromy matrix

$$\mathcal{M}(w) = \begin{pmatrix} \frac{w_0-w}{w_0+w} & 0 \\ 0 & \frac{w_0+w}{w_0-w} \end{pmatrix} \in SL(2, \mathbf{C}) \qquad (14.154)$$

and use

$$w - w_0 = -\frac{\rho}{4t_0}(t - t_0)\left(\frac{1}{t} - t_0\right) \qquad (14.155)$$

with the special value $w_0 = \frac{1}{2}$ and $t_0 \equiv t(x; w_0)$.

We use light cone coordinates, with the following notation to facilitate the comparison with the standard literature

$$u \equiv x^+, \quad v \equiv x^- \qquad (14.156)$$

then the remaining conformal invariance is entirely fixed by choosing the coordinates in such a way that

$$\rho_+(u) = \tfrac{1}{2}(1-2u^2), \quad \rho_-(v) = \tfrac{1}{2}(1-2v^2) \Rightarrow \rho(u, v) = 1-u^2-v^2 \quad (14.157)$$

where $\rho(u, v) > 0$ because the interaction region, where the waves collide, is $u^2 + v^2 < 1$. Substituting (14.155) into (14.154) and defining two particular solutions (14.146) in our gauge as

$$t_1(u, v) \equiv t\left(u, v; w = \frac{1}{2}\right) = \frac{\sqrt{1-u^2}-v}{\sqrt{1-u^2}+v} > 0 \qquad (14.158)$$

$$t_1(u, v) \equiv t\left(u, v; w = -\frac{1}{2}\right) = -\frac{\sqrt{1-v^2}+u}{\sqrt{1-v^2}-u} < 0 \qquad (14.159)$$

where the inequalities hold in the interaction region, we obtain in a straightforward way the desired factorization form for the monodromy matrix. Then it follows that

$$\hat{v}(u, v; t) = \begin{pmatrix} \sqrt{-\frac{t_2}{t_1}\frac{t-t_1}{t-t_2}} & 0 \\ 0 & \sqrt{-\frac{t_1}{t_2}\frac{t-t_2}{t-t_1}} \end{pmatrix}. \qquad (14.160)$$

Putting $t = 0$ we recover $v(u, v)$ in the triangular gauge, and then read directly the result for Δ by virtue of (14.69). We get

$$\Delta = -\frac{t_1}{t_2} = \frac{1-\xi}{1+\xi}, \quad B = 0 \qquad (14.161)$$

where the oblate spherical coordinates have been introduced

$$\xi \equiv u\sqrt{1-v^2} + v\sqrt{1-u^2} \qquad (14.162)$$

$$\eta \equiv u\sqrt{1-v^2} - v\sqrt{1-v^2}. \qquad (14.163)$$

From (14.161) we have $P_\pm^2 = Q_\pm = 0$, with $P_\pm^1 = \frac{1}{2}\Delta^{-1}\partial_\pm\Delta$. Putting equation (14.161) into this second relation, we gain

$$\frac{1}{2}\Delta^{-1}\partial_\pm\Delta = \frac{1}{2}t_1^{-1}\partial_\pm t_1 - \frac{1}{2}t_2^{-1}\partial_\pm t_2 \qquad (14.164)$$

which using the formulae

$$t^{-1}\partial_+ t = \rho^{-1}\partial_+\rho\frac{1-t}{1+t}, \qquad t^{-1}\partial_- t = \rho^{-1}\partial_-\rho\frac{1+t}{1-t} \qquad (14.165)$$

becomes

$$P_\pm^1 = \frac{1}{2}\rho^{-1}\partial_\pm\rho\left(\frac{1\mp t_1}{1\pm t_1} - \frac{1\mp t_2}{1\pm t_2}\right). \qquad (14.166)$$

Now we use the expression given here for P_+^1 to integrate the equation for the conformal factor. Some further calculations show that

$$\lambda^2 = 8uv\frac{(1-t_1t_2)^2}{(1-t_1^2)(1-t_2^2)} \qquad (14.167)$$

where the undetermined overall factor has been chosen for convenience. Then, this result yields the four-dimensional metric

$$ds^2 = (1+\xi)^2\left(\frac{d\xi^2}{1-\xi^2} - \frac{d\eta^2}{1-\eta^2}\right) - \rho^2\frac{1-\xi}{1+\xi}(dx^2)^2 - \frac{1+\xi}{1-\xi}(dx^3)^2. \qquad (14.168)$$

This is (a special case of) the so-called Ferrari–Ibanez colliding plane wave solution.

Acknowledgments

A Nagar wishes to thank Professor Pietro Fré and Professor Vittorio Gorini for their encouragement, and two students of Como University, Valentina Riva and Massimo Busetti, for their hospitality and the help given to him in arranging these lectures.

Further reading

Kramer D *et al* 1980 *Exact Solutions of Einstein Field Equations* (Cambridge: Cambridge University Press)

Hoensealers C and Dietz W (ed) 1984 *Solutions of Einstein's Equations: Techniques and Results* (Berlin: Springer)

Maison D 1978 *Phys. Rev. Lett.* **41** 521

Belinskii V A and Zakharov V E 1978 *Zh. Eksp. Teor. Fiz.* **48** 985

Breitenlohner P and Maison D 1987 *Ann. Inst. Poincaré* **46** 215

Julia B and Nicolai H 1996 *Nucl. Phys.* B **482** 431

Chapter 15

Gyroscopes and gravitational waves

Donato Bini[1] *and Fernando de Felice*[2]
[1] *Istituto per Applicazioni della Matematica, CNR, I-80131 Napoli, Italy and International Center for Relativistic Astrophysics, University of Rome, I-00185 Roma, Italy*
[2] *Dipartimento di Fisica 'G. Galilei', Università degli Studi di Padova, Via Marzolo, 8, I-35131, Padova, Italy and INFN, Sezione di Padova, Italy*

The behaviour of a gyroscope in geodesic motion is studied in the field of a plane gravitational wave. We find that, with respect to a special set of frames, the compass of inertia undergoes a precession which, to first order in the dimensionless amplitude h of the wave, is dominated by the cross-polarization alone. This suggests that a gyro might act as a filter of the polarization state of the wave.

15.1 Introduction

The (direct) detection of gravitational waves is still an open question, although indirect evidence for their existence has been obtained from the observation of the binary pulsar system PSR 1913+16 [1]. Besides the well-known bar antennae, there is a growing interest in laser interferometry detectors, like LIGO and VIRGO, which are sensitive to the low frequency ($\sim$10 Hz) gravitational waves which are emitted by sources like coalescing binaries.

The purpose of this paper is to study the behaviour of a test gyroscope which is acted upon by a plane gravitational wave with the purpose to see whether this interaction leads to observable effects. It is well known that in the absence of significant coupling between the background curvature and the multipole moments of the energy–momentum tensor of an extended body, the spin vector is Fermi–Walker transported along the body's own trajectory (see [2]

268

and references therein). The effects of a gravitational wave on a frame which is not Fermi–Walker transported, are best appreciated by studying the precession of a gyro at rest in that frame. Clearly it is essential to identify a class of frames which optimize the corresponding precession effect.

In section 15.2 we give a short review of the observer dependent spacetime splitting which enables one to describe in physical terms the motion of a test particle as well as that of a test gyroscope. In section 15.3 we discuss the spacetime of a plane gravitational wave and confine our attention to the family of static observers; we give an example of a tetrad frame adapted to these observers with respect to which the precession of a gyroscope is induced by the cross-polarization only. In section 15.4 we discuss the non-trivial problem of how to fix, in an operational and non-ambiguous way, a frame of reference which is not Fermi–Walker transported in the spacetime of a plane gravitational wave. Finally, in section 15.5 we calculate the precession of a gyroscope, in a general geodesic motion, with respect to the above frame.

In what follows, Greek indices run from 0–3, latin indices from 1–3.

15.2 Splitting formalism and test particle motion: a short review

A given family of test observers, namely a congruence of timelike lines with unit tangent vector field u (i.e. $u \cdot u = -1$) induces a splitting of the spacetime into space plus time through the orthogonal decomposition of the tangent space at each point into the local time direction along u and the local rest space LRS_u.

Projection of spacetime tensor fields onto LRS_u is accomplished using the projection operator

$$P_{(u)} = I + u \otimes u \tag{15.1}$$

and yields a family of spatial tensors (belonging to $\text{LRS}_u \otimes \cdots \otimes \text{LRS}_u$, i.e. for which any contraction with u vanishes). The collection of all the spatially projected tensor fields, associated to a given spacetime tensor field, will be referred to as the 'measure' of the spacetime tensor itself. For instance, the measure of the unit volume four-form $\eta_{\alpha\beta\mu\nu}$, gives only one non-trivial spatial field: $\epsilon_{(u)\alpha\beta\gamma} = u^\delta \eta_{\delta\alpha\beta\gamma}$ which can be used in turn to define the spatial cross product $\times_u$ in LRS_u.

One can also spatially project the various derivative operators so that the result of the derivative of any tensor field is itself a spatial tensor; examples are: the spatial Lie derivative, $\pounds_{(u)X} = P_{(u)} \pounds_X$ for any vector field X, the spatial covariant derivative $\nabla_{(u)\alpha} = P_{(u)} P_{(u)\alpha}^{\beta} \nabla_\beta$, the Lie temporal derivative, $\nabla_{(\text{lie},u)} = P_{(u)} \pounds_u$, the Fermi–Walker temporal derivative, $\nabla_{(\text{fw},u)} = P_{(u)} \nabla_u$ and several other natural derivatives for which a detailed discussion can be found in [2].

The measure of the covariant derivative of the four-velocity of the observers, gives rise to the kinematical coefficients of the observer congruence, namely the acceleration, vorticity expansion

$$a_{(u)}^\alpha = \nabla_{(fw,u)} u^\alpha,$$

$$\omega_{(u)\alpha\beta} = P_{(u)}{}^\gamma{}_\alpha P_{(u)}{}^\delta{}_\beta \nabla_{[\gamma} u_{\delta]},$$

$$\theta_{(u)\alpha\beta} = P_{(u)}{}^\gamma{}_\alpha P_{(u)}{}^\delta{}_\beta \nabla_{(\gamma} u_{\delta)}, \tag{15.2}$$

and the spatial dual of the vorticity field

$$\omega_{(u)}{}^\alpha = \tfrac{1}{2}\epsilon_{(u)}{}^{\alpha\beta\gamma} \omega_{(u)\beta\gamma}. \tag{15.3}$$

When dealing with different families of test observers, say u and U, the mixed projection map $P_{(U,u)} = P_{(u)} P_{(U)}$ from LRS_U to LRS_u (and the analogous compositions of two or more projectors) will be useful. Let ℓ_U be the world line of a nonzero rest mass test particle with U as its unit timelike tangent vector. The orthogonal decomposition of U relative to the family of test observers u, identifies its relative velocity $\nu_{(U,u)} = \nu\hat{\nu}_{(U,u)}$ where $\nu = \|\nu_{(U,u)}\| = \|\nu_{(u,U)}\|$ and $\hat{\nu}_{(U,u)}$ is the unit spatial vector, so that

$$U = \gamma[u + \nu\hat{\nu}_{(U,u)}]. \tag{15.4}$$

Here $\gamma = (1 - \nu^2)^{-1/2}$ is the local relative Lorentz factor. If the four acceleration of the particle

$$a_{(U)} = \nabla_U U = \frac{D}{d\tau_U} U$$

is non-vanishing, then its projection onto LRS_u, leads to the acceleration-equals-force equation:

$$P_{(U,u)} a_{(U)} \equiv \gamma F_{(U,u)}$$

where $F_{(U,u)}$ is the spatial *force* acting on the particle as *seen* by the observer u. In a similar way, one defines a *spatial gravitoinertial force*

$$F^{(G)}_{(fw,U,u)} = -\gamma^{-1} P_{(u)} \frac{Du}{d\tau_U}$$

$$= -P_{(u)} \frac{Du}{d\tau_{(U,u)}}$$

$$= \gamma[g_{(u)} + \nu(\tfrac{1}{2}\hat{\nu}_{(U,u)} \times_u H_{(u)} - \theta_{(u)} L\hat{\nu}_{(U,u)})], \tag{15.5}$$

where τ_U is a proper time parametrization for U and $\tau_{(U,u)} = \int_{\ell_U} \gamma\, d\tau_U$ is the corresponding Cattaneo relative standard time parametrization; $g_{(u)} = -a_{(u)}$ and $H_{(u)} = 2\omega_{(u)}$ are, respectively, the electric- and magnetic-like components of the gravitoinertial force. This terminology is justified by the Lorentz form of the gravitoinertial force which appears in the last of equations (15.5).

If we define $p_{(U,u)} = \gamma\nu_{(U,u)}$, $E_{(U,u)} = \gamma$ and

$$\frac{D_{(fw,U,u)}}{d\tau_{(U,u)}} = \gamma^{-1} P_{(u)} \frac{D}{d\tau_U} = \nabla_{(fw,u)} + \nu^\alpha_{(U,u)} \nabla(u)_\alpha \tag{15.6}$$

the latter being the measure of the (rescaled) absolute derivative along U, the $(3+1)$ version of the equation of motion of the particle and of the energy theorem, acquires the Newtonian form

$$\frac{D_{(fw,U,u)}}{d\tau_{(U,u)}} P_{(U,u)} = F_{(U,u)} + F^{(G)}_{(fw,U,u)},$$

$$\frac{dE_{(U,u)}}{d\tau_{(U,u)}} = [F_{(U,u)} + F^{(G)}_{(fw,U,u)}] \cdot_u \nu_{(U,u)}. \tag{15.7}$$

Let us now consider the motion of a test gyroscope. As it is well known, the spin vector $S_{(U)}$ of a gyroscope carried by an observer U, is Fermi–Walker transported along his worldline (i.e. $S_{(U)}$ does not precess with respect to spatial axes which are Fermi–Walker dragged along U), namely:

$$\frac{D_{(fw,U)}}{d\tau_U} S_{(U)} = P_{(U)} \frac{D}{d\tau_U} S_{(U)} = 0. \tag{15.8}$$

Suppose that we have chosen a spatial triad $\bar{e}(U)_{\hat{a}}$ which is adapted to the observer U and is *not* a Fermi–Walker frame. The observer U will then *see* the spin $S(U)$ of the gyroscope to precess with respect to these axes according to the law:

$$\left[\frac{dS^{\hat{a}}_{(U)}}{d\tau_U} - \epsilon^{\hat{a}}{}_{\hat{b}\hat{c}} \zeta^{\hat{b}}_{(fw,U,\bar{e}(U)_{\hat{a}})} S^{\hat{c}}_{(U)} \right] \bar{e}(U)_{\hat{a}} = 0 \tag{15.9}$$

where

$$\zeta^{\hat{a}}_{(fw,U,\bar{e}(U)_{\hat{a}})} \equiv \epsilon^{\hat{a}\hat{b}\hat{c}} \bar{e}(U)_{\hat{b}} \cdot \nabla_{(tw,U)} \bar{e}(U)_{\hat{c}} \tag{15.10}$$

is the precession rate vector.

However, we may want the gyroscope to be analysed by a different observer, u say, who is not comoving with the gyro's centre of mass. In this case we need a smooth family of these observers, each one intersecting the gyro's worldline at any of its spacetime points where he *measures* the instantaneous precession of the spin vector relative to a suitably defined frame, adapted to u. Of course, we require that the observer's u are synchronized so that their measurements can be compared. The results of these measurements are described by a smooth and at least once differentiable function of the proper-time of u.

Let $\{e(u)_{\hat{a}}\}$ be a field of spatial triads adapted to u; then the restriction of $\{e(u)_{\hat{a}}\}$ to the worldline ℓ_U of the gyroscope, allows one to define on ℓ_U a field of tetrad frames, adapted to U, given by $\{(U, \bar{e}(U)_{\hat{a}}\}$, where:

$$\bar{e}(U)_{\hat{a}} = B_{(lrs,u,U)} e(u)_{\hat{a}}, \tag{15.11}$$

$B_{(lrs,u,U)} = P_{(U)} B_{(u,U)} P_{(u)} : LRS_u \rightarrow LRS_U$ being the boost map between the rest spaces of the observers U and u; this map has been studied extensively in [2–4]. Since the boost is an isometry, the precession of $S_{(U)}$ with respect to the axes $\bar{e}(U)_{\hat{a}}$ is the same as the precession, with respect to the axes $e(u)_{\hat{a}}$, of the boosted spinvector $s_{(u)}$, which reads:

$$s_{(u)} = B_{(\text{lrs},U,u)}S_{(U)} = \left[P_{(u)} - \frac{\gamma}{\gamma+1}\nu_{(U,u)} \otimes \nu_{(U,u)}\right]P_{(U,u)}S_{(U)}. \quad (15.12)$$

Hence, from (15.9) and (15.11) and the acceleration-equals-force equation for U; we find:

$$\left[\frac{\mathrm{d}s_{(u)}^{\hat{a}}}{\mathrm{d}\tau_U} - \gamma\epsilon^{\hat{a}}{}_{\hat{b}\hat{c}}[\zeta_{(\text{fw},U,u)} + \zeta_{(\text{sc,fw},U,u)}]^{\hat{b}}s_{(u)}^{\hat{c}}\right]e(u)_{\hat{a}} = 0 \quad (15.13)$$

where[1]

$$\zeta_{(\text{fw},U,u)} = \frac{1}{\gamma+1}\nu_{(U,u)} \times_u [F^{(G)}_{(\text{fw},U,u)} - \gamma F_{(U,u)}]$$

$$\zeta_{(\text{sc,fw},U,u)} = \frac{1}{2}\delta^{\hat{a}\hat{b}}\frac{D_{(\text{fw},U,u)}}{\mathrm{d}\tau_{(U,u)}}e(u)_{\hat{a}} \times_u e(u)_{\hat{b}} \quad (15.14)$$

so that

$$\zeta_{(\text{fw},U,\bar{e}(U)_{\hat{a}})} = \gamma B_{(\text{lrs},u,U)}[\zeta_{(\text{fw},U,u)} + \zeta_{(\text{sc,fw},U,u)}]. \quad (15.15)$$

Finally, by rescaling equation (15.13) to the proper-time of u, one has

$$\frac{\mathrm{d}s_{(u)}^{\hat{a}}}{\mathrm{d}\tau_{(U,u)}} - \epsilon^{\hat{a}}{}_{\hat{b}\hat{c}}[\zeta_{(\text{fw},U,u)} + \zeta_{(\text{sc,fw},U,u)}]^{\hat{b}}s_{(u)}^{\hat{c}} = 0 \quad (15.16)$$

where

$$\zeta_{(\text{fw},U,u)} + \zeta_{(\text{sc,fw},U,u)} \equiv \tilde{\zeta}_{(\text{fw},u,e(u)_{\hat{a}})} = \gamma^{-1}B_{(\text{lrs},U,u)}\zeta_{(\text{fw},U,\bar{e}(U)_{\hat{a}})} \quad (15.17)$$

is the angular velocity precession of the gyroscope as *measured* by the observer u with respect to the axes $e(u)_{\hat{a}}$.

It is worth mentioning here that while the observer U, who is comoving with the gyro's centre of mass, measures the precession (15.10) along his own worldline, the observer's u can only compare the instantaneous measurements of $\tilde{\zeta}_{(\text{fw},u,e(u)_{\hat{a}})}$ in (15.17), made by each of them along the gyro's worldline. Evidently either type of measurements requires the tetrad frames to be operationally well defined. This will be discussed in the following section.

15.3 The spacetime metric

The metric of a plane monochromatic gravitational wave, elliptically polarized and propagating along a direction which we fix as the coordinate x direction, can be written in the 'TT' gauge as [5]:

$$\mathrm{d}s^2 = -\mathrm{d}t^2 + \mathrm{d}x^2 + (1 - h_{22})\,\mathrm{d}y^2 + (1 + h_{22})\,\mathrm{d}z^2 - 2h_{23}\,\mathrm{d}y\,\mathrm{d}z \quad (15.18)$$

---

[1] This notation for the Fermi–Walker relative angular velocity $\zeta_{(\text{fw},U,u)}$ and the Fermi–Walker space-curvature relative angular velocity $\zeta_{(\text{sc,fw},U,u)}$ has been introduced in [2].

with $h_{AB} = h_{AB}(t - x)$, $(A, B = 2, 3)$. Let u denote the tangent vector field of a family of geodesic, non-rotating and expanding observers defined by

$$u^b = -dt, \quad u = \partial_t. \tag{15.19}$$

One can adapt to these observers an infinite number of spatial frames, by rotating any given one arbitrarily. For example, consider the following u-frame $\{e_{\hat{\alpha}}\} = \{e_{\hat{0}} = u, e_{\hat{a}} = e(u)_{\hat{a}}\}$ with its dual $\{\omega^{\hat{\alpha}}\} = \{\omega^{\hat{0}} = -u^b, \omega^{\hat{a}} = \omega(u)^{\hat{a}}\}$

$$u = \partial_t$$

$$e(u)_{\hat{1}} = \partial_x$$

$$e(u)_{\hat{2}} = (1 - h_{22})^{-1/2}\partial_y \simeq (1 + \tfrac{1}{2}h_{22})\partial_y$$

$$e(u)_{\hat{3}} = (1 - h_{22}^2 - h_{23}^2)^{-1/2}[(1 - h_{22})^{-1/2}h_{23}\partial_y + (1 - h_{22})^{1/2}\partial_z]$$

$$\simeq h_{23}\partial_y + (1 - \tfrac{1}{2}h_{22})\partial_z \tag{15.20}$$

$$-u^b = dt$$

$$\omega(u)^{\hat{1}} = dx$$

$$\omega(u)^{\hat{2}} = (1 - h_{22})^{1/2}\left[dy - \frac{h_{23}}{1 - h_{22}}dz\right] \simeq \left(1 - \tfrac{1}{2}h_{22}\right)dy - h_{23}\,dz$$

$$\omega(u)^{\hat{3}} = \left(\frac{1 - h_{22}^2 - h_{23}^2}{1 - h_{22}}\right)^{1/2}dz \simeq \left(1 + \tfrac{1}{2}h_{22}\right)dz,$$

where $\simeq$ denotes the corresponding weak-field limit. Any other spatial frame $\{\tilde{e}(u)_{\hat{a}}\}$, adapted to the observers (15.19), can be obtained from this one by a spatial rotation R

$$\tilde{e}(u)_{\hat{a}} = e(u)_{\hat{b}}R^{\hat{b}}{}_{\hat{a}}. \tag{15.21}$$

Among all the possible frames, there exists only one with respect to which the local compass of inertia experiences no precession. This frame is Fermi–Walker transported along u, namely it satisfies the condition

$$P_{(u)}\frac{D}{d\tau_u}\tilde{e}(u)_{\hat{a}} \equiv \nabla_{(\text{fw},u)}\tilde{e}(u)_{\hat{a}} = 0. \tag{15.22}$$

A Fermi–Walker frame is the most natural of the u-frames; its spatial directions, in fact, are fixed by three mutually orthogonal axes of small size comoving gyroscopes. However, if a metric perturbation causes a dragging of the local compass of inertia, the only way to detect and measure it, is to select a frame which is *not* Fermi–Walker transported. In this frame, in fact, a gyroscope would be seen to precess and indeed its precession contains all the informations about gravitational dragging. Nonetheless, it is quite non-trivial to identify, in an operational way, a frame which is *not* Fermi–Walker transported along the observer's worldline when it is acted upon by a gravitational wave.

Frame (15.20) is clearly Fermi–Walker transported in the absence of gravitational waves (h_{22} and h_{23} being time independent), but it is not so when they are present. The Fermi rotation of the frame, in this case, is described by the (antisymmetric) angular velocity spatial tensor [6]:

$$C_{(u)\hat{b}\hat{a}} = e(u)_{\hat{b}} \cdot_u \nabla_{(fw,u)} e(u)_{\hat{a}}, \qquad (15.23)$$

hence, a gyroscope carried by the observer u will precess with respect to frame (15.20) with an angular velocity tensor which has only one independent nonzero component, namely:

$$C_{(u)\hat{3}\hat{2}} = -\frac{[h_{23,t}(1 - h_{22}) + h_{22,t}h_{23}]}{2(1 - h_{22})\sqrt{1 - h_{22}^2 - h_{23}^2}} \simeq -\frac{1}{2}h_{23,t}. \qquad (15.24)$$

However, frame $e(u)_{\hat{a}}$ cannot be operationally defined, so result (15.24) is of little physical significance although it shows the existence of frames which respond to one state of polarization only, at least to first order in h_{AB}. We are, therefore, motivated to search for 'frames' that can be fixed from a viable experimental set up.

15.4   Searching for an operational frame

Let us consider the timelike geodesics of the metric (15.18). These are well known [7]; the four-velocity of a general such geodesic can be written as

$$U_g = \frac{1}{2E}[(1 + f + E^2)\partial_t + (1 + f - E^2)\partial_x]$$
$$+ \frac{1}{1 - h_{22}^2 - h_{23}^2}\{[\alpha(1 + h_{22}) + \beta h_{23}]\partial_y + [\beta(1 - h_{22}) + \alpha h_{23}]\partial_z\},$$
$$(15.25)$$

where α, β and E are Killing constants and $f = g_{AB}U^A U^B$ is equal to

$$f = \frac{1}{1 - h_{22}^2 - h_{23}^2}[\alpha^2(1 + h_{22}) + \beta^2(1 - h_{22}) + 2\alpha\beta h_{23}]$$
$$\simeq \alpha^2(1 + h_{22}) + \beta^2(1 - h_{22}) + 2\alpha\beta h_{23}. \qquad (15.26)$$

If $u = \partial_t$ is the family of observers who make the mesurements and $\{e(u)_{\hat{a}}\}$ is an adapted spatial frame, then the relative velocity $v_{(U_g,u)}{}^{\hat{a}}$ of U_g with respect to u is defined by the relation

$$U_g = \gamma_{(U_g,u)}[u + v_{(U_g,u)}{}^{\hat{a}}e(u)_{\hat{a}}], \qquad (15.27)$$

where

$$\gamma_{(U_g,u)} = \frac{1 + f + E^2}{2E}$$

$$\simeq \frac{1}{2E}[1 + E^2 + \alpha^2(1 + h_{22}) + \beta^2(1 - h_{22}) + 2\alpha\beta h_{23}] \quad (15.28)$$

and the relative velocity components can be obtained by comparison of (15.28) and (15.25).

Hereafter, we restrict ourselves to the weak field approximation (first order in h_{AB}) and assume, without loss of generality that, in the absence of a gravitational wave, the spatial velocity $v_{(U_g,u)}$, was only in the y coordinate direction. This corresponds to the requirement that

$$\beta = 0, \quad E = \sqrt{1 + \alpha^2}. \quad (15.29)$$

In this case we find

$$v_{(U,u)}{}^{\hat{1}} = \frac{\alpha^2}{2(1 + \alpha^2)} h_{22},$$

$$v_{(U,u)}{}^{\hat{2}} = \frac{\alpha}{\sqrt{1 + \alpha^2}}\left[1 + \frac{h_{22}}{2(1 + \alpha^2)}\right], \quad (15.30)$$

$$v_{(U,u)}{}^{\hat{3}} = \frac{\alpha}{\sqrt{1 + \alpha^2}} h_{23}.$$

We now require that the four-velocity of a test gyroscope is $U = U_g$ as given by (15.25) with conditions (15.29). The assumption of geodesicity is justified if we consider the limit of zero size gyroscopes. In this case, in fact, not only can we neglect the multipole moments of the gyro's stress tensor higher than the dipole, but also ignore the tidal term which enters the Papapetrou–Dixon equations and arises from the coupling of the gyro's spin with the background curvature. This term is of the order of the ratio of the average size of the gyroscope and the gravitational-wave wavelength.

Let us consider the restriction of the vector field $u = \partial_t$ of stationary observers, to the gyroscope's worldline and require that these observers *monitor* the behaviour of the spin of the moving gyro, measuring the (instantaneous) precession vector $\tilde{\zeta}$ given by equation (15.17). As already mentioned, the spatial frame $e(u)_{\hat{a}}$ given by (15.20) is not operationally well defined, so we have to find one which is so.

To find a spatial frame which is suitable for actual experiments, note that the observer's u can unambiguously determine in their rest-frame, a spatial direction given by that of the relative velocity v of the gyroscope. Suppose they fix, by guessing if necessary, a direction of propagation of the gravitational wave and term this as the x-axis with unit vector $e(u)_{\hat{1}}$. Then from these two directions, namely that of the relative velocity of the gyro and of the wave propagation, it is possible to construct a spatial triad as follows

$$\lambda(u)_{\hat{1}} = e(u)_{\hat{1}} = \partial_x$$
$$\lambda(u)_{\hat{2}} = \hat{\nu}_{(U,u)} \times {}_u e(u)_{\hat{1}}$$
$$\lambda(u)_{\hat{3}} = \lambda(u)_{\hat{1}} \times {}_u \lambda(u)_{\hat{2}}. \qquad (15.31)$$

This frame can be operationally constructed apart from guessing the direction of propagation of the wave. Indeed such a guess is also required to fit data from bar antenna detectors, for example. Obviously this frame is not unique: any other spatial triad obtained from it after a rotation which depends at most on the (known) modulus of the relative velocity (or is constant) is equally useful.

The spatial triads $e(u)_{\hat{a}}$ in (15.20) and $\lambda(u)_{\hat{a}}$ in (15.31), differ by a rotation

$$\lambda(u)_{\hat{a}} = \mathcal{R}^{\hat{b}}{}_{\hat{a}} e(u)_{\hat{b}}.$$

In the weak field limit, the only non-trivial components of $\mathcal{R}^{\hat{b}}{}_{\hat{a}}$, are

$$\mathcal{R}^{\hat{1}}{}_{\hat{1}} = -\mathcal{R}^{\hat{2}}{}_{\hat{3}} = \mathcal{R}^{\hat{3}}{}_{\hat{2}} = 1, \quad \mathcal{R}^{\hat{2}}{}_{\hat{2}} = \mathcal{R}^{\hat{3}}{}_{\hat{3}} = h_{23}$$

so that

$$\lambda(u)_{\hat{1}} = \partial_x$$
$$\lambda(u)_{\hat{2}} \simeq h_{23} e(u)_{\hat{2}} - e(u)_{\hat{3}} \simeq -(1 - \tfrac{1}{2} h_{22}) \partial_z$$
$$\lambda(u)_{\hat{3}} \simeq e(u)_{\hat{2}} + h_{23} e(u)_{\hat{3}} \simeq (1 + \tfrac{1}{2} h_{22}) \partial_y + h_{23} \partial_z. \qquad (15.32)$$

15.5　Precession of a gyroscope in geodesic motion

The precession of the gyro which is measured by the observer's u all along its worldline, is the image of the precession measured by the comoving observer U, under the boost $B(U, u)$ as shown in (15.17). In order to study the spin precession seen by the observer comoving with the gyro, we must first decide with respect to what axes (non-Fermi–Walker transported but operationally well defined) the precession will be measured, as explained in section 15.2.

Since the observer's u intersect the worldline of the observer U carrying the gyro, the u-frames $\{\lambda(u)_{\hat{a}}\}$ in (15.31) form a smooth field of frames on it, so the observer U can identify spatial directions in his rest space simply by boosting the directions $\lambda(u)_{\hat{a}}$.

At each event along his worldline in fact he will see the axes $\lambda(u)_{\hat{a}}$ defined in (15.31) to be in relative motion, therefore the boost of these axes, namely

$$\bar{\lambda}(U)_{\hat{a}} = B_{(\mathrm{lrs}, u, U)} \lambda(u)_{\hat{a}} = \lambda(u)_{\hat{a}} + \frac{\gamma}{\gamma + 1} [\nu_{(U,u)} \cdot \lambda(u)_{\hat{a}}](u + U), \qquad (15.33)$$

from (15.11), will be *seen* by U as the corresponding axes with the same orientation which are 'momentarily at rest'. The orientation of the spin vector S with respect to the axes $\bar{\lambda}(U)_{\hat{a}}$ is also the orientation of S with respect to the moving axes $\lambda(u)_{\hat{a}}$.

The velocity of spin precession then corresponds to the spatial dual of the Fermi–Walker structure functions of $\bar{\lambda}(U)_{\hat{a}}$, namely $C_{(\text{fw})}(U, \bar{\lambda}(U)_{\hat{a}})_{\hat{b}\hat{a}}$, according to relation (15.23).

Confining our attention to the weak-field approximation, the components of the precession velocity with respect to the triad $\bar{\lambda}(U)_{\hat{a}}$ are

$$\zeta_{(\text{fw},U,\bar{\lambda}(U)_{\hat{a}})}{}^{\hat{1}} \simeq -\tfrac{1}{2}h_{23,t}$$

$$\zeta_{(\text{fw},U,\bar{\lambda}(U)_{\hat{a}})}{}^{\hat{2}} \simeq \alpha/2 h_{22,t}$$

$$\zeta_{(\text{fw},U,\bar{\lambda}(U)_{\hat{a}})}{}^{\hat{3}} \simeq \alpha/2 h_{23,t}. \tag{15.34}$$

We observe that in the limit of small linear momentum, $\alpha \ll 1$, the dominant precession is in the direction of wave propagation $e(u)_{\hat{1}}$ (to zeroth order, $\bar{\lambda}(U)_{\hat{1}} \simeq e(u)_{\hat{1}}$) and is induced by the cross-polarization only. (Note that the precession in the direction of propagation of the wave does not depend on α.) In this case we can conclude that the gyro can act as a polarization filter for gravitational waves.

In the opposite limit of large linear momentum, $\alpha \gg 1$, the precession vector lies mainly in the plane orthogonal to the propagation direction and is contributed likewise by both polarizations. Indeed the measurement of the precession induced by a plane gravitational wave, of a gyroscope set in relativistic motion, would enable one to identify the local direction of propagation of the wave. A similar situation will be encountered in the rest frame of u, where from (15.17), (15.28) and (15.29) we have:

$$\tilde{\zeta}_{(\text{fw},u,\lambda(u)_{\hat{a}})}^{\hat{1}} = \gamma_{(U,u)}^{-1} B_{(\text{lrs},U,u)} \zeta_{(\text{fw},U,\bar{\lambda}(U)_{\hat{a}})}{}^{\hat{1}} \simeq -\frac{1}{2}\frac{1}{\sqrt{1+\alpha^2}}h_{23,t}$$

$$\tilde{\zeta}_{(\text{fw},u,\lambda(u)_{\hat{a}})}^{\hat{2}} = \gamma_{(U,u)}^{-1} B_{(\text{lrs},U,u)} \zeta_{(\text{fw},U,\bar{\lambda}(U)_{\hat{a}})}{}^{\hat{2}} \simeq \frac{1}{2}\frac{\alpha}{\sqrt{1+\alpha^2}}h_{22,t}$$

$$\tilde{\zeta}_{(\text{fw},u,\lambda(u)_{\hat{a}})}^{\hat{3}} = \gamma_{(U,u)}^{-1} B_{(\text{lrs},U,u)} \zeta_{(\text{fw},U,\bar{\lambda}(U)_{\hat{a}})}{}^{\hat{3}} \simeq \frac{1}{2}\frac{\alpha}{\sqrt{1+\alpha^2}}h_{23,t}. \tag{15.35}$$

Finally, let us note that results (15.35) are only slightly modified after rotating the frame (15.31) by a constant angle ϕ around the propagation direction of the wave. In fact, in this case, the new spatial u-frame becomes

$$f(u)_{\hat{1}} = \lambda(u)_{\hat{1}}$$
$$f(u)_{\hat{2}} = \cos\phi\lambda(u)_{\hat{2}} + \sin\phi\lambda(u)_{\hat{3}}$$
$$f(u)_{\hat{3}} = -\sin\phi\lambda(u)_{\hat{2}} + \cos\phi\lambda(u)_{\hat{3}}; \tag{15.36}$$

when $\phi = 0$ it reduces to (15.31). Once the boosted frame $\bar{f}(U)_{\hat{a}} = B_{(\text{lrs},U,u)} f(u)_{\hat{a}}$ in LRS_U is obtained, the components of the precession velocity turn out to be

$$\zeta_{(\text{fw},U,\bar{f}(U)_{\hat{a}})}{}^{\hat{1}} \simeq -\tfrac{1}{2}h_{23,t}$$

$$\zeta_{(\text{fw},U,\bar{f}(U)_{\hat{a}})}{}^{\hat{2}} \simeq \alpha/2[\cos\phi h_{22,t} - \sin\phi h_{23,t}]$$

$$\zeta_{(\text{fw},U,\bar{f}(U)_{\hat{a}})}{}^{\hat{3}} \simeq \alpha/2[-\sin\phi h_{22,t} + \cos\phi h_{23,t}], \qquad (15.37)$$

again showing that, in the limit of small linear momentum α, the precession mainly occurs about the direction of propagation of the wave.

15.6 Conclusions

We have operationally defined a tetrad frame adapted to a family of static observers in the background of a plane gravitational wave. Then we have used this family to study the precession angular velocity of a gyroscope moving along a spacetime geodesic. The results show that, to first order in h and in the case of non-relativistic motion ($\alpha \ll 1$), the observed gyroscopic precession is mainly induced by the cross-polarization only so a gyro appears to behave as a polarization filter.

Assuming the form $h_{23} = h_{\times}\sin(\frac{2\pi c}{\lambda_{\text{GW}}}(t - x/c) + \psi)$ for the cross-polarization, with an obvious meaning for the symbols, the precession frequency (in conventional units) would be

$$\Omega_{(\text{gyro})}(t) \simeq -\frac{\pi c h_{\times}}{\lambda_{\text{GW}}}\cos\left(\frac{2\pi c}{\lambda_{\text{GW}}}(t - x/c) + \psi\right). \qquad (15.38)$$

The values of the precession frequency are very small, as expected. With a typical amplitude of 10^{-21} at the Earth and a frequency of 10^3 Hz, we could hope for a maximum precession of the order of 10^{-18} s^{-1}.

Clearly the precession effect is larger with high-frequency gravitational waves or when the gyroscope is close enough to the wave source to allow for a higher value of $h_{\times}$. This situation is encountered by a spinning neutron star, say, in a compact binary system.

The type of analysis we have considered, is most suitable to describe the interaction of a moving gyroscope with a *continuous* flow of plane gravitational waves with metric form as in (15.18). These are expected to be emitted by sources like compact binaries. If we have an impulsive source, like a supernova, then one expects a burst of gravitational radiation which is better described by a gravitational sandwich. This latter case is now under investigation.

References

[1] Taylor J H and Weisberg J M 1989 *Astrophys. J.* **345** 434
[2] Jantzen R T, Carini P and Bini D 1992 *Ann. Phys.* **215** 1
[3] Bini D, Carini P and Jantzen R T 1997a *Int. J. Mod. Phys.* D **6** 1
[4] Bini D, Carini P and Jantzen R T 1997b *Int. J. Mod. Phys.* D **6** 143
[5] Misner C W, Thorne K S and Wheeler J A 1973 *Gravitation* (New York: Freeman)
[6] de Felice F and Clarke C J S 1990 *Relativity on Curved Manifolds* (Cambridge: Cambridge University Press)
[7] de Felice F 1978 *J. Phys. A: Math. Gen.* **12** 1223

Chapter 16

Elementary introduction to pre-big bang cosmology and to the relic graviton background

Maurizio Gasperini
Dipartimento di Fisica, Università di Bari, Via G. Amendola 173,
70126 Bari, Italy and Istituto Nazionale di Fisica Nucleare,
Sezione di Bari, Bari, Italy

This is a contracted version of a series of lectures for graduate and undergraduate students given at the '*VI Seminario Nazionale di Fisica Teorica*' (Parma, September 1997), at the Second International Conference '*Around VIRGO*' (Pisa, September 1998), and at the Second *SIGRAV* School on '*Gravitational Waves in Astrophysics, Cosmology and String Theory*' (Center 'A. Volta', Como, April 1999). The aim is to provide an elementary, self-contained introduction to string cosmology and, in particular, to the background of relic cosmic gravitons predicted in the context of the so-called 'pre-big bang' scenario. No special preparation is required besides a basic knowledge of general relativity and of standard (inflationary) cosmology. All the essential computations are reported in full details either in the main text or in the appendices. For a deeper and more complete approach to the pre-big bang scenario the interested reader is referred to the updated collection of papers available at http://www.to.infn.it/~gasperin/.

16.1 Introduction

The purpose of these lectures is to provide an introduction to the background of relic gravitational waves expected in a string cosmology context, and to discuss its main properties. To this purpose, it seems to be appropriate to include a short presentation of string cosmology, in order to explain the basic ideas underlying the so-called pre-big bang scenario, which is one of the most promising scenarios for the production of a detectable graviton background of cosmological origin.

After a short, qualitative presentation of the pre-big bang models, we will concentrate on the details of the cosmic graviton spectrum: we will discuss the theoretical predictions for different models, and will compare the predictions with existing phenomenological constraints, and with the expected sensitivities of the present gravity-wave detectors. A consistent part of these lectures will thus be devoted to introducing the basic notions of cosmological perturbation theory, which are required to compute the graviton spectrum and to understand why the amplification of tensor metric perturbations, at high frequency, is more efficient in string cosmology than in the standard inflationary context. Let me start by noting that a qualitative, but effective representation of the main difference between string cosmology and standard, inflationary cosmology can be obtained by plotting the curvature scale of the Universe versus time, as illustrated in figure 16.1.

According to the cosmological solutions of the so-called 'standard' scenario [1], the spacetime curvature decreases in time. As we go back in time the curvature grows monotonically, and blows up at the initial 'big bang' singularity, as illustrated in the top part of figure 16.1 (a similar plot, in the standard scenario, also describes the behaviour of the temperature and of the energy density of the gravitational sources).

According to the standard inflationary scenario [2], in contrast, the Universe in the past is expected to enter a de Sitter, or 'almost' de Sitter phase, during which the curvature tends to stay frozen at a nearly constant value. From a classical point of view, however, this scenario has a problem, since a phase of expansion at constant curvature cannot be extended back in time for ever [3], for reasons of geodesic completeness. This point was clearly stressed also in Alan Guth's recent survey of inflationary cosmology [4]:

> ... *Nevertheless, since inflation appears to be eternal only into the future, but not to the past, an important question remains open. How did it all start? Although eternal inflation pushes this question far into the past, and well beyond the range of observational tests, the question does not disappear.*

A possible anwer to this question, in a quantum cosmology context, is that the universe emerges in a de Sitter state 'from nothing' [5] (or from some unspecified 'vacuum'), through a process of quantum tunnelling. We will not discuss the quantum approach in these lectures, but let us note that the computation of the transition probability requires an appropriate choice of the boundary conditions [6], which in the context of standard inflation are imposed *ad hoc* when the universe is in an unknown state, deeply inside the non-perturbative, quantum gravity regime. In a string cosmology context, in contrast, the initial conditions are referrred asymptotically to a low-energy, classical state which is known, and well controlled by the low-energy string effective action [7].

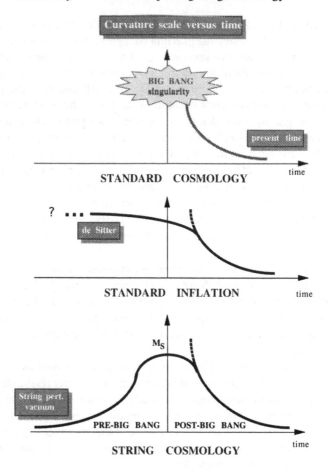

Figure 16.1. Time evolution of the curvature scale in the standard cosmological scenario, in the conventional inflationary scenario, and in the string cosmology scenario.

From a classical point of view, however, the answer to the above question—what happens to the universe before the phase of constant curvature, which cannot last for ever—is very simple, as we are left with only two possibilities. Either the curvature starts growing again, at some point in the past (but in this case the singularity problems remain, it is simply shifted back in time), or the curvature starts decreasing.

In this second case we are just led to the string cosmology scenario, illustrated in the bottom part of figure 16.1. String theory suggests indeed for the curvature a specular behaviour (or better a 'dual' behaviour, as we shall see in a moment) around the time axis. As we go back in time the curvature grows, reaches a maximum controlled by the string scale, and then decreases towards a state which is asymptotically flat and with negligible interactions (vanishing coupling constants), the so-called 'string perturbative vacuum'. In this scenario the phase of high, but finite (nearly Planckian) curvature is what replaces the big bang singularity of the standard scenario. It thus comes naturally, in a string cosmology context, to call 'pre-big bang' [8] the initial phase with growing curvature, in contrast to the subsequent, standard, 'post-big bang' phase with decreasing curvature.

At this point, a number of questions may arise naturally. In particular:

- *Motivations*: why such a cosmological scenario, characterized by a 'bell-like' shape of the curvature, seems to emerge in a string cosmology context and not, for instance, in the context of standard cosmology based on the Einstein equations?
- *Kinematics*: in spite of the differences, is the kinematic of the pre-big bang phase still appropriate to solve the well-known problems (horizon, flatness …) of the standard scenario? After all, we do not want to lose the main achievements of the conventional inflationary models.
- *Phenomenology*: are there phenomenological consequences that can discriminate between string cosmology models and other inflationary models? Are such effects observable, at least in principle?

In the following sections we will present a quick discussion of the three points listed above.

16.2 Motivations: duality symmetry

There are various motivations, in the context of string theory, suggesting a cosmological scenario like that illustrated in figure 16.1. All the motivations are however related, more or less directly, to an important property of string theory, the duality symmetry of the effective action.

To illustrate this point, let us start by recalling that in general relativity the solutions of the standard Einstein action,

$$S = -\frac{1}{2\lambda_p^{d-1}} \int d^{d+1}x \sqrt{|g|} R \tag{16.1}$$

(d is the number of spatial dimensions, and $\lambda_p = M_p^{-1}$ is the Planck length scale), are invariant under 'time-reversal' transformations. Consider, for instance, a homogeneous and isotropic solution of the cosmological equations, represented by a scale factor $a(t)$:

$$ds^2 = dt^2 - a^2(t) dx_i^2. \tag{16.2}$$

If $a(t)$ is a solution, then also $a(-t)$ is a solution. On the other hand, when t goes into $-t$, the Hubble parameter $H = \dot{a}/a$ changes sign,

$$a(t) \to a(-t), \quad H = \dot{a}/a \to -H. \tag{16.3}$$

To any standard cosmological solution $H(t)$, describing decelerated expansion and decreasing curvature $(H > 0, \dot{H} < 0)$, is thus associated with a 'reflected' solution, $H(-t)$, describing a contracting universe because H is negative.

This is the situation in general relativity. In string theory the action, in addition to the metric, contains at least another fundamental field, the scalar dilaton ϕ. At the tree-level, namely to lowest order in the string coupling and in the higher-derivatives (α') string corrections, the effective action which guarantees the absence of conformal anomalies for the motion of strings in curved backgrounds (see apppendix A) can be written as:

$$S = -\frac{1}{2\lambda_s^{d-1}} \int d^{d+1}x \sqrt{|g|} e^{-\phi} [R + (\partial_\mu \phi)^2] \tag{16.4}$$

$(\lambda_s = M_s^{-1}$ is the fundamental string length scale; see appendix B for notations and sign conventions). In addition to the invariance under time-reversal, the above action is also invariant under the 'dual' inversion of the scale factor, accompanied by an appropriate transformation of the dilaton (see [9] and the first paper of [8]). More precisely, if $a(t)$ is a solution for the cosmological background (16.2), then $a^{-1}(t)$ is also a solution, provided the dilaton transforms as:

$$a \to \tilde{a} = a^{-1}, \quad \phi \to \tilde{\phi} = \phi - 2d \ln a \tag{16.5}$$

(this transformation implements a particular case of T-duality symmetry, usually called 'scale factor duality', see appendix B).

When a goes into a^{-1}, the Hubble parameter H again goes into $-H$ so that, to each one of the two solutions related by time reversal, $H(t)$ and $H(-t)$, is associated a dual solution, $\tilde{H}(t)$ and $\tilde{H}(-t)$, respectively (see figure 16.2). The space of solutions is thus reached in a string cosmology context. Indeed, because of the combined invariance under the transformations (16.3) and (16.5), a cosmological solution has in general four branches: two branches describe expansion (positive H), two branches describe contraction (negative H). Also, as illustrated in figure 16.2, for two branches the curvature scale $(\sim H^2)$ grows in time, with a typical 'pre-big bang' behaviour, while for the other two branches the curvature scale decreases, with a typical 'post-big bang' behaviour.

It follows, in this context, that to any given decelerated expanding solution, $H(t) > 0$, with decreasing curvature, $\dot{H}(t) < 0$ (typical of the standard cosmological scenario), is always associated a 'dual partner' describing accelerated expansion, $\tilde{H}(-t) > 0$, and growing curvature, $\dot{\tilde{H}}(-t) > 0$. This doubling of solutions has no analogue in the context of Einstein cosmology, where there is no dilaton, and the duality symmetry cannot be implemented.

It should be stressed, before proceeding further, that the duality symmetry is not restricted to the case of homogeneous and isotropic backgrounds like (16.2), but is expected to be a general property of the solutions of the string effective action (possibly valid at all orders [10], with the appropriate generalizations). The inversion of the scale factor, in particular, and the associated transformation

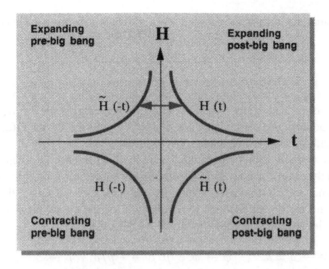

Figure 16.2. The four branches of a low-energy string cosmology background.

(16.5), is only a special case of a more general, $O(d, d)$ symmetry of the string effective action, which is manifest already at the lowest order. In fact, the tree-level action in general contains, besides the metric and the dilaton, also a second rank antisymmetric tensor $B_{\mu\nu}$, the so-called Kalb–Ramond 'universal' axion:

$$S = -\frac{1}{2\lambda_s^{d-1}} \int d^{d+1}x \sqrt{|g|} e^{-\phi} \left[R + (\partial_\mu \phi)^2 - \frac{1}{12}(\partial_{[\mu} B_{\nu\alpha]})^2 \right]. \quad (16.6)$$

Given a background, even anisotropic, but with d Abelian isometries, this action is invariant under a global, pseudo-orthogonal group of $O(d, d)$ transformations which mix in a non-trivial way the components of the metric and of the antisymmetric tensor, leaving invariant the so-called 'shifted' dilaton $\bar{\phi}$:

$$\bar{\phi} = \phi - \ln \sqrt{|\det g_{ij}|}. \quad (16.7)$$

In the particular, 'cosmological' case in which we are interested, the d isometries correspond to spatial translations (namely, we are in the case of a homogeneous, Bianchi I type metric background). For this background, the action (16.6) can be rewritten in terms of the $(2d \times 2d)$ symmetric matrix M, defined by the spatial components of the metric, g_{ij}, and of the axion, B_{ij}, as:

$$B \equiv B_{ij}, \quad G \equiv g_{ij}, \quad M = \begin{pmatrix} G^{-1} & -G^{-1}B \\ BG^{-1} & G - BG^{-1}B \end{pmatrix}. \quad (16.8)$$

In the cosmic time gauge, the action takes the form (see appendix B)

$$S = -\frac{\lambda_s}{2} \int dt\, e^{-\bar{\phi}} \left[(\dot{\bar{\phi}})^2 + \frac{1}{8} \text{Tr}\, \dot{M}(M^{-1})^{\cdot} \right], \quad (16.9)$$

and is manifestly invariant under the set of global transformations [11]:

$$\overline{\phi} \to \overline{\phi}, \quad M \to \Lambda^T M \Lambda, \quad \Lambda^T \eta \Lambda = \eta, \quad \eta = \begin{pmatrix} 0 & I \\ I & 0 \end{pmatrix}, \qquad (16.10)$$

where I is the d-dimensional unit matrix, and η is the $O(d,d)$ metric in off-diagonal form. The transformation (16.5), representing scale factor duality, is now reproduced as a particular case of (16.10) with the trivial $O(d,d)$ matrix $\Lambda = \eta$, and for an isotropic background with $B_{\mu\nu} = 0$.

This $O(d,d)$ symmetry holds even in the presence of matter sources, provided they transform according to the string equations of motion in the given background [12]. In the perfect fluid approximation, for instance, the inversion of the scale factor corresponds to a reflection of the equation of state, which preserves however the 'shifted' energy $\overline{\rho} = \rho |\det g_{ij}|^{1/2}$:

$$a \to \tilde{a} = a^{-1}, \quad \overline{\phi} \to \overline{\phi}, \quad p/\rho \to -p/\rho, \quad \overline{\rho} \to \overline{\rho}. \qquad (16.11)$$

A detailed discussion of the duality symmetry is outside the purpose of these lectures. What is important, in our context, is the simultaneous presence of duality and time-reversal symmetry: by combining these two symmetries, in fact, it is possible in principle to obtain cosmological solutions of the 'self-dual' type, characterized by the conditions

$$a(t) = a^{-1}(-t), \quad \overline{\phi}(t) = \overline{\phi}(-t). \qquad (16.12)$$

They are important, as they connect in a smooth way the phase of growing and decreasing curvature, and also describe a smooth evolution from the string perturbative vacuum (i.e. the asymptotic no-interaction state in which $\phi \to -\infty$ and the string coupling is vanishing, $g_s = \exp(\phi/2) \to 0$), to the present cosmological phase in which the dilaton is frozen, with an expectation value [13] $\langle g_s \rangle = M_s/M_p \sim 0.3\text{--}0.03$ (see figure 16.3).

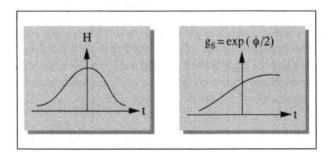

Figure 16.3. Time evolution of the curvature scale H and of the string coupling $g_s = \exp(\phi/2) \simeq M_s/M_p$, for a typical self-dual solution of the string cosmology equations.

The explicit occurrence of self-dual solutions and, more generally, of solutions describing a complete and smooth transition between the phase of pre- and post-big bang evolution, seems to require in general the presence of higher order (higher loop and/or higher derivative) corrections to the string effective action [14] (see, however, [15, 16]). So, in order to give only a simple example of combined {duality $\oplus$ time-reversal} transformation, let us consider here the low-energy, asymptotic regimes, which are well described by the lowest order effective action.

By adding matter sources, in the perfect fluid form, to the action (16.4), the string cosmology equations for a $d = 3$, homogeneous, isotropic and conformally flat background can be written as (see appendix C, equations (16.200), (16.202), (16.199), respectively):

$$\dot{\phi}^2 - 6H\dot{\phi} + 6H^2 = e^{\phi}\rho,$$
$$\dot{H} - H\dot{\phi} + 3H^2 = \tfrac{1}{2}e^{\phi}p,$$
$$2\ddot{\phi} + 6H\dot{\phi} - \dot{\phi}^2 - 6\dot{H} - 12H^2 = 0. \qquad (16.13)$$

For $p = \rho/3$, in particular, they are exactly solved by the standard solution with constant dilaton (see equations (16.216) and (16.217)),

$$a \sim t^{1/2}, \quad \rho = 3p \sim a^{-4}, \quad \phi = \text{constant}, \quad t \to +\infty, \qquad (16.14)$$

describing decelerated expansion and decreasing curvature scale:

$$\dot{a} > 0, \quad \ddot{a} < 0, \quad \dot{H} < 0. \qquad (16.15)$$

This is exactly the radiation-dominated solution of the standard cosmological scenario, based on the Einstein equations. In string cosmology, however, to this solution is associated a 'dual complement', i.e. an additional solution which can be obtained by applying on the background (16.14) a time-reversal transformation $t \to -t$, and the duality transformation (16.11):

$$a \sim (-t)^{-1/2}, \quad \phi \sim -3\ln(-t), \quad \rho = -3p \sim a^{-2}, \quad t \to -\infty. \quad (16.16)$$

This is still an exact solution of equations (16.13) (see appendix C), describing however accelerated (i.e. inflationary) expansion, with growing dilaton and growing curvature scale:

$$\dot{a} > 0, \quad \ddot{a} > 0, \quad \dot{H} > 0. \qquad (16.17)$$

We note, for future reference, that accelerated expansion with growing curvature is usually called 'superinflation' [17], or 'pole-inflation', to distinguish it from the more conventional power-inflation, with decreasing curvature.

The two solutions (16.14) and (16.16) provide a particular, explicit representation of the scenario illustrated in figure 16.3, in the two asymptotic

Table 16.1. Analogy between supersymmetry and duality.

	Supersymmetry	Duality + Time-reversal
Pair of partners	{bosons, fermions}	{growing curvature, decreasing curvature}
Known states	photons, gravitons, ...	decelerated, standard post-big bang
Predicted	photinos, gravitinos, ...	accelerated, inflationary pre-big bang

regimes of t large and positive, and t large and negative, respectively. The duality symmetry seems thus to provide an important motivation for the pre-big bang scenario, as it leads naturally to introduce a phase of growing curvature, and is a crucial ingredient for the 'bell-like' scenario of figure 16.3.

It should be noted that pure scale factor duality, by itself, is not enough to convert a phase of decreasing into growing curvature (see for instance figure 16.2, where it is clearly shown that H and $\tilde{H}$, in the same temporal range, lead to the same evolution of the curvature scale, $H^2 \sim \tilde{H}^2$). Time reflection is thus necessarily required, if we want to invert the curvature behaviour. From this point of view, time-reversal symmetry is more important than duality.

In a thermodynamic context, however, duality by itself is able to suggest the existence of a primordial cosmological phase with 'specular' properties with respect to the present, standard cosmological phase [18]. It must be stressed, in addition, that it is typically in the cosmology of extended objects that the phase of growing curvature may describe accelerated expansion instead of contraction, and that the growth of the curvature may be regularized, instead of blowing up to a singularity. For instance, it is with the string dilaton [8], or with a network of strings self-consistently coupled to the background [19], that we are naturally lead to superinflation. Also, in quantum theories of extended objects, it is the minimal, fundamental length scale of the theory that is expected to bound the curvature, and to drive superinflation to a phase of constant, limiting curvature [20] asymptotically approaching de Sitter, as explicitly checked in a string theory context [21]. Duality symmetries, on the other hand, are typical of extended objects (and of strings, in particular), so that it is certainly justified to think of duality as of a fundamental motivation and ingredient of the pre-big bang scenario.

Duality is an important symmetry of modern theoretical physics, and to conclude this section we would like to present an analogy with another very important symmetry, namely supersymmetry (see table 16.1).

According to supersymmetry, to any *bosonic state* is associated a *fermionic partner*, and *vice versa*. From the existence of bosons that we know to be present in nature, if we believe in supersymmetry, we can predict the existence of fermions not yet observed, like the photino, the gravitino, and so on.

In the same way, according to duality and time-reversal, to any *geometrical state* with decreasing curvature is associated a *dual partner* with growing curvature. On the other hand, our universe, at present, is in the standard post-big bang phase, with decreasing curvature. If we believe that duality has to be implemented, even approximately, in the course of the cosmological evolution, we can then predict the existence of a phase, in the past, characterized by growing curvature and by a typical pre-big bang evolution.

16.3   Kinematics: shrinking horizons

If we accept, at least as a working hypothesis, the possibility that our universe had in the past a 'dual' complement, with growing curvature, we are led to the second of the three questions listed in section 16.1: is the kinematics of the pre-big bang phase still appropriate to solve the problems of the standard inflationary scenario? The answer is positive, but in a non-trivial way.

Consider, for instance, the present cosmological phase. Today the dilaton is expected to be constant, and the universe should be appropriately described by Einstein equations. The gravitational part of such equations contains two types of terms: terms controlling the geometric curvature of a space-like section, evolving in time like a^{-2}, and terms controlling the gravitational kinetic energy, i.e. the spacetime curvature scale, evolving like H^2. According to present observations the spatial curvature term is non-dominant, i.e.

$$r = \frac{a^{-2}}{H^2} \sim \frac{\text{spatial curvature}}{\text{spacetime curvature}} \lesssim 1. \tag{16.18}$$

According to the standard cosmological solutions, on the other hand, the above ratio must grow in time. In fact, by putting $a \sim t^{\beta}$,

$$r \sim \dot{a}^{-2} \sim t^{2(1-\beta)}, \tag{16.19}$$

so that r keeps growing both in the matter-dominated ($\beta = 2/3$) and in the radiation-dominated ($\beta = 1/2$) era. Thus, as we go back in time, r becomes smaller and smaller, and when we set initial conditions (for instance, at the Planck scale) we have to impose an enormous fine tuning of the spatial curvature term, with respect to the other terms of the cosmological equations. This is the so-called flatness problem.

The problem can be solved if we introduce in the past a phase (usually called inflation), during which the value of r was decreasing, for a time long enough to compensate the subsequent growth during the phase of standard evolution. It is important to stress that this requirement, in general, can be implemented by two physically different classes of backgrounds.

Consider for simplicity a power-law evolution of the scale factor in cosmic time, with a power β, so that the time-dependence of r is the one given in equation (16.19). The two possible classes of backgrounds corresponding to a decreasing r are then the following:

- *Class I*: $a \sim t^{\beta}$, $\beta > 1$, $t \to +\infty$. This class of backgrounds corresponds to what is conventionally called 'power inflation', describing a phase of accelerated expansion and decreasing curvature scale, $\dot{a} > 0$, $\ddot{a} > 0$, $\dot{H} < 0$. This class contains, as a limiting case, the standard de Sitter inflation, $\beta \to \infty$, $a \sim e^{Ht}$, $\dot{H} = 0$, i.e. accelerated exponential expansion at constant curvature.
- *Class II*: $a \sim (-t)^{\beta}$, $\beta < 1$, $t \to 0_-$. This is the class of backgrounds corresponding to the string cosmology scenario. There are two possible subclasses:

  IIa: $\beta < 0$, describing superinflation, i.e. accelerated expansion with growing curvature scale, $\dot{a} > 0$, $\ddot{a} > 0$, $\dot{H} > 0$;

  IIb: $0 < \beta < 1$, describing accelerated contraction and growing curvature scale, $\dot{a} < 0$, $\ddot{a} < 0$, $\dot{H} < 0$.

A phase of growing curvature, if accelerated like in the pre-big bang scenario, can thus provide an unconventional, but acceptable, inflationary solution of the flatness problem (the same is true for the other standard kinematical problems, see [8]). It is important to stress, in particular, that the two subclasses *IIa*, *IIb*, do not correspond to different models, as they are simply different kinematical representations of the *same* scenario in two *different frames*, the string frame (S-frame), in which the effective action takes the form (16.4),

$$S(g, \phi) = - \int d^{d+1}x \sqrt{|g|}\, e^{-\phi}[R + g^{\mu\nu}\partial_{\mu}\phi\partial_{\nu}\phi], \qquad (16.20)$$

and the Einstein frame (E-frame), in which the dilaton is minimally coupled to the metric, and has a canonical kinetic term:

$$S(\tilde{g}, \tilde{\phi}) = - \int d^{d+1}x \sqrt{|\tilde{g}|}[\tilde{R} - \tfrac{1}{2}\tilde{g}^{\mu\nu}\partial_{\mu}\tilde{\phi}\partial_{\nu}\tilde{\phi}]. \qquad (16.21)$$

In order to illustrate this point, we shall proceed in two steps. First, we will show that, through a field redefinition $g = g(\tilde{g}, \tilde{\phi})$, $\phi = \phi(\tilde{g}, \tilde{\phi})$, it is always possible to move from the S-frame to the E-frame; second, we will show that, by applying such a redefinition, a superinflationary solution obtained in the S-frame becomes an accelerated contraction in the E-frame, and *vice versa*.

We shall consider, for simplicity, an isotropic, spatially flat background with d spatial dimensions, and we set:

$$g_{\mu\nu} = \text{diag}(N^2, -a^2\delta_{ij}), \quad \phi = \phi(t), \qquad (16.22)$$

where $g_{00} = N^2$ is to be fixed by an arbitrary choice of gauge. For this background we get:

$$\Gamma_{0i}{}^{j} = H\delta_i^j, \quad \Gamma_{ij}{}^{0} = \frac{a\dot{a}}{N^2}\delta_{ij}, \quad \Gamma_{00}^{0} = \frac{\dot{N}}{N} = F$$

$$R = \frac{1}{N^2}[2dFH - 2d\dot{H} - d(d+1)H^2], \qquad (16.23)$$

and the S-frame action (16.20) becomes

$$S(g, \phi) = - \int d^{d+1}x \frac{a^d e^{-\phi}}{N} [2dFH - 2d\dot{H} - d(d+1)H^2 + \dot{\phi}^2]. \quad (16.24)$$

Modulo a total derivative, we can eliminate the first two terms, and the action takes the quadratic form

$$S(g, \phi) = - \int d^{d+1}x \frac{a^d e^{-\phi}}{N} [\dot{\phi}^2 - 2dH\dot{\phi} + d(d-1)H^2]. \quad (16.25)$$

where, as expected, N plays the role of a Lagrange multiplier (no kinetic term in the action).

In the E-frame the variables are $\tilde{N}, \tilde{a}, \tilde{\phi}$, and the action (16.21), after integration by parts, takes the canonical form

$$S(\tilde{g}, \tilde{\phi}) = - \int d^{d+1}x \frac{\tilde{a}^d}{\tilde{N}} \left[-\frac{1}{2}\dot{\tilde{\phi}}^2 + d(d-1)H^2 \right]. \quad (16.26)$$

A quick comparison with equation (16.25) finally leads to the field redefinition (no coordinate transformation!) connecting the Einstein and string frame:

$$\tilde{a} = a e^{-\phi/(d-1)}, \quad \tilde{N} = N e^{-\phi/(d-1)}, \quad \tilde{\phi} = \phi \sqrt{\frac{2}{d-1}}. \quad (16.27)$$

In fact, the above transformation gives

$$\tilde{H} = H - \frac{\dot{\phi}}{d-1} \quad (16.28)$$

and, when inserted into equation (16.26), exactly reproduces the S-frame action (16.25).

Consider now a superinflationary, pre-big bang solution obtained in the S-frame, for instance the isotropic, d-dimensional vacuum solution

$$a = (-t)^{-1/\sqrt{d}}, \quad e^{\phi} = (-t)^{-(\sqrt{d}+1)}, \quad t < 0, \ t \to 0_- \quad (16.29)$$

(see appendix B, equations (16.160) and (16.161)), and look for the corresponding E-frame solution. The above solution is valid in the synchronous gauge, $N = 1$, and if we choose, for instance, the synchronous gauge also in the E-frame, we can fix $\tilde{N}$ by the condition:

$$\tilde{N} \, dt \equiv N e^{-\phi/(d-1)} \, dt = d\tilde{t}, \quad (16.30)$$

which defines the E-frame cosmic time, $\tilde{t}$, as:

$$d\tilde{t} = e^{-\phi/(d-1)} \, dt. \quad (16.31)$$

After integration

$$t \sim \tilde{t}^{\frac{d-1}{d+\sqrt{d}}}, \qquad (16.32)$$

and the transformed solution takes the form:

$$\tilde{a} = (-\tilde{t})^{1/d}, \quad e^{\tilde{\phi}} = (-\tilde{t})^{-\sqrt{\frac{2(d-1)}{d}}}, \quad \tilde{t} < 0, \; \tilde{t} \to 0_-. \qquad (16.33)$$

One can easily check that this solution describes accelerated contraction with growing dilaton and growing curvature scale:

$$\frac{d\tilde{a}}{d\tilde{t}} < 0, \quad \frac{d^2\tilde{a}}{d\tilde{t}^2} < 0, \quad \frac{d\tilde{H}}{d\tilde{t}} < 0, \quad \frac{d\tilde{\phi}}{d\tilde{t}} > 0. \qquad (16.34)$$

The same result applies if we transform other isotropic solutions from the string to the Einstein frame, for instance the perfect fluid solution of appendix C, equation (16.216). We leave this simple exercise to the interested reader.

Having discussed the 'dynamical' equivalence (in spite of the kinematical differences) of the two classes of string cosmology metrics, *IIa* and *IIb*, it seems appropriate at this point to stress the main dynamical difference between standard inflation, *class I* metrics, and pre-big bang inflation, *class II* metrics. Such a difference can be conveniently illustrated in terms of the proper size of the event horizon, relative to a given comoving observer.

Consider in fact the proper distance, $d_e(t)$, between the surface of the event horizon and a comoving observer, at rest at the origin of an isotropic, conformally flat background [22]:

$$d_e(t) = a(t) \int_t^{t_M} dt' \, a^{-1}(t'). \qquad (16.35)$$

Here t_M is the maximal allowed extension, towards the future, of the cosmic time coordinate for the given background manifold. The above integral converges for all the above classes of accelerated (expanding or contracting) scale factors. In the case of *class I* metrics we have, in particular,

$$d_e(t) = t^\beta \int_t^\infty dt' \, t'^{-\beta} = \frac{t}{\beta - 1} \sim H^{-1}(t) \qquad (16.36)$$

for power-law inflation ($\beta > 1, t > 0$), and

$$d_e(t) = e^{Ht} \int_t^\infty dt' \, e^{-Ht'} = H^{-1} \qquad (16.37)$$

for de Sitter inflation. For *class II* metrics ($\beta < 1, t < 0$) we have instead

$$d_e(t) = (-t)^\beta \int_t^0 dt' \, (-t')^{-\beta} = \frac{(-t)}{1 - \beta} \sim H^{-1}(t). \qquad (16.38)$$

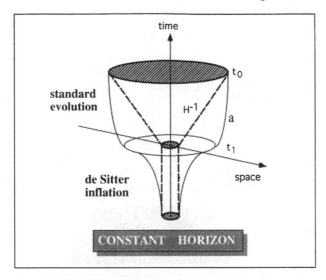

Figure 16.4. Qualitative evolution of the Hubble horizon (broken curve) and of the scale factor (full curve) in the standard inflationary scenario.

In all cases the proper size $d_e(t)$ evolves in time like the so-called Hubble horizon (i.e. the inverse of the Hubble parameter), and then like the inverse of the curvature scale. The size of the horizon is thus constant or growing in standard inflation (*class I*), decreasing in pre-big bang inflation (*class II*), both in the S-frame and in the E-frame.

Such an important difference is clearly illustrated in figures 16.4 and 16.5, where the broken curves represent the evolution of the horizon, the full curves the evolution of the scale factor. The shaded area at time t_0 represents the portion of universe inside our present Hubble radius. As we go back in time, according to the standard scenario, the horizon shrinks linearly, ($H^{-1} \sim t$), but the decrease of the scale factor is slower so that, at the beginning of the phase of standard evolution ($t = t_1$), we end up with a causal horizon much smaller than the portion of universe that we presently observe. This is the well-known 'horizon problem' of the standard scenario.

In figure 16.4 the phase of standard evolution is preceded in time by a phase of standard de Sitter inflation. Going back in time, for $t < t_1$, the scale factor keeps shrinking, and our portion of universe 're-enters' inside the Hubble radius during a phase of constant (or slightly growing in time) horizon.

In figure 16.5 the standard evolution is preceded in time by a phase of pre-big bang inflation, with growing curvature. The universe 're-enters' the Hubble radius during a phase of shrinking horizon. To emphasize the difference, we have plotted the evolution of the scale factor both in the expanding S-frame, $a(t)$, and in the contracting E-frame, $\tilde{a}(t)$. Unlike in standard inflation, the proper size of

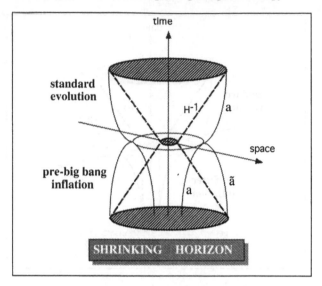

Figure 16.5. Qualitative evolution of the Hubble horizon (broken curve) and of the scale factor (full curve) in the pre-big bang inflationary scenario, in the S-frame, $a(t)$, and in the E-frame, $\tilde{a}(t)$.

the initial portion of the universe may be very large in strings (or Planck) units, *but not larger than the initial horizon* itself [23], as emphasized in the picture. The initial horizon is large because the initial curvature scale is small, in string units, $H_i \ll 1/\lambda_s$.

This is a basic consequence of the choice of the initial state which, in the pre-big bang scenario, approaches the flat, cold and empty string perturbative vacuum [8], and which is to be contrasted to the extremely curved, hot and dense initial state of the standard scenario, characterizing a universe which starts inflating at the Planck scale, $H_i \sim 1/\lambda_p$.

16.4   Open problems and phenomenological consequences

In order to give an honest presentation of the pre-big bang scenario, it is fair to say that the string cosmology models are not free from various (more or less important) difficulties, and that many aspects of the scenario are still unclear. A detailed discussion of such aspects is outside the purpose of this paper, but we would like to mention here at least three important open problems. Presented in 'time-ordered' form (from the beginning to the end of inflation) they are the following.

- The first concerns the initial conditions, and in particular the decay of the string perturbative vacuum. The question is whether or not the 'switching on' of a long inflationary phase requires fine tuning. Originally raised in [24], this problem was recently reproposed as a fundamental difficulty of the pre-big bang scenario [25] (see [23, 26–28]).

- The second concerns the transition from the pre- to the post-big bang phase, which is expected to occur in the high curvature and strong coupling regime. There is a quantum cosmology approach, based on the scattering of the Wheeler–De Witt wavefunction in minisuperspace [7], but the problem seems to require, in general, the introduction of higher derivative (α') and quantum loop corrections [21, 29] in the string effective action (see [15, 16]).

- The third problem concerns the final matching to the standard Friedman–Robertson–Walker phase, with a transition from the dilaton-dominated to the radiation-dominated regime, and all the associated problems of dilaton oscillations, reheating, preheating, particle production, entropy production [30], and so on.

All these problems are under active investigation, and further work is certainly needed for a final answer. However, even assuming that all the problems will be solved in a satisfactory way, we are left eventually with a further question, the third one listed in the introduction, which is the basic question (in our opinion). Are there phenomenological consequences that can discriminate string cosmology from the other inflationary scenarios? and, in particular, are such consequences observable (at least in principle)?

The answer is positive. There are many phenomenological differences, even if all the differences seem to have the same 'common denominator', i.e. the fact that the quantum fluctuations of the background fields are amplified in different models with different spectra. The spectrum, in particular, tends to follow the behaviour of the curvature scale during the phase of inflation. In the standard scenario the curvature is constant or decreasing, so that the spectrum tends to be flat, or decreasing with frequency. In string cosmology the curvature is growing, and the spectrum tends to grow with frequency.

In the following sections we will discuss in detail this effect for the case of tensor metric perturbations. Here we would like to note that the phenomenological consequences of the pre-big bang scenario can be classified into three different types, depending on the possibility of their observation: *type I* effects, referring to observations to be performed in a not so far future (20–30 years?); *type II* effects, referring to observations to be performed in a near future (a few years); *type III* effects, referring to observations already (in part) performed. To conclude this very quick presentation of the pre-big bang scenario, let me give one example for each type of phenomenological effect.

- *Type I*: the production of a relic graviton background that, in the frequency range of conventional detectors ($\sim 10^2$–10^3 Hz), is much higher (by 8–9 orders of magnitude) than the background expected in conventional inflation [8, 31–33]. The sensitivity of the presently operating gravitational antennae is not enough to detect it, however, and we have to wait for the advanced, second generation interferometric detectors (LIGO [34], VIRGO [35]), or for interferometers in space (LISA [36]).

- *Type II*: the large-scale CMB anisotropy 'seeded' by the inhomogeneous fluctuations of a massless [37] or massive [38] axion background. Metric fluctuations are indeed too small, on the horizon scale, to be responsible for the temperature anisotropies detected by COBE [39]; the axion spectrum, in contrast, can be sufficiently flat [40] for that purpose. Such a different origin of the anisotropy may lead to non-Gaussianity, or to differences (with respect to the standard inflationary scenario) in the height and position of the first Doppler peak of the spectrum [41]. Such differences could be soon confirmed, or disproved, by the planned satellite observations (MAP [42], PLANCK [43], ...).

- *Type III*: the production of primordial magnetic fields strong enough to 'seed' the galactic dynamo, and to explain the origin of the cosmic magnetic fields observed on a large (galactic, intergalactic) scale [44]. In the standard inflationary scenario, in fact, the amplification of the vacuum fluctuations of the electromagnetic field is not efficient enough [45], because of the conformal invariance of the Maxwell equations. In string cosmology, in contrast, the electromagnetic field is also coupled to the dilaton, and the fluctuations are amplified by the accelerated growth of the dilaton during the phase of pre-big bang evolution.

Finally, we wish to mention a further important phenomenogical effect, typical of string cosmology (and that we do not know how to classify within the tree types defined above, however): dilaton production, i.e. the amplification of the dilatonic fluctuations of the vacuum, and the formation of a cosmic background of relic dilatons [46].

The possibility of detecting such a background is strongly dependent on the value of the dilaton mass, that we do not know, at present. If dilatons are massless [47], then the amplitude and the spectrum of the relic background should be very similar to those of the graviton background, and the relic dilatons could be possibly detected, in the future, by gravitational antennae able to respond to scalar modes, unless their coupling to bulk matter is too small [47], of course.

If dilatons are massive, the mass has to be large enough to be compatible with existing tests of the equivalence principle and of macroscopic gravitational forces. In addition, there is a rich phenomenology of cosmological bounds, which leaves open only two possible mass windows [46]. Interestingly enough, however, in the allowed light mass sector the dilaton lifetime is longer than the present age of the universe, and the dilaton fraction of critical energy density ranges from 0.01–1: in this context, the dilaton becomes a new, interesting dark matter candidate (see [48] for a detailed discussion of the allowed mass windows, and of the possibility that light but non-relativistic dilatons could represent today a significant fraction of dark matter on a cosmological scale). We have no idea, however, of how to detect directly such a massive dilaton background, because the mass is light, but it is heavy enough ($\gtrsim 10^{-4}$ eV) to be far outside the sensitivity range of resonant gravitational detectors.

The rest of this lecture will be devoted to discussing various theoretical and phenomenological aspects of graviton production, in a general cosmological context and, in particular, in the context of the pre-big bang scenario. Let us start by recalling some basic notions of cosmological perturbation theory, which are required for the computation of the graviton spectrum.

16.5 Cosmological perturbation theory

The standard approach to cosmological perturbation theory is to start with a set of non-perturbed equations, for instance the Einstein or the string cosmology equations,

$$G_{\mu\nu} = T_{\mu\nu}, \tag{16.39}$$

to expand the metric and the matter fields around a given background solution,

$$g_{\mu\nu} \to g_{\mu\nu}^{(0)} + \delta^{(1)}g_{\mu\nu}, \quad T_{\mu\nu} \to T_{\mu\nu}^{(0)} + \delta^{(1)}T_{\mu\nu}, \quad G_{\mu\nu}^{(0)} = T_{\mu\nu}^{(0)}, \tag{16.40}$$

and to obtain, to first order, a linearized set of equations describing the classical evolution of perturbations,

$$\delta^{(1)}G_{\mu\nu} = \delta^{(1)}T_{\mu\nu}. \tag{16.41}$$

In principle, the procedure is simple and straightforward. In practice, however, we have to go through a series of formal steps, that we list here in 'chronological' order:

* choice of the 'frame';
* choice of the 'gauge';
* normalization of the amplitude;
* computation of the spectrum.

16.5.1 Choice of the frame

The choice of the frame is that of the basic set of fields (metric included) used to parametrize the action. The action, in general, can be expressed in terms of different fields. In string cosmology, for instance, there is a preferred frame, the S-frame, in which the lowest order gravidilaton action takes the form of (16.20). It is preferred because the metric appearing in the action is the same as the σ-model metric to which test strings are minimally coupled (see appendix A): with respect to this metric, the motion of free strings is then geodesics. It is always possible, however, through the field redefinition

$$\tilde{g}_{\mu\nu} = g_{\mu\nu}e^{-2\phi/(d-1)}, \quad \tilde{\phi} = \sqrt{\frac{2}{d-1}}\phi, \tag{16.42}$$

to introduce the more conventional E-frame (16.21) in which the dilaton is minimally coupled to the metric, with a canonical kinetic term.

In the two frames the field equations are different, and the perturbation equations are also different. This seems to raise a potential problem: which frame is to be used to evaluate the physical effects of the cosmological perturbations?

The problem is only apparent, however, because physical observables (like the perturbation spectrum) are the same in both frames. The reason is that there is a compensation between the different perturbation equations and the different background solution around which we expand. A general proof of this result can be given by using the notion of canonical variable (see subsection 16.5.3). Here I will give only an explicit example for tensor perturbations in a $d = 3$, isotropic and spatially flat background.

Let us start in the E-frame, with the background equations:

$$R_{\mu\nu} = \tfrac{1}{2}\partial_\mu\phi\partial_\nu\phi, \qquad (16.43)$$

referring to the 'tilded' variables of (16.42) (we will omit the 'tilde', for simplicity, and will explicitly reinsert it at the end of the computation). Considering the transverse-traceless part of metric perturbations:

$$\delta^{(1)}\phi = 0, \quad \delta^{(1)}g_{\mu\nu} = h_{\mu\nu}, \quad \delta^{(1)}g^{\mu\nu} = -h^{\mu\nu}, \quad \nabla_\nu h_\mu{}^\nu = 0 = h_\mu{}^\nu \quad (16.44)$$

(∇_μ denotes covariant differentiation with respect to the unperturbed metric g, and the indices of h are also raised and lowered with g). The perturbation of the background equations gives:

$$\delta^{(1)}R_\mu{}^\nu = 0. \qquad (16.45)$$

We can then work in the synchronous gauge, where

$$g_{00} = 1, \quad g_{0i} = 0, \quad g_{ij} = -a^2\delta_{ij},$$
$$h_{00} = 0, \quad h_{0i} = 0, \quad g^{ij}h_{ij} = 0, \quad \partial_j h_i{}^j = 0. \qquad (16.46)$$

To first order in h we get

$$\delta^{(1)}\Gamma_{0i}{}^j = \tfrac{1}{2}\dot{h}_i{}^j, \quad \delta^{(1)}\Gamma_{ij}{}^0 = -\tfrac{1}{2}\dot{h}_{ij},$$
$$\delta^{(1)}\Gamma_{ij}{}^k = \tfrac{1}{2}(\partial_i h_j{}^k + \partial_j h_i{}^k - \partial^k h_{ij}). \qquad (16.47)$$

The $(0, 0)$ component of equation (16.45) is trivially satisfied (as well as the perturbation of the scalar field equation); the (i, j) components, by using the identities (see for instance [54])

$$g^{jk}\dot{h}_{ik} = \dot{h}_i{}^j + 2H h_i{}^j,$$
$$g^{jk}\ddot{h}_{ik} = \ddot{h}_i{}^j + 2\dot{H}h_i{}^j + 4H\dot{h}_i{}^j + 4H^2 h_i{}^j, \qquad (16.48)$$

give

$$\delta^{(1)}R_i{}^j = -\frac{1}{2}\left(\ddot{h}_i{}^j + 3H\dot{h}_i{}^j - \frac{\nabla^2}{a^2}h_i{}^j\right) \equiv -\frac{1}{2}\Box h_i{}^j = 0. \qquad (16.49)$$

In terms of the conformal time coordinate, $d\eta = dt/a$, this wave equation can be finally rewritten, for each polarization mode, as

$$\tilde{h}'' + 2\frac{\tilde{a}'}{\tilde{a}}\tilde{h}' - \nabla^2\tilde{h} = 0, \qquad (16.50)$$

(where we have explicitly reinserted the tilde, and where a prime denotes differentiation with respect to the conformal time, which is the same in the Einstein and in the string frame, according to equations (16.27) and (16.31)).

Let us now repeat the computation in the S-frame, where the background equations for the metric (equation (16.196) with no contribution from $H_{\mu\nu\alpha}$, $T_{\mu\nu}$, V and σ) can be written explicitly as

$$R_\mu{}^\nu + g^{\nu\alpha}(\partial_\mu\partial_\alpha\phi - \Gamma_{\mu\alpha}{}^\rho\partial_\rho\phi) = 0. \qquad (16.51)$$

Perturbing to first order,

$$\delta^{(1)}R_\mu{}^\nu - (\delta^{(1)}g^{\nu\alpha}\Gamma_{\mu\alpha}{}^0 + g^{\nu\alpha}\delta^{(1)}\Gamma_{\mu\alpha}{}^0)\dot\phi = 0. \qquad (16.52)$$

The $(0,0)$ component, as well as the perturbation of the dilaton equation, are trivially satisfied. The (i, j) components, using again the identities (16.48), lead to [31]:

$$\Box h_i{}^j - \dot\phi \dot{h}_i{}^j = 0. \qquad (16.53)$$

In conformal time, and for each polarization component,

$$h'' + \left(2\frac{a'}{a} - \phi'\right)h' - \nabla^2 h = 0. \qquad (16.54)$$

This last equation seems to be different from the E-frame equation (16.50). Recalling, however, the relation (16.27) between a and $\tilde{a}$, we have

$$2\frac{\tilde{a}'}{\tilde{a}} = 2\frac{a'}{a} - \phi', \qquad (16.55)$$

so that we have the same equation for h and $\tilde{h}$, the same solution, and the same spectrum when the solution is expanded in Fourier modes. The perturbation analysis is thus *frame-independent*, and we can safely choose the more convenient frame to compute the spectrum.

16.5.2 Choice of the gauge

The second step is the choice of the gauge, i.e. the choice of the coordinate system within a given frame. The perturbation spectrum is, of course, gauge-independent, but the the perturbative analysis *is not*, in general. It is possible, in fact, that the validity of the linear approximation is broken in a given gauge, but still valid in a different, more appropriate gauge.

Since this effect is particularly important, let me give, in short, an explicit example for the scalar perturbations of the metric tensor in a $d = 3$, isotropic and conformally flat background, in the E-frame (we will omit the tilde, for simplicity). The perturbed metric, in the so-called longitudinal gauge, depends on the two Bardeen potentials Φ and Ψ as [49]:

$$ds^2 = a^2[(1 + 2\Phi)\,d\eta^2 - (1 - 2\Psi)\,dx_i^2]. \tag{16.56}$$

By perturbing the Einstein equations (16.43), the dilaton equation, and combining the results for the various components, one obtains to first order that $\Phi = \Psi$, and that the metric fluctuations satisfy the equation:

$$\Phi'' + 6\frac{a'}{a}\Phi' - \nabla^2\Phi = 0. \tag{16.57}$$

We now consider the particular, exact solution of the vacuum string cosmology equations in the E-frame,

$$a(\eta) = |\eta|^{1/2}, \quad \phi(\eta) = -\sqrt{3}\ln|\eta|, \quad \eta \to 0_-, \tag{16.58}$$

corresponding to a phase of accelerated contraction and growing dilaton (i.e. the pre-big bang solution (16.33), written in conformal time, for $d = 3$). For this background, the perturbation equation (16.57) becomes a Bessel equation for the Fourier modes Φ_k,

$$\Phi_k'' + \frac{3}{\eta}\Phi_k' + k^2\Phi_k = 0, \quad \nabla^2\Phi_k = -k^2\Phi_k, \tag{16.59}$$

and the asymptotic solution, for modes well outside the horizon ($|k\eta| \ll 1$),

$$\varphi_k = A_k \ln|k\eta| + B_k|k\eta|^{-2} \tag{16.60}$$

contains a growing part which blows up ($\sim \eta^{-2}$) as the background approaches the high curvature regime ($\eta \to 0_-$). In this limit the linear approximation breaks down, so that the longitudinal gauge is not, in general, consistent with the perturbative expansion around a homogeneous, inflationary pre-big bang background, as scalar inhomogeneities may become too large.

In the same background (16.58) the problem is absent, however, for tensor perturbations, since their growth outside the horizon is only logarithmic. From equation (16.50) we have in fact the asymptotic solution

$$h_k = A_k + B_k \ln|k\eta|, \quad |k\eta| \ll 1. \tag{16.61}$$

This may suggest that the breakdown of the linear approximation, for scalar perturbations, is an artefact of the longitudinal gauge. This is indeed confirmed by the fact that, in a more appropriate off-diagonal (also called 'uniform curvature' [50]) gauge,

$$ds^2 = a^2[(1 + 2\varphi)\,d\eta^2 - dx_i^2 - 2\partial_i B\,dx^i\,d\eta], \tag{16.62}$$

Table 16.2. The four classes of accelerated backgrounds.

$\alpha < -1$	Power-inflation	$\dot{a} > 0,$	$\ddot{a} > 0,$	$\dot{H} < 0$
$\alpha = -1$	de Sitter	$\dot{a} > 0,$	$\ddot{a} > 0,$	$\dot{H} = 0$
$-1 < \alpha < 0$	Super-inflation	$\dot{a} > 0,$	$\ddot{a} > 0,$	$\dot{H} > 0$
$\alpha > 0$	Accelerated contraction	$\dot{a} < 0,$	$\ddot{a} < 0,$	$\dot{H} < 0$

the growing mode is 'gauged down', i.e. it is suppressed enough to restore the validity of the linear approximation [51] (the off-diagonal part of the metric fluctuations remains growing, but the growth is suppressed in such a way that the amplitude, normalized to the vacuum fluctuations, keeps smaller than one for all scales k, provided the curvature is smaller than one in string units). This result is also confirmed by a covariant and gauge invariant computation of the spectrum, according to the formalism developed by Bruni and Ellis [52].

It should be stressed, however, that the presence of a growing mode, and the need for choosing an appropriate gauge, is a problem typical of the pre-big bang scenario. In fact, let us come back to tensor perturbations, in the E-frame: for a generic accelerated background the scale factor can be parametrized in conformal time with a power α, as follows:

$$a = (-\eta)^{\alpha}, \quad \eta \to 0_-, \tag{16.63}$$

and the perturbation equation (16.50) gives, for each Fourier mode, the Bessel equation

$$h_k'' + \frac{2\alpha}{\eta} h_k' + k^2 h_k = 0, \tag{16.64}$$

with asymptotic solution, outside the horizon ($|k\eta| \ll 1$):

$$h_k = A_k + B_k \int^{\eta} \frac{\mathrm{d}\eta'}{a^2(\eta')} = A_k + B_k |\eta|^{1-2\alpha}. \tag{16.65}$$

The solution tends to be constant for $\alpha < 1/2$, while it tends to grow for $\alpha > 1/2$. It is now an easy exercise to re-express the scale factor (16.63) in cosmic time,

$$\mathrm{d}t = a\mathrm{d}\eta, \quad a(t) \sim |t|^{\alpha/(1+\alpha)}, \tag{16.66}$$

and to check that, by varying α, we can parametrize all types of accelerated backgrounds introduced in section 16.3: accelerated expansion (with decreasing, constant and growing curvature), and accelerated contraction, with growing curvature (see table 16.2).

In the standard, inflationary scenario the metric is expanding, $\alpha < 0$, so that the amplitude h_k is frozen outside the horizon. In the pre-big bang scenario, in contrast, the metric is contracting in the E-frame, so that h_k may grow if the contraction is fast enough, i.e. $\alpha > 1/2$ (in fact, the growing mode problem was first pointed out in the context of Kaluza–Klein inflation and dynamical dimensional reduction [53], where the internal dimensions are contracting). For the low-energy string cosmology background (16.58) we have $\alpha = 1/2$, the growth is simply logarithmic (see (16.61)), and the linear approximation can be applied consistently, provided the curvature remains bounded by the string scale [51]. However, for $\alpha > 1/2$ the growth of the amplitude may require a different gauge for a consistent linearized description.

16.5.3 Normalization of the amplitude

The linearized equations describing the classical evolution of perturbations can be obtained in two ways:

- by perturbing directly the background equations of motion;
- by perturbing the metric and the matter fields to first order, by expanding the action up to terms quadratic in the first order fluctuations,

$$g \to g + \delta^{(1)}g, \qquad \delta^{(2)}S \equiv S[(\delta^{(1)}g)^2], \qquad (16.67)$$

and then by varying the action with respect to the fluctuations.

The advantage of the second method is to define the so-called 'normal modes' for the oscillation of the system {gravity + matter sources}, namely the variables which diagonalize the kinetic terms in the perturbed action, and satisfy canonical commutation relations when the fluctuations are quantized. Such canonical variables are required, in particular, to normalize perturbations to a spectrum of quantum, zero-point fluctuations, and to study their amplification from the vacuum state up to the present state of the universe.

Let us apply such a procedure to tensor perturbations, in the S-frame, for a $d = 3$ isotropic background. In the syncronous gauge, the transverse-traceless, first-order metric perturbations $h_{\mu\nu} = \delta^{(1)}g_{\mu\nu}$ satisfy equation (16.46). We expand all terms of the low-energy gravidilaton action (16.20) up to order h^2:

$$\delta^{(1)}g^{\mu\nu} = -h^{\mu\nu}, \qquad \delta^{(2)}g^{\mu\nu} = h^{\mu\alpha}h_\alpha{}^\nu,$$
$$\delta^{(1)}\sqrt{-g} = 0, \qquad \delta^{(2)}\sqrt{-g} = -\tfrac{1}{4}\sqrt{-g}\,h_{\mu\nu}h^{\mu\nu}, \qquad (16.68)$$

and so on for $\delta^{(1)}R_{\mu\nu}$, $\delta^{(2)}R_{\mu\nu}$ (see, for instance, [54]). By using the background equations, and integrating by part, we finally arrive at the quadratic action

$$\delta^{(2)}S = \frac{1}{4} \int d^4x\, a^3 e^{-\phi} \left(\dot{h}^j_i \dot{h}^i_j + h^j_i \frac{\nabla}{a^2} h^i_j \right). \qquad (16.69)$$

By separating the two physical polarization modes, i.e. the standard 'cross' and 'plus' gravity wave components,

$$h^j_i h^i_j = 2(h^2_+ + h^2_\times), \qquad (16.70)$$

we get, for each mode (now generically denoted with h), the effective scalar action

$$\delta^{(2)} S = \frac{1}{2} \int d^4x \, a^3 e^{-\phi} \left(\dot{h}^2 + h \frac{\nabla}{a^2} h \right),$$
(16.71)

which can be rewritten, using conformal time, as

$$\delta^{(2)} S = \frac{1}{2} \int d^3x \, d\eta \, a^2 e^{-\phi} (h'^2 + h\nabla h).$$
(16.72)

The variation with respect to h gives finally equation (16.54), i.e. the same equation obtained by perturbing directly the background equations in the S-frame.

The above action describes a scalar field h, non-minimally coupled to a time-dependent external field, $a^2 e^{-\phi}$ (also called 'pump field'). In order to impose the correct quantum normalization to vacuum fluctuations, we introduce now the so-called 'canonical variable' ψ, defined in terms of the pump field as

$$\psi = zh, \quad z = ae^{-\phi/2}.$$
(16.73)

With such a definition the kinetic term for ψ appears in the standard canonical form: for each mode k, in fact, we get the action

$$\delta^{(2)} S_k = \frac{1}{2} \int d\eta \left(\psi_k'^2 - k^2 \psi_k^2 + \frac{z''}{z} \psi_k^2 \right),$$
(16.74)

and the corresponding canonical evolution equation:

$$\psi_k'' + [k^2 - V(\eta)]\psi_k = 0, \quad V(\eta) = \frac{z''}{z},$$
(16.75)

which has the form of a Schrodinger-like equation, with an effective potential depending on the external pump field. This form of the canonical equation, by the way, is the same for all types of perturbations (with different potentials, of course). What is important, in our context, is that for an accelerated inflationary background $V(z) \to 0$ as $\eta \to -\infty$. This means that, asymptotically, the canonical variable satisfies the free-field oscillating equation

$$\eta \to -\infty, \quad \psi_k'' + k^2 \psi_k = 0,$$
(16.76)

and can be normalized to an initial vacuum fluctuation spectrum,

$$\eta \to -\infty, \quad \psi_k = \frac{1}{\sqrt{2k}} e^{-ik\eta},$$
(16.77)

in such a way as to satisfy the free field canonical commutation relations, $[\psi_k, \psi_j^{*'}] = i\delta_{kj}$. The normalization of ψ_k then fixes the normalization of the metric variable $h_k = \psi_k/z$.

It is important to stress that there is no need to introduce the canonical variable to study the classical evolution of perturbations, but that such variable is needed for the initial normalization to a vacuum fluctuation spectrum. We can also normalize perturbation in a different way, of course but in that case we are studying the amplification not of the vacuum fluctuations, but of a different spectrum [55].

At this point, two remarks are in order. The first concerns the frame-independence of the spectrum. The above procedure can also be applied in the E-frame, to define a canonical variable $\tilde{\psi}$: one then obatins for $\tilde{\psi}_k$ the canonical equation (16.75), with a pump field that depends only on the metric, $\tilde{z} = \tilde{a}$. However, by using the conformal transformation connecting the two frames, it turns out that the two pump fields are the same, $\tilde{z} = \tilde{a} = ae^{-\phi/2} = z$, so that for ψ and $\tilde{\psi}$ we have the same potential, the same evolution equation, the same solution, and thus the same spectrum.

The second remark is that the canonical procedure can be applied to any action, and in particular to the string effective action including higher curvature corrections of order α', which can be written as [21]:

$$S = \int d^4x \sqrt{-g} e^{-\phi} \left\{ -R - \partial_\mu \phi \partial^\mu \phi + \frac{\alpha'}{4} [R_{GB}^2 - (\partial_\mu \phi \partial^\mu \phi)^2] \right\} \quad (16.78)$$

where $R_{GB}^2 \equiv R_{\mu\nu\alpha\beta}^2 - 4R_{\mu\nu}^2 + R^2$ is the Gauss–Bonnet invariant (we have chosen a convenient field redefinition that removes terms with higher-than-second derivatives from the equations of motion, see appendix A). From the quadratic perturbed action we obtain α' corrections to the pump fields. The canonical equation turns out to be the same as before, but with a k-dependent effective potential [54], and such an equation can be used to estimate the effects of the higher curvature corrections on the amplification of tensor perturbations. A numerical integration [54], in which the metric fluctuations are expanded around the high-curvature background solution of [21], leads in particular to the results illustrated in figure 16.6.

The qualitative behaviour is similar, both with and without α' corrections in the perturbed equations: the fluctuations are oscillating inside the horizon and frozen outside the horizon, as usual. However, the final amplitude is enhanced when α' corrections are included, and this suggests that the energy spectrum of the gravitational radiation, computed with the low-energy perturbation equation, may represent a sort of lower bound on the total amount of produced gravitons.

16.5.4 Computation of the spectrum

The final, amplified perturbation spectrum is to be obtained from the solutions of the canonical equation (16.75). In order to solve such an equation we need explicitly the effective potential $V[z(\eta)]$ which, in general, vanishes asymptotically at large positive and negative values of the conformal time. Consider, for instance, the tensor perturbation equation in the E-frame, so that

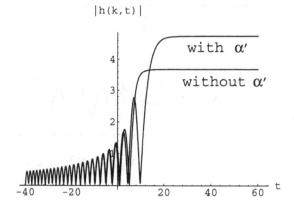

Figure 16.6. Amplification of tensor fluctuations, with and without the higher-curvature corrections included in the canonical perturbation equation.

the pump field is simply the scale factor. The typical cosmological background in which we are interested in should describe a transition fom an initial accelerated, inflationary evolution,

$$\eta \to -\infty, \quad a \sim |\eta|^{\alpha}, \quad V \sim \eta^{-2}, \tag{16.79}$$

to a final standard, radiation-dominated phase,

$$\eta \to +\infty, \quad a \sim \eta, \quad V = 0. \tag{16.80}$$

In this context, the evolution of fluctuations, initially normalized as in equation (16.77), can be described as a scattering of the canonical variable by an effective potential, according to the Schrödinger-like perturbation equation (16.75) (see figure 16.7).

However, the differential variable in equation (16.75) is (conformal) time, not space. As a consequence, the eigenfrequencies represent (comoving) energies, not momenta. Thus, even normalizing the initial state to a positive frequency mode, as in equation (16.77), the final state is, in general, a mixture of positive and negative frequency modes, i.e. of positive and negative energy states,

$$\eta \to +\infty, \quad \psi_{\text{out}} \sim c_+ e^{-ik\eta} + c_- e^{+ik\eta}. \tag{16.81}$$

In a quantum field theory context, such a mixing represents a process of pair production from the vacuum. The coefficients $c_\pm$ are the so-called Bogoliubov coefficients, parametrizing a unitary transformation between $|\text{in}\rangle$ and $|\text{out}\rangle$ states. In matrix form, they connect the set of $|\text{in}\rangle$ annihilation and creation operators, $\{\psi_{\text{in}}, b_k, b_k^\dagger\}$ to the out ones $\{\psi_{\text{out}}, a_k, a_k^\dagger\}$, as follows:

$$a_k = c_+ b_k + c_-^* b_{-k}^\dagger, \quad a_{-k}^\dagger = c_- b_k + c_+^* b_{-k}^\dagger. \tag{16.82}$$

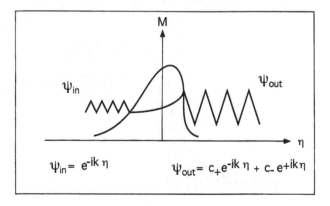

Figure 16.7. Scattering and amplification of the canonical variable.

Thus, even starting from the vacuum,

$$\bar{n}_{in} = \langle 0|b^{\dagger}b|0\rangle = 0, \tag{16.83}$$

we end up with a final number of produced pairs which is nonzero, in general, and is controlled by the Bogoliubov coefficient c_- as

$$\bar{n}_{out} = \langle 0|a^{\dagger}a|0\rangle = |c_-|^2 \neq 0. \tag{16.84}$$

In a second quantization approach, the amplification of perturbations can thus be seen as a process of pair production from the vacuum (or from any otherwise specified initial state), under the action of a time-dependent external field (the gravidilaton background, in the string cosmology case). Equivalently, the process can be described as a 'squeezing' of the initial state [56] (this description is useful to evaluate the associated entropy production [57]), or, in a semiclassical language, as a 'parametric amplification' [58] of the wavefunction ψ_k, which is scattered by an effective potential barrier through an 'antitunnelling' process [59]. Quite independently of the adopted language, the differential energy density of the produced radiation, for each mode k, depends on the number of produced pairs, and can be written as

$$d\rho_k = 2k\bar{n}_k \frac{d^3k}{(2\pi)^3}, \quad \bar{n}_k = |c_-(k)|^2. \tag{16.85}$$

The computation of the so-called energy spectrum, defined as the spectral energy density per logarithmic interval of frequency,

$$\frac{d\rho_k}{d\ln k} \equiv k\frac{d\rho_k}{dk} = \frac{k^4}{\pi^2}|c_-(k)|^2, \tag{16.86}$$

thus requires the computation of $c_-(k)$, and then the knowledge of the asymptotic solution of the canonical pertubation equation at large positive times.

To give an explicit example we shall consider here a very simple model consisting of two cosmological phases, an initial accelerated evolution up to the time η_1, and a subsequent radiation-dominated evolution for $\eta > \eta_1$:

$$a \sim (-\eta)^\alpha, \quad \eta < \eta_1,$$
$$a \sim \eta, \quad \eta > \eta_1. \tag{16.87}$$

The effective potential for tensor perturbations in the E-frame, $|a''/a|$, starts from zero at $-\infty$, grows like η^{-2}, reaches a maximum $\sim \eta_1^{-2}$, and vanishes in the radiation phase. We must solve the canonical perturbation equation for $\eta < \eta_1$ and $\eta > \eta_1$. In the first phase the equation reduces to a Bessel equation:

$$\psi_k'' + \left[k^2 - \frac{\alpha(\alpha - 1)}{\eta^2} \right] \psi_k = 0, \tag{16.88}$$

with general solution [60]

$$\psi_k = |\eta|^{1/2} [A H_\nu^{(2)}(|k\eta|) + B H_\nu^{(1)}(|k\eta|)], \quad \nu = |\alpha - 1/2|, \tag{16.89}$$

where $H_\nu^{(1,2)}$ are the first- and second-kind Hankel functions, of argument $k\eta$ and index $\nu = |\alpha - 1/2|$ determined by the kinematics of the background. By using the large argument limit of the Hankel functions for $\eta \to -\infty$,

$$H_\nu^{(2)}(k\eta) \sim \frac{1}{\sqrt{k\eta}} e^{-ik\eta}, \quad H_\nu^{(1)}(k\eta) \sim \frac{1}{\sqrt{k\eta}} e^{+ik\eta}, \tag{16.90}$$

we choose initially a positive frequency mode, normalizing the solution to a vacuum fluctuation spectrum,

$$A = 1/2, \quad B = 0. \tag{16.91}$$

In the second phase $V = 0$, and we have the free oscillating solution:

$$\psi_k = \frac{1}{\sqrt{k}} (c_+ e^{-ik\eta} + c_- e^{+ik\eta}). \tag{16.92}$$

The matching of ψ and ψ' at $\eta = \eta_1$ gives now the coefficients $c_\pm$ (more precisely, the matching would require the continuity of the perturbed metric projected on a spacelike hypersurface containing η_1, and the continuity of the extrinsic curvature of that hypersurface [61]; but in many cases these conditions are equivalent to the continuity of the canonical variable ψ, and of its first time derivative).

For an approximate determination of the spectrum, which is often sufficient for practical purposes, it is convenient to distinguish two regimes, in which the comoving frequency k is much higher or much lower than the frequency associated to the top of the effective potential barrier, $|V(\eta_1)|^{1/2} \simeq \eta_1^{-1}$. In the first case, $k \gg 1/|\eta_1| \equiv k_1$, we can approximate the Hankel functions with their large argument limit, and we find that there is no particle production,

$$|c_+| \simeq 1, \quad |c_-| \simeq 0. \tag{16.93}$$

In practice, c_- is not exactly zero, but is exponentially suppressed as a function of the frequency, just like the quantum reflection probability for a wave with a frequency well above the top of a potential step. We will neglect such an effect here, as we are mainly interested in a qualitative estimate of the perturbation spectrum.

In the second case, $k \ll 1/|\eta_1| \equiv k_1$, we can use the small argument limit of the Hankel functions,

$$H_\nu^{(2)} \sim a(k\eta_1)^\nu - ib(k\eta_1)^{-\nu}, \quad H_\nu^{(1)} \sim a(k\eta_1)^\nu + ib(k\eta_1)^{-\nu}, \tag{16.94}$$

and we find

$$|c_+| \simeq |c_-| \simeq |k\eta_1|^{-\nu-1/2}, \tag{16.95}$$

corresponding to a power-law spectrum:

$$\frac{d\rho_k}{d\ln k} = \frac{k^4}{\pi^2}|c_-(k)|^2 \simeq \frac{k_1^4}{\pi^2}\left(\frac{k}{k_1}\right)^{3-2\nu}, \quad k < k_1, \tag{16.96}$$

with a cut-off frequency $k_1 = \eta_1^{-1}$ controlled by the height of the effective potential.

For a comparison with present observations, it is finally convenient to express the spectrum in terms of the proper frequency, $\omega(t) = k/a(t)$, and in units of critical energy density, $\rho_c(t) = 3M_p^2 H^2(t)/8\pi$. We then obtain the dimensionless spectral distribution,

$$\Omega(\omega, t) = \frac{\omega}{\rho_c(t)}\frac{d\rho(\omega)}{d\omega} \simeq \frac{8}{3\pi}\frac{\omega_1^4}{M_p^2 H^2}\left(\frac{\omega}{\omega_1}\right)^{3-2\nu}$$

$$\simeq g_1^2 \Omega_\gamma(t)\left(\frac{\omega}{\omega_1}\right)^{3-2\nu}, \quad \omega < \omega_1, \tag{16.97}$$

where

$$\omega_1 = \frac{k_1}{a} = \frac{1}{a\eta_1} \simeq \frac{H_1 a_1}{a} \tag{16.98}$$

is the maximal amplified proper frequency, $g_1 = H_1/M_p$, and

$$\Omega_\gamma(t) = \frac{\rho_\gamma}{\rho_c} = \left(\frac{H_1}{H}\right)^2\left(\frac{a_1}{a}\right)^4 \tag{16.99}$$

is the energy density (in critical units) of the radiation that becomes dominant at $t = t_1$, rescaled down at a generic time t (today, $\Omega_\gamma(t_0) \sim 10^{-4}$).

Table 16.3. Slope of the graviton spectrum.

	Scale factor	Bessel index	Spectrum
de Sitter, constant curvature	$\alpha = -1$	$3 - 2\nu = 0$	flat
Power-inflation, decreasing curvature	$\alpha < -1$	$3 - 2\nu < 0$	decreasing
Pre-big bang inflation, growing curvature	$\alpha > -1$	$3 - 2\nu > 0$	increasing

It is important to stress that the amplitude of the spectrum is controlled by $g_1 = H_1/M_p$, i.e. by the curvature scale in Planck units at the time of the transition t_1 (a fundamental parameter of the given inflationary model). The slope of the spectrum, $3 - 2\nu$, is instead controlled by the kinematics of the background. In fact, it depends on the Bessel index ν which, in its turn, depends on α, the power of the scale factor (equation (16.89). The behaviour in frequency of the graviton spectrum, in particular, tends to follow the behaviour of the curvature scale during the epoch of accelerated evolution, see table 16.3.

The standard inflationary scenario is thus characterized by a flat or decreasing graviton spectrum; in string cosmology, instead, we must expect a growing spectrum. This has important phenomenological implications, that will be discussed in the following section.

16.6 The relic graviton background

As discussed in the previous section, one of the most firm predictions of all inflationary models is the amplification of the traceless-transverse part of the quantum fluctuations of the metric tensor, and the formation of a primordial, stochastic background of relic gravitational waves, distributed over a quite large range of frequencies (see [62] for a discussion of the stochastic properties of such a background, and [63] for a possible detection of the associated 'squeezing' [64]).

In a string cosmology context, the expected graviton background has been already discussed in a number of detailed review papers [59, 65, 66]. Here we will summarize the main properties of the background predicted in the context of the pre-big bang scenario.

For a phenomelogical discussion of the spetrum, it is convenient to consider the plane $\{\Omega_G, \omega\}$. In this plane there are three main phenomenological constraints:

- A first constraint comes from the large scale isotropy of the CMB radiation. The degree of anisotropy measured by COBE imposes a bound on the energy density of the graviton background at the scale of the present Hubble radius [67],

$$\Omega_G(\omega_0) \lesssim 10^{-10}, \quad \omega_0 \sim 10^{-18} \text{ Hz}. \tag{16.100}$$

- A second constraint comes from the absence of distortion of the pulsar timing-data [68], and gives the bound

$$\Omega_G(\omega_p) \lesssim 10^{-8}, \quad \omega_p \sim 10^{-8} \text{ Hz.} \tag{16.101}$$

- A third constraint comes from nucleosynthesis [69], which implies that the total graviton energy density, integrated over all modes, cannot exceed the energy density of one massless degree of freedom in thermal equilibrium, evaluated at the nucleosynthesis epoch. This gives a bound for the peak value of the spectrum [33],

$$h_{100} \int d\ln\omega\, \Omega_G(\omega, t_0) \lesssim 0.5 \times 10^{-5},$$

$$h_{100} = H_0/(100 \text{ km s}^{-1} \text{ Mpc}^{-1}), \tag{16.102}$$

which applies to all scales.

A further bound can be obtained by considering the production of primordial black holes [70]. The production of gravitons, in fact, could be associated with the formation of black holes, whose possible evaporation, at the present epoch, is constrained by a number of astrophysical observations. The absence of evaporation imposes an indirect upper limit on the graviton background. In a string cosmology context, however, and in the frequency range of interest for observations, this upper limit is roughly of the same order as the nucleosynthesis bound [70].

For flat or decreasing spectra it is now evident that the more constraining bound is the low-frequency one, obtained from the COBE data. In the standard inflationary scenario, characterized by flat or decreasing spectra (see table 16.3), the maximal allowed graviton background can thus be plotted as in figure 16.8. The flat spectrum corresponds to de Sitter inflation, the decreasing spectra to power inflation. The breakdown in the spectrum, around $\omega_{eq} \sim 10^{-16}$ Hz, is due to the transition from the radiation-dominated to the matter-dominated phase, which only affects the low-frequency part of the spectrum, namely those modes re-entering the horizon in the matter-dominated era. For such modes there is an additional potential barrier in the canonical perturbation equation, which induces an additional amplification $\sim(\omega_{eq}/\omega)^2 > 1$, with respect to the flat de Sitter spectrum.

However, this break of the spectrum is not important for our purposes. What is important is the fact that the observed anisotropy constrains the maximal amplitude of the spectrum. But the amplitude depends on the inflation scale, as stressed in the previous section. From the COBE bound (16.100), imposed on the modified de Sitter spectrum at the Hubble scale,

$$\Omega_G(\omega_0, t_0) = g_1^2 \Omega_\gamma(t)(\omega_{eq}/\omega)^2 \lesssim 10^{-10}, \tag{16.103}$$

we thus obtain a direct constraint on the inflation scale:

$$H_1/M_p \lesssim 10^{-5} \tag{16.104}$$

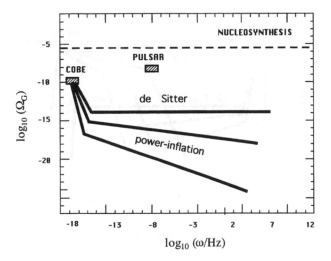

Figure 16.8. Graviton spectra in the standard inflationary scenario.

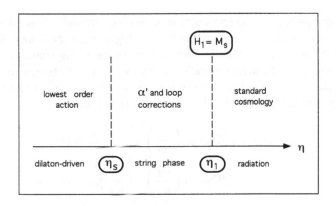

Figure 16.9. A minimal model of the pre-big bang background.

(for power-inflation the bound is even stronger [31]).

This bound applies to all models characterized by a flat or decreasing spectrum. The bound can be evaded, however, if the spectrum is growing, like in the string cosmology context. To illustrate this point, let us consider the simplest class of the so-called 'minimal' pre-big bang models, characterized by three main kinematic phases [32, 44]: an initial low-energy, dilaton-driven phase, an intermediate high-energy 'string' phase, in which α' and loop corrections become important, and a final standard, radiation-dominated phase (see figure 16.9). The timescale η_s marks the transition to the high-curvature phase, and the timescale η_1, characterized by a final curvature of order one in string units, marks the transition to the radiation-dominated cosmology.

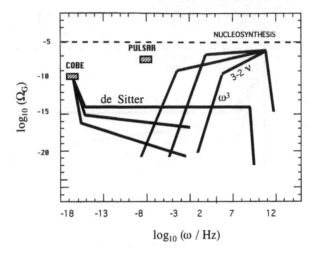

$$\log_{10}(\omega / \text{Hz})$$

Figure 16.10. Graviton spectra in minimal pre-big bang models.

By computing graviton production in this background [32] we find that the spectrum is characterized by two branches: a high-frequency branch, for modes crossing the horizon (or 'hitting' the barrier) in the string phase, $\omega > \omega_s = (a\eta_s)^{-1}$; and a low-frequency branch, for modes crossing the horizon in the initial dilaton phase, $\omega < \omega_s$. The slope is cubic at low frequency, and flatter at high frequency, and the spectrum can be parametrized as follows:

$$\Omega_G(\omega, t_0) \simeq g_1^2 \Omega_\gamma(t_0) \left(\frac{\omega}{\omega_1}\right)^{3-2\nu}, \quad \omega_s < \omega < \omega_1,$$

$$\simeq g_1^2 \Omega_\gamma(t_0) \left(\frac{\omega_1}{\omega_s}\right)^{2\nu} \left(\frac{\omega}{\omega_1}\right)^3, \quad \omega < \omega_s \qquad (16.105)$$

(modulo logarithmic corrections). There are two main parameters: the transition frequency ω_s, and the Bessel index ν, for the high-frequency part of the spectrum. These parameters represent our present ignorance about the duration and the kinematic details of the high curvature phase.

In spite of this uncertainty, however, there is a rather precise prediction for the height and the position of the peak of the spectrum, which turns out to be fixed in terms of the fundamental ratio M_s/M_p as:

$$\Omega_G(\omega_1) \sim 10^{-4}(M_s/M_p)^2, \quad \omega_1 \sim 10^{11}(M_s/M_p)^{1/2} \text{ Hz.} \qquad (16.106)$$

The behaviour of the spectrum, in this class of models, is illustrated in figure 16.10. A precise computation [33] shows that, given the maximal expected value of the string scale [13] ($M_s/M_p \simeq 0.1$), the peak value is automatically compatible with the nucleosynthesis bound, as well as with bounds from the production of primordial black holes.

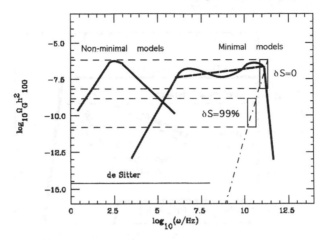

Figure 16.11. Peak and end point of the spectrum in minimal and non-minimal models.

In the minimal models the position of the peak is fixed. Actually, what is really fixed, in string cosmology, is the maximal height of the peak, but not necessarily the position in frequency, and it is not impossible, in more complicated non-minimal models, to shift the peak at lower frequencies.

In minimal models, in fact, the beginning of the radiation phase coincides with the end of the string phase. We may also consider models, however, in which the dilaton coupling $g_s^2 = \exp(\phi)$ is still small at the end of the high curvature phase, and the radiation era begins much later, when $g_s^2 \sim 1$, after an intermediate dilaton-dominated regime. The main difference between the two cases is that in the second case the effective potential which amplifies tensor perturbations is non-monotonic [59], so that there are high-frequency modes re-entering the horizon before the radiation era. As a consequence, the perturbation spectrum is also non-monotonic, and the peak does not coincide any longer, in general, with the end point of the spectrum (see also [71]), as illustrated in figure 16.11.

For minimal models the peak is around 100 GHz, for non-minimal models it could be at lower frequencies. These are good news from an experimental point of view, of course, but non-minimal models seem to be less natural, at least from a theoretical point of view. The boxes around the peak, appearing in figure 16.11, represent the uncertainty in the position of the peak due to our present ignorance about the precise value of the ratio M_s/M_p (for the illustrative purpose of figure 16.11, the ratio is assumed to vary in the range 0.1–0.01). The wavy line, in the high frequency branch of the spectrum, represents the fact that the spectrum associated with the string phase could be monotonic on average, but locally oscillating [72]. Finally, the lower strip, labelled by $\delta s = 99\%$, represents the fact that even the height of the peak could be lower than expected, if the produced gravitons have been diluted by some additional reheating phase, occurring well below the string scale, during the standard evolution.

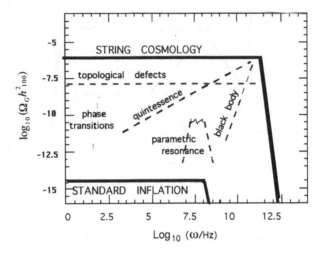

Figure 16.12. Allowed region for the spectrum of vacuum fluctuations in string cosmology and in standard inflation, compared with other relic spectra of primordial origin.

This last effect can be parametrized in terms of δs, which is the fraction of present entropy density in radiation, due to such additional, low-scale reheating. The position of the peak then depends on δs as [33]:

$$\omega_1(t_0) \simeq T_0 \left(\frac{M_s}{M_p}\right)^{1/2} (1 - \delta s)^{1/3},$$

$$\Omega_G(\omega_1, t_0) \simeq 7 \times 10^{-5} h_{100}^{-2} \left(\frac{M_s}{M_p}\right)^2 (1 - \delta s)^{4/3}, \qquad (16.107)$$

where $T_0 = 2.7$ K $\simeq 3.6 \times 10^{11}$ Hz. Such a dependence is not dramatic, however, because even for $\delta s = 99\%$ the peak keeps well above the standard inflationary prediction, represented by the line labelled 'de Sitter' in figure 16.11.

Given the various theoretical uncertainties, the best we can do, at present, is to define the maximal allowed region for the expected graviton background, i.e. the region spanned by the spectrum when all its parameters are varied. Such a region is illustrated in figure 16.12, for the phenomenologically interesting high-frequency range. The figure emphasizes the possible, large enhancement (of about eight orders of magnitude) of the intensity of the background in string cosmology, with respect to the standard inflationary scenario.

It may be useful to stress again the reason of such enhancement. In the standard inflationary scenario the graviton spectrum is decreasing, the normalization is imposed at low frequency, and the peak value is controlled by the anisotropy of the CMB radiation, $\delta T / T \lesssim 10^{-5}$. Thus, at low frequency,

$$\Omega_{\mathrm{G}}(t_0) \lesssim \Omega_\gamma(t_0) \left(\frac{\Delta T}{T}\right)^2_{\mathrm{COBE}} \sim 10^{-14}. \tag{16.108}$$

In string cosmology the spectrum is growing, the normalization is imposed at high frequency, and the peak value is controlled by the fundamental ratio $M_{\mathrm{s}}/M_{\mathrm{p}} \lesssim 0.1$. Thus

$$\Omega_{\mathrm{G}}(t_0) \lesssim \Omega_\gamma(t_0) \left(\frac{M_{\mathrm{s}}}{M_{\mathrm{p}}}\right)^2 \lesssim 10^{-6}. \tag{16.109}$$

The graviton background obtained from the amplification of the vacuum fluctuations, in string cosmology and in standard inflation, is compared in figure 16.12 with other, more unconventional graviton spectra. In particular, the graviton spectrum obtained from cosmic strings and topological defects [73], from bubble collision at the end of a first order phase transition [74], and from a phase of parametric resonance of the inflaton oscillations [75]. Also shown in figure 16.12 is the spectrum from models of quintessential inflation [76], and a thermal black body spectrum for a temperature of about one kelvin. All these cosmological backgrounds are higher than the background expected from the vacuum fluctuations in standard inflation, but not in a string cosmology context.

It may be interesting, at this point, to recall the expected sensitivities of the present, and near future, gravitational antennae, referred to the plots of figure 16.12.

At present, the best direct, experimental upper bound on the energy of a stochastic graviton background comes from the cross-correlation of the data of the two resonant bars NAUTILUS and EXPLORER [77]:

$$\Omega_{\mathrm{G}} h_{100} \lesssim 60, \quad \nu \simeq 907 \,\mathrm{Hz} \tag{16.110}$$

(similar sensitivities are also reached by AURIGA [78]). Unfortunately, the bound is too high to be significant for the plots of figure 16.12. However, a much better sensitivity, $\Omega_{\mathrm{G}} \sim 10^{-4}$ around $\nu \sim 10^3$ Hz, is expected from the present resonant bar detectors, if the integration time of the data is extended to about one year. A similar, or slightly better sensitivity, $\Omega_{\mathrm{G}} \sim 10^{-5}$ around $\nu \sim 10^2$ Hz, is expected from the first operating version of the interferometric detectors, such as LIGO and VIRGO. At high frequency, from the kilohertz to the megahertz range, a promising possibility seems to be the use of resonant electromagnetic cavities as gravity-wave detectors [79]. Work is in progress [80] to attempt to improve their sensitivity.

The present, and near future, available sensitivities of resonant bars and interferometers, therefore, are still outside the allowed region of figure 16.12, determined by the border line

$$\Omega_{\mathrm{G}} h_{100} \simeq 10^{-6}. \tag{16.111}$$

Such sensitivities are not so far from the border, after all, but to get really inside we have to wait for the cross-correlation of two spherical resonant-mass detectors [81], expected to reach $\Omega_G \sim 10^{-7}$ in the kilohertz range, or for the advanced interferometers, expected to reach $\Omega_G \sim 10^{-10}$ in the range of 10^2 Hz. At lower frequencies, around 10^{-2}–10^{-3} Hz, the space interferometer LISA [36] seems to be able to reach very high sensitivities, up to $\Omega_G \sim 10^{-11}$. Work is in progress, however, for a more precise computation of their sensitivity to a cosmic stochastic background [82].

Detectors able to reach, and to cross the limiting sensitivity (16.111), could explore for the first time the parameter space of string cosmology and of Planck scale physics. The detection of a signal from a pre-big bang background, extrapolated to the gigahertz range, could give a first experimental indication on the value of the fundamental ratio M_s/M_p. Even the absence of a signal, inside the allowed region, would be significant, as we could exclude some portion of parameter space of the string cosmology models, obtaining in such a way direct experimental information about processes occuring at (or very near to) the string scale.

16.7   Conclusion

The conclusion of these lectures is very short and simple.

There is a rich structure of stochastic, gravitational-wave backgrounds, of cosmological origin, in the frequency range of present (or planned for the future) gravity-wave detectors.

Among such backgrounds, the stronger one seems to be the background possibly predicted in the context of the pre-big bang scenario, in a string cosmology context, originating at (or very near to) the fundamental string scale. Also, the maximal predicted intensity of the background seems to be accessible to the sensitivity of the future advanced detectors.

If this is the case, the future gravity wave detectors will be able to test string theory models, or perhaps models referring to some more fundamental unified theory, such as D-brane theory, M-theory, and so on. In any case, such detectors will give direct experimental information on Planck scale physics.

Acknowledgments

It is a pleasure to thank the organizers of the *Second SIGRAV School*, and all the staff of the Center 'A. Volta', in Villa Olmo, for their hospitality and perfect organization of this interesting School.

Appendix A. The string effective action

The motion of a point particle in an external gravitational field, $g_{\mu\nu}$, is governed by the action

$$S = -\frac{m}{2} \int d\tau\, \dot{x}^\mu \dot{x}^\nu g_{\mu\nu}(x), \qquad (16.112)$$

where $x^\mu(\tau)$ are the spacetime coordinates of the particle, and τ is an affine parameter along the particle worldline.

The time evolution of a one-dimensional object like a string describes a world-surface, or 'world-sheet', instead of a worldline, and the action governing its motion is given by the surface integral

$$S = -\frac{M_s^2}{2} \int d\tau\, d\sigma\, \sqrt{-\gamma}\, \gamma^{ij} \partial_i x^\mu \partial_j x^\nu g_{\mu\nu}(x), \qquad (16.113)$$

where $\partial_i \equiv \partial/\partial\xi^i$ and $\xi^i = (\tau, \sigma)$ are, respectively, the timelike and spacelike coordinates on the string world-sheet ($i, j = 1, 2$). The coordinates $x^\mu(\tau, \sigma)$ are the fields governing the embedding of the string world-sheet in the external (also called 'target') space. The parameter M_s^2 defines (in units $h/2\pi = 1 = c$) the so-called string tension (the mass per unit length), and its inverse defines the fundamental length scale of the theory (often called, for historical reasons, the α' parameter):

$$M_s^2 \equiv \frac{1}{\lambda_s^2} \equiv \frac{1}{2\pi\alpha'}. \qquad (16.114)$$

In a curved metric background $g_{\mu\nu}$ depends on x^μ, and the nonlinear action (16.113) represents what is called a 'σ-model' defined on the string world sheet.

For the point particle action (16.112) the variation with respect to x^μ leads to the well known geodesic equations of motion,

$$\ddot{x}^\mu + \Gamma_{\alpha\beta}{}^\mu \dot{x}^\alpha \dot{x}^\beta = 0. \qquad (16.115)$$

The string equations of motion are similarly obtained by varying with respect to x^μ the action (16.113): we get then the Euler–Lagrange equations

$$\partial_i \frac{\partial L}{\partial(\partial_i x^\mu)} = \frac{\partial L}{\partial x^\mu}, \qquad L = \sqrt{-\gamma}\, \gamma^{ij} \partial_i x^\mu \partial_j x^\nu g_{\mu\nu}, \qquad (16.116)$$

which can be written explicitly as

$$\Box x^\mu + \gamma^{ij} \partial_i x^\alpha \partial_j x^\beta \Gamma_{\alpha\beta}{}^\mu = 0, \qquad \Box \equiv \frac{1}{\sqrt{-\gamma}} \partial_i \sqrt{-\gamma}\, \gamma^{ij} \partial_j. \qquad (16.117)$$

These equations describe the geodesic evolution of a test string in a given external metric. The variation with respect to γ_{ik} imposes the so-called 'constraints', i.e. the vanishing of the world sheet stress tensor T_{ik},

$$T_{ij} = \frac{2}{\sqrt{-\gamma}} \frac{\delta S}{\delta \gamma^{ij}} = \partial_i x^\mu \partial_j x^\nu g_{\mu\nu} - \frac{1}{2} \gamma_{ij} \partial_k x^\mu \partial^k x_\mu = 0. \qquad (16.118)$$

It is important to note, at this point, that for a classical string it is always possible to impose the so-called 'conformal gauge' in which the world sheet metric is flat, $\gamma_{ij} = \eta_{ij}$. In fact, in an appropriate basis, the two-dimensional metric tensor can always be set in a diagonal form, $\gamma_{ij} = \text{diag}(a, b)$, and then, by using reparametrization invariance on the world sheet, $b^2 \, d\sigma^2 = a^2 \, d\sigma'^2$, the metric can be set in a conformally flat form, $\gamma_{ij} = a^2 \eta_{ij}$. Since the action (16.113) is invariant under the conformal (or Weyl) transformation $\gamma_{ij} \rightarrow \Omega^2(\xi_k)\gamma_{ij}$,

$$\sqrt{-\gamma}\gamma^{ij} \rightarrow \sqrt{\Omega^4}\,\Omega^{-2}\sqrt{-\gamma}\gamma^{ij}, \qquad (16.119)$$

we can always eliminate the conformal factor a^2 in front of the Minkowski metric, by choosing $\Omega = a^{-1}$. In the conformal gauge the equations of motion (16.117) reduce to

$$\ddot{x}^\mu - x''^\mu + \Gamma_{\alpha\beta}{}^\mu (\dot{x}^\alpha + x'^\alpha)(\dot{x}^\beta - x'^\beta) = 0, \qquad (16.120)$$

where $\dot{x} = dx/d\tau$, $x' = dx/d\sigma$, and the constraints (16.118) become

$$g_{\mu\nu}(\dot{x}^\mu \dot{x}^\nu + x'^\mu x'^\nu) = 0, \qquad g_{\mu\nu}\dot{x}^\mu x'^\nu = 0. \qquad (16.121)$$

We now come to the crucial observation which leads to the effective action governing the motion of the background fields. The conformal transformation (16.119) is an invariance of classical theory. Let us require that there are no 'anomalies', i.e. no quantum violations of this classical symmetry. By imposing such a constraint, we will obtain a set of differential equations to be satisfied by the background fields coupled to the string. Thus, unlike a point particle which does not impose any constraint on the external geometry in which it is moving, the consistent quantization of a string gives constraints for the external fields. The background geometry cannot be chosen arbitrarily, but must satisfy the set of equations (also called β-function equations) which guarantee the absence of conformal anomalies. The string effective action used in this paper is the action which reproduces such a set of equations for the background fields, and in particular for the metric.

The derivation of the background equations of motion and of the effective action, from the σ-model action (16.113), can be performed order by order by using a perturbative expansion in powers of α' (indeed, in the limit $\alpha' \rightarrow 0$ the action becomes very large in natural units, so that the quantum corrections are expected to become smaller and smaller). Such a procedure, however, is in general long and complicated, even to lowest order, and a detailed derivation of the background equations is outside the scope of these lectures. Let us sketch here the procedure for the simplest case in which the only external field coupled to the string is the metric tensor $g_{\mu\nu}$.

In the conformal gauge, the action (16.113) becomes:

$$S = -\frac{1}{4\pi\alpha'} \int d^2\xi \, \partial_i x^\mu \partial^i x^\nu g_{\mu\nu}. \qquad (16.122)$$

Let us formally assume a deformation of the number of world sheet dimensions, from 2 to $2 + \epsilon$, and perform the conformal transformation: $\eta_{ij} \to \eta_{ij} \exp(\rho)$. Expanding we get, for small ϵ,

$$
-\frac{1}{4\pi\alpha'} \int d^{2+\epsilon}\xi \, e^{\rho\epsilon/2} \partial_i x^\mu \partial^i x^\nu g_{\mu\nu}
$$
$$
= -\frac{1}{4\pi\alpha'} \int d^{2+\epsilon}\xi \, \partial_i x^\mu \partial^i x^\nu g_{\mu\nu} \left(1 + \frac{\epsilon}{2}\rho + \cdots\right). \qquad (16.123)
$$

For $\epsilon \to 0$ the ρ dependence disappears and the classical action is conformally invariant. In order to preserve this invariance also for the quantum theory, at the one-loop level, let us treat the σ-model as a quantum field theory for $x^\mu(\sigma, \tau)$, and let us consider the quantum fluctuations $\hat{x}^\mu$ around a given expectation value x_0^μ. For the general reparametrization invariance of the theory we can always choose for x_0 a locally inertial frame, such that $g_{\mu\nu}(x_0) = \eta_{\mu\nu}$. By expanding the metric around x_0, the leading corrections are of second order in the fluctuations, because in a locally inertial frame the first derivatives of the metric (and then the Cristoffel connection) can always be set to zero (but not the curvature). With an appropriate choice of coordinates, called Riemann normal coordinates, the metric can thus be expanded as:

$$
g_{\mu\nu}(x) = \eta_{\mu\nu} - \tfrac{1}{3} R_{\mu\nu\alpha\beta}(x_0) \hat{x}^\alpha \hat{x}^\beta + \cdots \qquad (16.124)
$$

and the action for the quantum fluctuations becomes, to lowest order in the curvature,

$$
S = -\frac{1}{4\pi\alpha'} \int d^{2+\epsilon}\xi \left[\partial_i \hat{x}^\mu \partial^i \hat{x}_\mu \left(1 + \frac{\epsilon}{2}\rho\right) \right.
$$
$$
\left. -\frac{1}{3} \partial_i \hat{x}^\mu \partial^i \hat{x}^\nu R_{\mu\nu\alpha\beta}(x_0) \hat{x}^\alpha \hat{x}^\beta \left(1 + \frac{\epsilon}{2}\rho\right) + \cdots \right]. \qquad (16.125)
$$

It must be noted that, at the quantum level, the dependence of ρ does not disappear in general from the action in the limit $\epsilon \to 0$, since there are one-loop terms that diverge like ϵ^{-1}, just to cancel the ϵ dependence and to give a contribution proportional to ρ to the effective action. By evaluating, for instance, the two-point function for the quantum operator $\hat{x}^\alpha \hat{x}^\beta$, in the coincidence limit $\sigma \to \sigma'$ (the tadpole graph), one obtains [83]

$$
\langle \hat{x}^\alpha(\sigma) \hat{x}^\beta(\sigma') \rangle_{\sigma \to \sigma'} \sim \eta^{\alpha\beta} \lim_{\sigma \to \sigma'} \int d^{2+\epsilon}k \, \frac{e^{ik\cdot(\sigma-\sigma')}}{k^2} \sim \eta^{\alpha\beta}\epsilon^{-1}, \qquad (16.126)
$$

which gives the one-loop contribution to the action

$$
\Delta S \sim \int d^{2+\epsilon}\xi \, \partial_i \hat{x}^\mu \partial^i \hat{x}^\nu R_{\mu\nu} \rho. \qquad (16.127)
$$

This term violates, at one loop, the conformal invariance, unless we restrict to a background geometry satisfying the condition

$$R_{\mu\nu} = 0, \tag{16.128}$$

which are just the usual Einstein equations in vacuum.

A similar procedure can be applied if the string moves in a richer external background (not only pure gravity). Indeed, pure gravity is not enough, as a consistent quantum theory for closed bosonic strings, for instance, must contain at least three massless states (besides the unphysical tachyon, removed by supersymmetry) in the lowest energy level: the graviton, the scalar dilaton and the pseudoscalar Kalb–Ramond axion. The σ-model describing the propagation of a string in such a background must thus contain the coupling to the metric, to the dilaton ϕ, and to the two-form $B_{\mu\nu} = -B_{\nu\mu}$:

$$S = -\frac{1}{4\pi\alpha'} \int d^2\xi \, \partial_i x^\mu \partial_j x^\nu (\sqrt{-\gamma}\gamma^{ij} g_{\mu\nu} + \epsilon^{ij} B_{\mu\nu})$$
$$- \frac{1}{4\pi} \int d^2\xi \, \sqrt{-\gamma} \frac{\phi}{2} R^{(2)}(\gamma), \tag{16.129}$$

where ϵ^{ij} is the two-dimensional Levi-Civita tensor density, $\epsilon_{12} = -\epsilon_{21} = 1$, and $R^{(2)}(\gamma)$ is the two-dimensional scalar curvature for the world sheet metric γ. The condition of conformal invariance, at the one-loop level, leads to the equations

$$R_{\mu\nu} + \nabla_\mu\nabla_\nu - \tfrac{1}{4}H_{\mu\alpha\beta}H_\nu{}^{\alpha\beta} = 0, \quad H_{\mu\nu\alpha} = \partial_\mu B_{\nu\alpha} + \partial_\nu B_{\alpha\mu} + \partial_\alpha B_{\mu\nu},$$
$$R + 2\nabla^2\phi - (\nabla\phi)^2 - \tfrac{1}{12}H^2_{\mu\nu\alpha} = 0,$$
$$\nabla_\mu(e^{-\phi}H^{\mu\nu\alpha}) = 0, \tag{16.130}$$

which can be obtained by extremizing the effective action

$$S = -\frac{1}{2\lambda_s^{d-1}} \int d^{d+1}x \, \sqrt{|g|}e^{-\phi}\left[R + (\nabla\phi)^2 - \frac{1}{12}H^2_{\mu\nu\alpha}\right] \tag{16.131}$$

(see appendix C).

It should be noted that the inclusion of the dilaton in the condition of conformal invariance cannot be avoided, since the dilaton coupling in the action (16.129) breaks conformal invariance already at the classical level ($\sqrt{-\gamma}R^{(2)}$ is not invariant under a Weyl rescaling of γ). However, the dilaton term is of order α' with respect to the other terms of the action (for dimensional reasons), so that it is correct to sum up the classical dilaton contribution to the quantum, one-loop effects, as they are all of the same order in α'. Without the dilaton, however, the world sheet curvature density $\sqrt{-\gamma}R^{(2)}$ does not contribute to the string equations of motion, as it is a pure Eulero two-form in two dimensions (just like the Gauss–Bonnet term in four dimensions).

Let me note, finally, that the expansion around x_0 can be continued to higher orders,

$$g(x) = \eta + R\hat{x}\hat{x} + \partial R\hat{x}\hat{x}\hat{x} + R^2\hat{x}\hat{x}\hat{x}\hat{x} + \cdots, \qquad (16.132)$$

thus introducing higher curvature terms, and higher powers of α', in the effective action:

$$S = -\frac{1}{2\lambda_s^{d-1}} \int d^{d+1}x \sqrt{|g|}e^{-\phi} \left[R + (\nabla\phi)^2 - \frac{\alpha'}{4}R_{\mu\nu\alpha\beta}^2 + \cdots \right]. \qquad (16.133)$$

At any given order, unfortunately, there is an intrinsic ambiguity in the action due to the fact that, with an appropriate field redefinition of order α',

$$g_{\mu\nu} \rightarrow g_{\mu\nu} + \alpha'(R_{\mu\nu} + \partial_\mu\phi\partial_\nu\phi + \cdots)$$
$$\phi \rightarrow \phi + \alpha'(R + \nabla^2\phi + \cdots), \qquad (16.134)$$

we obtain a number of different actions, again of the same order in α' (see, for instance, [84]). This ambiguity cannot be eliminated until we limit to an effective action truncated to a given finite order.

The higher curvature (or higher derivative) expansion of the effective action is typical of string theory: it is controlled by the fundamental, minimal length parameter $\lambda_s = (2\pi\alpha')^{1/2}$, in such a way that the higher order corrections disappear in the point-particle limit $\lambda_s \rightarrow 0$. At any given order in α', however, there is also the more conventional expansion in power of the coupling constant g_s (i.e. the loop expansion of quantum field theory: tree-level $\sim g^{-2}$, one-loop $\sim g^0$, two-loop $\sim g^2$, ...). The important observation is that, in a string theory context, the effective coupling constant is controlled by the dilaton. Consider, for instance, a process of graviton scattering, in four dimensions. Comparing the action of (16.4) with the standard, gravitational Einstein action (16.1), it follows that the effective coupling constant, to lowest order, is

$$\sqrt{8\pi G} = \lambda_p = \lambda_s e^{\phi/2} \qquad (16.135)$$

(G is the usual Newton constant). Each loop adds an integer power of the square of the dimensionless coupling constant, which is controlled by the dilaton as

$$g_s^2 = (\lambda_p/\lambda_s)^2 = (M_s/M_p)^2 = e^\phi. \qquad (16.136)$$

We may thus expect, for the loop expansion of the action, the following general scheme:

$$S = -\int e^{-\phi}\sqrt{-g}(R + \nabla\phi^2 + \alpha'R^2 + \cdots) \quad \text{tree level}$$

$$-\int \sqrt{-g}(R + \nabla\phi^2 + \alpha'R^2 + \cdots) \quad \text{one-loop}$$

$$-\int e^{+\phi}\sqrt{-g}(R + \nabla\phi^2 + \alpha'R^2 + \cdots) \quad \text{two-loop}$$

$$\vdots \qquad (16.137)$$

Unfortunately, each term in the action, at each loop order, is multiplied by a dilaton 'form factor' which is different in general for *different fields* and for *different orders*. This difference can lead to an effective violation of the universality of the gravitational interactions [85] in the low-energy, macroscopic regime, and this violation can be reconciled with the present tests of the equivalence principle only if the dilaton is massive enough, to make short enough the range of the non-universal dilatonic interactions.

The tree-level relation (16.136) is valid also for a higher-dimensional effective action, provided e^ϕ represents the shifted four-dimensional dilaton which includes the volume of the extra-dimensional, compact internal space, and which controls the grand-unification gauge coupling, α_{GUT}, as [13]

$$\alpha_{GUT} = \exp\langle\phi\rangle = (M_s/M_p)^2 \sim 0.1\text{--}0.001. \qquad (16.138)$$

However, the relation (16.136) is no longer valid, in general, if the gauge interactions are confined in four dimensions and only gravity propagates in the extra dimensions. In that case the relation depends on the volume of the extra dimensions, whose size may be allowed to be large in Planck units [86]. For internal dimensions of volume V_n the relation becomes, in particular,

$$M_p^2 = M_s^{2+n} V_n e^{-\Phi}, \qquad (16.139)$$

where Φ is the dilaton in $d = 3 + n$ dimensions. In this case, the string mass parameter could be much smaller than the value expected from equation (16.138), provided the internal volume is correspondingly larger.

Appendix B. Duality symmetry

Notations and conventions

In this paper we use the metric signature $(+---)$, and we define the Riemann and Ricci tensor as follows:

$$R_{\mu\nu\alpha}{}^\beta = \partial_\mu \Gamma_{\nu\alpha}{}^\beta + \Gamma_{\mu\rho}{}^\beta \Gamma_{\nu\alpha}{}^\rho - (\mu \leftrightarrow \nu),$$
$$R_{\nu\alpha} = R_{\mu\nu\alpha}{}^\mu.$$

Consider the gravidilaton effective action, in the S-frame, to lowest order in α' and in the quantum loop expansion:

$$S = -\frac{1}{2\lambda_s^{d-1}} \int d^{d+1}x \sqrt{|g|} e^{-\phi} [R + (\nabla\phi)^2]. \qquad (16.140)$$

For a homogeneous, but anisotropic, Bianchi I type metric background:

$$\phi = \phi(t), \quad g_{00} = N^2(t), \quad g_{ij} = -a_i^2(t)\delta_{ij}, \qquad (16.141)$$

we have ($H_i = \dot{a}_i/a_i$):

$$(\nabla\phi)^2 = \frac{\dot{\phi}^2}{N^2}, \quad \sqrt{-g} = N\prod_{i=1}^{d} a_i,$$

$$\Gamma_{00}{}^0 = \frac{\dot{N}}{N} \equiv F, \quad \Gamma_{0i}{}^j = H_i\delta_i^j, \quad \Gamma_{ij}{}^0 = \frac{a_i\dot{a}_i}{N^2}\delta_{ij},$$

$$R = \frac{1}{N^2}\left[2F\sum_i H_i - 2\sum_i \dot{H}_i - \sum_i H_i^2 - \left(\sum_i H_i\right)^2\right]. \quad (16.142)$$

By noting that

$$\frac{\mathrm{d}}{\mathrm{d}t}\left[2\frac{\mathrm{e}^{-\phi}}{N}\left(\prod_{k=1}^{d} a_k\right)\left(\sum_i H_i\right)\right]$$

$$= \frac{\mathrm{e}^{-\phi}}{N}\left(\prod_{k=1}^{d} a_k\right)\left[2\sum_i \dot{H}_i - 2F\sum_i H_i - 2\dot{\phi}\sum_i H_i + 2\left(\sum_i H_i\right)^2\right], \quad (16.143)$$

the action (16.140), modulo a total derivative, can be rewritten as:

$$S = -\frac{1}{2\lambda_{\mathrm{s}}^{d-1}}\int \mathrm{d}^d x\,\mathrm{d}t \prod_{i=1}^{d} a_i \frac{\mathrm{e}^{-\phi}}{N}\left[\dot{\phi}^2 - \sum_i H_i^2 + \left(\sum_i H_i\right)^2 - 2\dot{\phi}\sum_i H_i\right]. \quad (16.144)$$

We now introduce the so-called shifted dilaton $\bar{\phi}$, defined by

$$\mathrm{e}^{-\bar{\phi}} = \int \frac{\mathrm{d}^d x}{\lambda_{\mathrm{s}}^d}\prod_{i=1}^{d} a_i \mathrm{e}^{-\phi}, \quad (16.145)$$

from which

$$\phi = \bar{\phi} + \sum_i \ln a_i, \quad \dot{\phi} = \dot{\bar{\phi}} + \sum_i H_i \quad (16.146)$$

(by assuming spatial sections of finite volume, ($\int \mathrm{d}^d x\,\sqrt{|g|})_{t=\mathrm{constant}} < \infty$, we have absorbed into ϕ the constant shift $-\ln(\lambda_{\mathrm{s}}^{-d}\int \mathrm{d}^d x)$, required to secure the scalar behaviour of $\bar{\phi}$ under coordinate reparametrizations preserving the comoving gauge). The action becomes:

$$S = -\frac{\lambda_{\mathrm{s}}}{2}\int \mathrm{d}t\,\frac{\mathrm{e}^{-\bar{\phi}}}{N}\left(\dot{\bar{\phi}}^2 - \sum_i H_i^2\right). \quad (16.147)$$

By inverting one of the d scale factors the corresponding Hubble parameter changes sign,

$$a_i \to \tilde{a}_i = a_i^{-1}, \quad H_i \to \tilde{H}_i = \frac{\dot{\tilde{a}}_i}{\tilde{a}_i} = a_i\frac{\mathrm{d}a_i^{-1}}{\mathrm{d}t} = -H_i, \quad (16.148)$$

so that the quadratic action (16.147) is clearly invariant under the inversion of any scale factor preserving the shifted dilaton,

$$a_i \to \tilde{a}_i = a_i^{-1}, \quad \bar{\phi} \to \bar{\phi} \qquad (16.149)$$

(for 'scale factor duality', see [9] and the first paper of [8]).

In order to derive the field equations, it is convenient to use the variables $\beta_i = \ln a_i$, so that $H_i = \dot{\beta}_i$, $\dot{H}_i = \ddot{\beta}_i$, and the action (16.147) is cyclic in β_i. By varying with respect to N, β_i and $\bar{\phi}$, and subsequently fixing the cosmic time gauge $N = 1$, we obtain, respectively,

$$\dot{\bar{\phi}}^2 - \sum_i H_i^2 = 0, \qquad (16.150)$$

$$\dot{H}_i - H_i \dot{\bar{\phi}} = 0, \qquad (16.151)$$

$$2\ddot{\bar{\phi}} - \dot{\bar{\phi}}^2 - \sum_i H_i^2 = 0. \qquad (16.152)$$

This is a system of $(d+2)$ equations for the $(d+1)$ variables $\{a_i, \phi\}$. However, only $(d+1)$ equations are independent (see, for instance, [21]: equation (16.150) represents a constraint on the set of initial data).

The above equations are invariant under a time reversal transformation

$$t \to -t, \quad H \to -H, \quad \dot{\bar{\phi}} \to -\dot{\bar{\phi}}, \qquad (16.153)$$

and also under the duality transformation (16.149). If we invert $k \leq d$ scale factors, $\tilde{a}_1 = a_1^{-1}, \ldots, \tilde{a}_k = a_k^{-1}$, the shifted dilaton is preserved, $\bar{\phi} = \tilde{\bar{\phi}}$, provided

$$\bar{\phi} = \phi - \sum_{i=1}^{d} \ln a_i = \tilde{\phi} - \sum_{i=1}^{k} \ln \tilde{a}_i - \sum_{i=k+1}^{d} \ln a_i, \qquad (16.154)$$

from which:

$$\tilde{\phi} = \phi - 2 \sum_{i=1}^{k} \ln a_i. \qquad (16.155)$$

Given an exact solution, represented by the set of variables

$$\{a_1, \ldots, a_d, \phi\}, \qquad (16.156)$$

the inversion of $k \leq d$ scale factors defines then a new exact solution, represented by the set of variables

$$\{a_1^{-1}, \ldots, a_k^{-1}, a_{k+1}, \ldots, a_d, \phi - 2\ln a_1, \ldots, -2\ln a_k\}. \qquad (16.157)$$

By inverting all the scale factors we obtain the transformation

$$\{a_i, \phi\} \rightarrow \{a_i^{-1}, \phi - 2 \sum_{i=1}^{d} \ln a_i\} \tag{16.158}$$

which, in the isotropic case, corresponds in particular to the duality transformation (16.5).

As a simple example, we consider here the particular isotropic solution

$$a = t^{1/\sqrt{d}}, \quad \overline{\phi} = -\ln t, \tag{16.159}$$

which satisfies identically the set of equations (16.150)–(16.152). By applying a duality *and* a time-reversal transformation we obtain the four different exact solutions

$$\{a_\pm(t) = t^{\pm 1/\sqrt{d}}, \overline{\phi}(t) = -\ln t\},$$
$$\{a_\pm(-t) = (-t)^{\pm 1/\sqrt{d}}, \overline{\phi}(-t) = -\ln(-t)\}, \tag{16.160}$$

corresponding to the four branches illustrated in figure 16.2, and describing decelerated expansion, $a_+(t)$, decelerated contraction, $a_-(t)$, accelerated contraction, $a_+(-t)$, accelerated expansion, $a_-(-t)$. The solution describes expansion or contraction if the sign of $\dot{a}$ is positive or negative, repectively, and the solution is accelerated or decelerated if $\dot{a}$ and $\ddot{a}$ have the same or the opposite sign, respectively.

It is important to consider also the dilaton behaviour. According to equation (16.146):

$$\phi_\pm(\pm t) = \overline{\phi}(\pm t) + d \ln a_\pm(\pm t) = (\pm \sqrt{d} - 1) \ln(\pm t). \tag{16.161}$$

It follows that, in a phase of growing curvature ($t < 0, t \rightarrow 0_-$), the dilaton is growing only for an expanding metric, $a_-(-t)$. This means that, in the isotropic case, there are only expanding pre-big bang solutions, i.e. solutions evolving from the string perturbative vacuum ($H \rightarrow 0, \phi \rightarrow -\infty$), and then characterized by a growing string coupling, $\dot{g}_s = (\exp \phi / 2)^{\cdot} > 0$.

In the more general, anisotropic case, and in the presence of contracting dimensions, a growing curvature solution is associated to a growing dilaton only for a large enough number of contracting dimensions. To make this point more precisely, consider the particular, exact solution of equations (16.150)–(16.152) with d expanding and n contracting dimensions, and scale factors $a(t)$ and $b(t)$, respectively:

$$a = (-t)^{-1/\sqrt{d+n}}, \quad b = (-t)^{1/\sqrt{d+n}}, \quad \overline{\phi} = -\ln(-t), \quad t \rightarrow 0_-. \tag{16.162}$$

This gives, for the dilaton,

$$\phi = \overline{\phi} + d \ln a + n \ln b = \frac{n - d - \sqrt{d+n}}{\sqrt{d+n}} \ln(-t), \tag{16.163}$$

so that the dilaton is growing if

$$d + \sqrt{d+n} > n. \tag{16.164}$$

For $n = 6$, in particular, this condition requires $d > 3$. This could represent a potential difficulty for the pre-big bang scenario, which might be solved, however, by quantum cosmology effects [87].

The scale factor duality of the action (16.147) is, in general, broken by the addition of a non-trivial dilaton potential (unless the potential depends on the dilaton through $\bar\phi$, of course). When the antisymmetric tensor $B_{\mu\nu}$ is included in the action, however, the scale factor duality can be lifted to a larger group of global symmetry transformations. To illustrate this important aspect of the string cosmology equations, we will consider here a set of cosmological background fields $\{\phi, g_{\mu\nu}, B_{\mu\nu}\}$, for which a synchronous frame exists where $g_{00} = 1$, $g_{0i} = 0$, $B_{0\mu} = 0$, and all the components ϕ, g_{ij}, B_{ij} do not depend on the spatial coordinates.

Let us write the action

$$S = -\frac{1}{2\lambda_s^{d-1}} \int d^{d+1}x \, \sqrt{|g|} e^{-\phi} \left[R + (\nabla\phi)^2 - \frac{1}{12} H_{\mu\nu\alpha}^2 \right] \tag{16.165}$$

directly in the synchronous gauge, as we are not interested in the field equations, but only in the symmetries of the action. We set $g_{ij} = -\gamma_{ij}$ and we find, in this gauge,

$$\Gamma_{ij}{}^0 = \tfrac{1}{2}\dot\gamma_{ij}, \quad \Gamma_{0i}{}^j = \tfrac{1}{2}g^{jk}\dot g_{ik} = \tfrac{1}{2}(g^{-1}\dot g)_i{}^j = (\gamma^{-1}\dot\gamma)_i{}^j$$
$$R_0{}^0 = -\tfrac{1}{4}\mathrm{Tr}(\gamma^{-1}\dot\gamma)^2 - \tfrac{1}{2}\mathrm{Tr}(\gamma^{-1}\ddot\gamma) - \tfrac{1}{2}\mathrm{Tr}(\dot\gamma^{-1}\dot\gamma),$$
$$R_i{}^j = -\tfrac{1}{2}(\gamma^{-1}\ddot\gamma)_i{}^j - \tfrac{1}{4}(\gamma^{-1}\dot\gamma)_i{}^j \mathrm{Tr}(\gamma^{-1}\dot\gamma) + \tfrac{1}{2}(\gamma^{-1}\dot\gamma\gamma^{-1}\dot\gamma)_i{}^j, \tag{16.166}$$

where

$$\mathrm{Tr}(\gamma^{-1}\dot\gamma) = (\gamma^{-1})^{ij}\dot\gamma_{ji} = g^{ij}\dot g_{ji}, \tag{16.167}$$

and so on (note also that $\dot\gamma^{-1}$ means $(\gamma^{-1})'$). Similarly we find, for the antisymmetric tensor,

$$H_{0ij} = \dot B_{ij}, \quad H^{0ij} = g^{ik}g^{jl}\dot B_{kl} = (\gamma^{-1}\dot B\gamma^{-1})^{ij},$$
$$H_{\mu\nu\alpha}H^{\mu\nu\alpha} = 3H_{0ij}H^{0ij} = -3\mathrm{Tr}(\gamma^{-1}\dot B)^2. \tag{16.168}$$

Let us introduce the shifted dilaton, by absorbing into ϕ the spatial volume, as before:

$$\sqrt{|\det g_{ij}|} e^{-\phi} = e^{-\bar\phi}, \tag{16.169}$$

from which

$$\dot{\bar\phi} = \dot\phi - \frac{1}{2}\frac{d}{dt}\ln(\det\gamma) = \dot\phi - \frac{1}{2}\mathrm{Tr}(\gamma^{-1}\dot\gamma). \tag{16.170}$$

By collecting the various contributions from ϕ, R and H^2, the action (16.165) can be rewritten as:

$$S = -\frac{\lambda_s}{2} \int dt \, e^{-\bar{\phi}} [\dot{\bar{\phi}}^2 + \tfrac{1}{4} \text{Tr}(\gamma^{-1}\dot{\gamma})^2 - \text{Tr}(\gamma^{-1}\ddot{\gamma})$$
$$- \tfrac{1}{2} \text{Tr}(\dot{\gamma}^{-1}\dot{\gamma}) + \dot{\bar{\phi}} \, \text{Tr}(\gamma^{-1}\dot{\gamma}) + \tfrac{1}{4} \text{Tr}(\gamma^{-1}\dot{B})^2]. \quad (16.171)$$

We can now eliminate the second derivatives, and the mixed term $(\sim \dot{\bar{\phi}}\dot{\gamma})$, by noting that

$$\frac{d}{dt}[e^{-\bar{\phi}} \, \text{Tr}(\gamma^{-1}\dot{\gamma})] = e^{-\bar{\phi}}[\text{Tr}(\gamma^{-1}\ddot{\gamma}) + \text{Tr}(\dot{\gamma}^{-1}\dot{\gamma}) - \dot{\bar{\phi}} \, \text{Tr}(\gamma^{-1}\dot{\gamma})]. \quad (16.172)$$

Finally, by using the identity,

$$(\gamma^{-1})^{\cdot} = -\gamma^{-1}\dot{\gamma}\gamma^{-1} \quad (16.173)$$

(following from $g^{-1}g = \gamma^{-1}\gamma = I$), we can rewrite the action in quadratic form, modulo a total derivative, as

$$S = -\frac{\lambda_s}{2} \int dt \, e^{-\bar{\phi}} \left[\dot{\bar{\phi}}^2 - \frac{1}{4} \text{Tr}(\gamma^{-1}\dot{\gamma})^2 + \frac{1}{4} \text{Tr}(\gamma^{-1}\dot{B})^2 \right]. \quad (16.174)$$

This action can be set into a more compact form by using the $(2d \times 2d)$ matrix M, defined in terms of the spatial components of the metric and of the antisymmetric tensor,

$$M = \begin{pmatrix} G^{-1} & -G^{-1}B \\ BG^{-1} & G - BG^{-1}B \end{pmatrix},$$
$$G = g_{ij} \equiv -\gamma_{ij}, \quad G^{-1} = g^{ij}, \quad B \equiv B_{ij}, \quad (16.175)$$

and using also the matrix η, representing the invariant metric of the $O(d,d)$ group in the off-diagonal representation,

$$\eta = \begin{pmatrix} 0 & I \\ I & 0 \end{pmatrix} \quad (16.176)$$

(I is the unit d-dimensional matrix). By computing $M\eta$, $\dot{M}\eta$ and $(\dot{M}\eta)^2$ we find, in fact,

$$\text{Tr}(\dot{M}\eta)^2 = 2\,\text{Tr}[\dot{\gamma}^{-1}\dot{\gamma} + (\gamma^{-1}\dot{B})^2] = 2\,\text{Tr}[-(\gamma^{-1}\dot{\gamma})^2 + (\gamma^{-1}\dot{B})^2], \quad (16.177)$$

and the action becomes

$$S = -\frac{\lambda_s}{2} \int dt \, e^{-\bar{\phi}} \left[\dot{\bar{\phi}}^2 + \frac{1}{8} \text{Tr}(\dot{M}\eta)^2 \right]. \quad (16.178)$$

We may note, at this point, that M is a symmetric matrix of the pseudo-orthogonal $O(d, d)$ group. In fact,

$$M^{T}\eta M = \eta, \quad M = M^{T} \tag{16.179}$$

for any B and G. Therefore,

$$M\eta = \eta M^{-1}, \quad (\dot{M}\eta)^2 = \eta(M^{-1})^{\cdot}M\eta, \tag{16.180}$$

and the action can be finally rewritten as

$$S = -\frac{\lambda_s}{2}\int dt\, e^{-\bar{\phi}}\left[\dot{\bar{\phi}}^2 + \frac{1}{8}\operatorname{Tr}\dot{M}(M^{-1})^{\cdot}\right]. \tag{16.181}$$

This form is explicitly invariant under the global $O(d, d)$ transformations (16.10), preserving the shifted dilaton:

$$\bar{\phi} \to \bar{\phi}, \quad M \to \Lambda^{T}M\Lambda, \quad \Lambda^{T}\eta\Lambda = \eta. \tag{16.182}$$

In fact

$$\operatorname{Tr}\dot{\tilde{M}}(\tilde{M}^{-1})^{\cdot} = \operatorname{Tr}[\Lambda^{T}\dot{M}\Lambda\Lambda^{-1}(M^{-1})^{\cdot}(\Lambda^{T})^{-1}] = \operatorname{Tr}\dot{M}(M^{-1})^{\cdot}. \tag{16.183}$$

In the absence of the antisymmetric tensor M is diagonal, and the special $O(d, d)$ transformation with $\Lambda = \eta$ corresponds to an inversion of the metric tensor:

$$M = \operatorname{diag}(G^{-1}, G),$$
$$\tilde{M} = \Lambda^{T}M\Lambda = \eta M\eta = \operatorname{diag}(G, G^{-1}) \Longrightarrow \tilde{G} = G^{-1}. \tag{16.184}$$

For a diagonal metric $G = a^2 I$, and the invariance under the scale factor duality transformation (16.5) is recovered as a particular case of the global $O(d, d)$ symmetry of the low-energy effective action.

Appendix C. The string cosmology equations

In order to derive the cosmological equations let us include in the action, for completeness, the antisymmetric tensor $B_{\mu\nu}$, a dilaton potential $V(\phi)$, and also the possible contribution of other matter sources represented by a Lagrangian density L_m:

$$S = -\frac{1}{2\lambda_s^{d-1}}\int d^{d+1}x\,\sqrt{|g|}e^{-\phi}\left[R + (\nabla\phi)^2 - \frac{1}{12}H_{\mu\nu\alpha}^2 + V(\phi)\right]$$
$$+ \int d^{d+1}x\,\sqrt{|g|}L_m. \tag{16.185}$$

In a scalar–tensor model of gravity, especially in the presence of higher derivative interactions, it is often convenient to write the action in the language of exterior differential forms, as this may simplify the variational procedure (see, for instance, [88]). Here, we will follow however the more traditional approach, by varying the action with respect to $g_{\mu\nu}$, ϕ and $B_{\mu\nu}$. We shall take into account the dynamical stress tensor $T_{\mu\nu}$ of the matter sources (defined in the usual way), as well as the scalar source σ representing a possible direct coupling of the dilaton to the matter fields:

$$\delta_g(\sqrt{-g}L_{\rm m}) = \tfrac{1}{2}\sqrt{-g}T_{\mu\nu}\delta g^{\mu\nu}, \quad \delta_\phi(\sqrt{-g}L_{\rm m}) = \sqrt{-g}\sigma\,\delta\phi. \qquad (16.186)$$

We start performing the variation with respect to the metric, using the standard, general relativistic results:

$$\delta\sqrt{-g} = -\tfrac{1}{2}\sqrt{-g}g_{\mu\nu}\delta g^{\mu\nu},$$

$$\delta(\sqrt{-g}R) = \sqrt{-g}(G_{\mu\nu}\delta g^{\mu\nu} + g_{\mu\nu}\nabla^2\delta g^{\mu\nu} - \nabla_\mu\nabla_\nu\delta g^{\mu\nu}), \qquad (16.187)$$

where $G_{\mu\nu}$ is the usual Einstein tensor. It must be noted, however, that the second covariant derivatives of $\delta g^{\mu\nu}$, when integrated by parts, are no longer equivalent to a divergence (and then to a surface integral), because of the dilaton factor $\exp(-\phi)$ in front of the Einstein action, which adds dilatonic gradients to the full variation. By performing a first integration by part, and using the metricity condition $\nabla_\alpha g_{\mu\nu} = 0$, we get in fact:

$$\delta_g S = \frac{1}{2}\int d^{d+1}x\,\sqrt{|g|}T_{\mu\nu}\delta g^{\mu\nu} - \frac{1}{2\lambda_{\rm s}^{d-1}}\int d^{d+1}x\,\sqrt{|g|}e^{-\phi}$$

$$\times\,[G_{\mu\nu} + \nabla_\alpha\phi g_{\mu\nu}\nabla^\alpha - \nabla_\mu\phi\nabla_\nu + \nabla_\mu\phi\nabla_\nu\phi - \tfrac{1}{2}g_{\mu\nu}(\nabla\phi)^2$$

$$-\,\tfrac{1}{2}g_{\mu\nu}V(\phi) + \tfrac{1}{2}g_{\mu\nu}\tfrac{1}{12}H_{\alpha\beta\gamma}^2 - \tfrac{3}{12}H_{\mu\alpha\beta}H_\nu{}^{\alpha\beta}]\delta g^{\mu\nu}$$

$$-\,\frac{1}{2\lambda_{\rm s}^{d-1}}\int d^{d+1}x\,\sqrt{|g|}\nabla_\alpha[e^{-\phi}g_{\mu\nu}\nabla^\alpha\delta g^{\mu\nu} - e^{-\phi}\nabla_\nu\delta g^{\nu\alpha}] = 0.$$

$$(16.188)$$

A second integration by parts of $\nabla\delta g^{\mu\nu}$ cancels the bilinear term $\nabla_\mu\phi\nabla_\nu\phi$, and leads to the field equations:

$$G_{\mu\nu} + \nabla_\mu\nabla_\nu\phi + \tfrac{1}{2}g_{\mu\nu}[(\nabla\phi)^2 - 2\nabla^2\phi - V(\phi) + \tfrac{1}{12}H_{\alpha\beta\gamma}^2] - \tfrac{1}{4}H_{\mu\alpha\beta}H_\nu{}^{\alpha\beta}$$

$$= \tfrac{1}{2}e^\phi T_{\mu\nu}. \qquad (16.189)$$

We have chosen units such that $2\lambda_{\rm s}^{d-1} = 1$, so that e^ϕ represents the $(d+1)$-dimensional gravitational constant (see appendix A). Also, we have implicitly added to the action the boundary term

$$\frac{1}{2\lambda_{\rm s}^{d-1}}\int_{\partial\Omega}\sqrt{|g|}e^{-\phi}K^\alpha\,d\Sigma_\alpha, \qquad (16.190)$$

whose variation with respect to g exactly cancels the contribution of the total divergence appearing in the last integral of equation (16.188):

$$\delta_g \int \sqrt{|g|}e^{-\phi} K^{\alpha} \, d\Sigma_{\alpha} = \int \sqrt{|g|}e^{-\phi} (g_{\mu\nu}\nabla^{\alpha}\delta g^{\mu\nu} - \nabla_{\nu}\delta g^{\nu\alpha}) \, d\Sigma_{\alpha}. \quad (16.191)$$

Here K^{α} is a geometric term representing the so-called extrinsic curvature on the d-dimensional closed hypersurface, of infinitesimal area $d\Sigma_{\alpha}$, bounding the total spacetime volume over which we are varying the action. Note that the integral (16.190) differs from the usual boundary term, used in general relativity [89] to derive the Einstein equations, only by the presence of the tree-level dilaton coupling $e^{-\phi}$ to the extrinsic curvature.

Let us now perform the variation with respect to the dilaton, again in units $2\lambda_s^{d-1} = 1$. We get the Euler–Lagrange equations:

$$\partial_{\mu}[-2\sqrt{-g}e^{-\phi}\partial^{\mu}\phi] = e^{-\phi}\sqrt{-g}[R + (\nabla\phi)^2 - \tfrac{1}{12}H^2 + V]$$
$$- e^{-\phi}\sqrt{-g}V' + \sqrt{-g}\sigma \quad (16.192)$$

(where $V' = \partial V/\partial\phi$), from which

$$R + 2\nabla^2\phi - (\nabla\phi)^2 + V - V' - \tfrac{1}{12}H^2 + e^{\phi}\sigma = 0. \quad (16.193)$$

The variation with respect to $B_{\mu\nu}$,

$$\delta_B \int d^{d+1}x \sqrt{|g|}e^{-\phi}(\partial_{\mu}B_{\nu\alpha})H^{\mu\nu\alpha} = 0, \quad (16.194)$$

gives finally

$$\partial_{\mu}(\sqrt{|g|}e^{-\phi}H^{\mu\nu\alpha}) = 0 = \nabla_{\mu}(e^{-\phi}H^{\mu\nu\alpha}). \quad (16.195)$$

Equations (16.189), (16.193) and (16.195) are the equations governing the evolution of the string cosmology background, at low energy. Note that equation (16.189) can also be given in a simplified form: if we eliminate the scalar curvature present inside the Einstein tensor, by using the dilaton equation (16.193), we obtain:

$$R_{\mu}{}^{\nu} + \nabla_{\mu}\nabla^{\nu}\phi - \tfrac{1}{2}\delta_{\mu}^{\nu}V' - \tfrac{1}{4}H_{\mu\alpha\beta}H^{\nu\alpha\beta} = \tfrac{1}{2}e^{\phi}(T_{\mu}{}^{\nu} - \delta_{\mu}^{\nu}\sigma). \quad (16.196)$$

For the purpose of these lectures, it will be enough to derive some simple solution of the string cosmology equations in the absence of the potential ($V = 0$), of the antisymmetric tensor ($B = 0$), and with a perfect fluid, minimally coupled to the dilaton ($\sigma = 0$), as the matter sources. Assuming for the background a Bianchi I type metric, we can work in the synchronous gauge, by setting

$$g_{\mu\nu} = \text{diag}(1, -a_i^2\delta_{ij}), \quad a_i = a_i(t), \quad \phi = \phi(t),$$
$$T_{\mu}{}^{\nu} = \text{diag}(\rho, -p_i^2\delta_i^j), \quad p_i/\rho = \gamma_i = \text{constant}, \quad \rho = \rho(t). \quad (16.197)$$

For this background:

$$\Gamma_{0i}{}^j = H_i \delta_i^j, \quad \Gamma_{ij}{}^0 = a_i \dot{a}_i \delta_{ij}, \quad R_0{}^0 = -\sum_i (\dot{H}_i + H_i^2),$$

$$R_i{}^j = -\dot{H}_i \delta_i^j - H_i \delta_i^j \sum_k H_k, \quad R = -\sum_i (2\dot{H}_i + H_i^2) - \left(\sum_i H_i\right)^2,$$

$$(\nabla\phi)^2 = \dot{\phi}^2, \quad \nabla^2\phi = \ddot{\phi} + \sum_i H_i \dot{\phi}, \quad \nabla_0\nabla^0\phi = \ddot{\phi},$$

$$\nabla_i\nabla^j\phi = H_i\dot{\phi}\delta_i^j. \tag{16.198}$$

The dilaton equation (16.193) gives then

$$2\ddot{\phi} + 2\dot{\phi}\sum_i H_i - \dot{\phi}^2 - \sum_i (2\dot{H}_i + H_i^2) - \left(\sum_i H_i\right)^2 = 0. \tag{16.199}$$

The (00) component of the equation (16.189) gives

$$\dot{\phi}^2 - 2\dot{\phi}\sum_i H_i - \sum_i H_i^2 + \left(\sum_i H_i\right)^2 = e^\phi \rho. \tag{16.200}$$

The diagonal, spatial components (i, i) of equation (16.189) (the off-diagonal components are trivially satisfied) give

$$\dot{H}_i + H_i \sum_k H_k - H_i\dot{\phi} - \frac{1}{2}\sum_i (2\dot{H}_i + H_i^2)$$

$$-\frac{1}{2}\left(\sum_i H_i\right)^2 - \frac{1}{2}\dot{\phi}^2 + \ddot{\phi} + \dot{\phi}\sum_i H_i = \frac{1}{2}e^\phi p_i. \tag{16.201}$$

The last five terms on the left-hand side add to zero because of the dilaton equation (16.199), and the spatial equations reduce to

$$\dot{H}_i - H_i\left(\dot{\phi} - \sum_k H_k\right) = \frac{1}{2}e^\phi p_i. \tag{16.202}$$

The above equations are clearly invariant under time-reversal, $t \to -t$. In order to make explicit also their duality invariance, let us introduce again the shifted dilaton (see equation (16.146)), such that

$$e^{\bar{\phi}} = e^\phi/\sqrt{-g}, \quad \dot{\bar{\phi}} = \dot{\phi} - \sum_i H_i, \tag{16.203}$$

and define

$$\bar{\rho} = \rho\sqrt{-g} = \rho \prod_i a_i, \quad \bar{p} = p\sqrt{-g} = p \prod_i a_i. \tag{16.204}$$

In terms of these variables, the time and space equations (16.200) and (16.202), and the dilaton equation (16.199), become, respectively:

$$\dot{\bar{\phi}}^2 - \sum_i H_i^2 = e^{\bar{\phi}} \bar{\rho},$$
(16.205)

$$\dot{H_i} - H_i \dot{\bar{\phi}} = \tfrac{1}{2} e^{\bar{\phi}} \bar{p}_i,$$
(16.206)

$$2\ddot{\bar{\phi}} - \dot{\bar{\phi}}^2 - \sum_i H_i^2 = 0.$$
(16.207)

They are explicitly invariant under the scale-factor duality transformation:

$$a_i \rightarrow a_i^{-1}, \quad \bar{\phi} \rightarrow \bar{\phi}, \quad \bar{\rho} \rightarrow \bar{\rho}, \quad \bar{p} \rightarrow -\bar{p}$$
(16.208)

which implies, for a perfect fluid source, a 'reflection' of the equation of state, $\gamma = p/\rho = \bar{p}/\bar{\rho} \rightarrow -\bar{p}/\bar{\rho} = -\gamma$ (see the first paper in [8]). A general $O(d, d)$ transformation changes, however, the equation of state in a more drastic way (see [12]), introducing also shear and bulk viscosity.

The above $(d + 2)$ equations are a system of independent equations for the $(d + 2)$ variables $\{a_i, \phi, \rho\}$. Their combination implies the usual covariant conservation of the energy density. By differentiating equation (16.205), and using (16.206) and (16.207) to eliminate $\dot{H_i}$, $\ddot{\bar{\phi}}$, respectively, we get in fact

$$\dot{\bar{\rho}} + \sum_i H_i \bar{p}_i = 0,$$
(16.209)

which, using the definitions (16.204), is equivalent to

$$\dot{\rho} + \sum_i H_i (\rho + p_i) = 0.$$
(16.210)

In order to obtain exact solutions, it is convenient to include this energy conservation equation in the full system of independent equations.

In these lectures we will present only a particular example of the matter-dominated solution by considering a d-dimensional, isotropic background characterized by a power-law evolution,

$$a \sim t^\alpha, \quad \bar{\phi} \sim -\beta \ln t, \quad p = \gamma \rho.$$
(16.211)

We use (16.205), (16.207) and (16.209) as independent equations. The integration of equation (16.209) gives immediately

$$\bar{\rho} = \rho_0 a^{-d\gamma};$$
(16.212)

equation (16.205) is then satisfied provided

$$d\gamma\alpha + \beta = 2.$$
(16.213)

Finally, equation (16.207) provides the constraint

$$2\beta - \beta^2 - d\alpha^2 = 0. \tag{16.214}$$

We then have a system of two equations for the two parameters α, β (note that, if α is a solution for a given γ, then also $-\alpha$ is a solution, associated to $-\gamma$). We have, in general, two solutions. The trivial flat space solution $\beta = 2, \alpha = 0$, corresponds to dust matter ($\gamma = 0$) according to equation (16.206). For $\gamma \neq 0$ we obtain instead

$$\alpha = \frac{2\gamma}{1 + d\gamma^2}, \quad \beta = \frac{2}{1 + d\gamma^2}, \tag{16.215}$$

which fixes the time evolution of a and $\overline{\phi}$:

$$a \sim t^{\frac{2\gamma}{1+d\gamma^2}}, \quad \overline{\phi} = -\frac{2}{1 + d\gamma^2} \ln t, \tag{16.216}$$

and also of the more conventional variables ρ, ϕ:

$$\rho = \overline{\rho} a^{-d} = \rho_0 a^{-d(1+\gamma)}, \quad \phi = \overline{\phi} + d \ln a = \frac{2(d\gamma - 1)}{1 + d\gamma^2} \ln t. \tag{16.217}$$

This particular solution reproduces the small curvature limit of the general solution with perfect fluid sources (see the last two papers of [8]), sufficiently far from the singularity. As in the vacuum solution (16.159) there are four branches, related by time-reversal and by the duality transformation (16.208), and characterized by the scale factors

$$a_\perp(\pm t) \sim (\pm t)^{\pm 2\gamma/(1+d\gamma^2)}. \tag{16.218}$$

The duality transformation that preserves $\overline{\phi}$ and $\overline{\rho}$, and inverts the scale factor, in this case is simply represented by the transformation $\gamma \to -\gamma$. Consider for instance the standard radiation-dominated solution, corresponding to $d = 3$, $\gamma = 1/3$, and $t > 0$, and associated to a constant dilaton, according to equation (16.217). A duality transformation gives a new solution with $\gamma = -1/3$, namely (from (16.216) and (16.217)):

$$a \sim t^{-1/2}, \quad \rho \sim a^{-2}, \quad \phi \sim -3 \ln t. \tag{16.219}$$

By performing an additional time reflection we then obtain the pre-big bang solution 'dual to radiation', already reported in eq. (16.16).

References

[1] Weinberg S 1972 *Gravitation and Cosmology* (New York: Wiley)
[2] Kolb E W and Turner M S 1990 *The Early Universe* (Redwood City, CA: Addison-Wesley)

Linde A D 1990 *Particle Physics and Inflationary Cosmology* (Chur: Harwood)
[3] Vilenkin A 1992 *Phys. Rev.* D **46** 2355
Borde A and Vilenkin A 1994 *Phys. Rev. Lett.* **72** 3305
[4] Guth A 1998 *The inflationary Universe* (London: Vintage) p 271
[5] Vilenkin A 1984 *Phys. Rev.* D **30** 509
Linde A D 1984 *Sov. Phys.–JEPT* **60** 211
Zel'dovich Y and Starobinski A A 1984 *Sov. Astron. Lett.* **10** 135
Rubakov V A 1984 *Phys. Lett.* B **148** 280
[6] Hartle J B and Hawking S W 1983 *Phys. Rev.* D **28** 2960
Hawking S W 1984 *Nucl. Phys.* B **239** 257
Hawking S W and Page D N 1986 *Nucl. Phys.* B **264** 185
[7] Gasperini M, Maharana J and Veneziano G 1996 *Nucl. Phys.* B **472** 394
Gasperini M and Veneziano G 1996 *Gen. Rel. Grav.* **28** 1301
Gasperini M 1998 *Int. J. Mod. Phys.* A **13** 4779
Cavagliá M and Ungarelli C 1999 *Class. Quantum Grav.* **16** 1401
[8] Veneziano G 1991 *Phys. Lett.* B **265** 287
Gasperini M and Veneziano G 1993 *Astropart. Phys.* **1** 317
Gasperini M and Veneziano G 1993 *Mod. Phys. Lett.* A **8** 3701
Gasperini M and Veneziano G 1994 *Phys. Rev.* D **50** 2519
An updated collection of papers on the pre-big bang scenario is available at
   http://www.to.infn.it/~gasperin
[9] Tseytlin A A 1991 *Mod. Phys. Lett.* A **6** 1721
[10] Sen A 1991 *Phys. Lett.* B **271** 295
Hassan S F and Sen A 1992 *Nucl. Phys.* B **375** 103
[11] Meissner K A and Veneziano G 1991 *Phys. Lett.* B **267** 33
Meissner K A and Veneziano G 1991 *Mod. Phys. Lett.* A **6** 3397
[12] Gasperini M and Veneziano G 1992 *Phys. Lett.* B **277** 256
[13] Kaplunovsky V 1985 *Phys. Rev. Lett.* **55** 1036
[14] Brustein R and Veneziano G 1994 *Phys. Lett.* B **329** 429
Kaloper N, Madden R and Olive K A 1995 *Nucl. Phys.* B **452** 677
Easther R, Maeda K and Wands D 1996 *Phys. Rev.* D **53** 4247
[15] Gasperini M, Maharana J and Veneziano G 1991 *Phys. Lett.* B **272** 277
[16] Gasperini M 1999 *Mod. Phys. Lett.* A **14** 1059
[17] Shadev D 1984 *Phys. Lett.* B **317** 155
Abbott R B, Barr S M and Ellis S D 1984 *Phys. Rev.* D **30** 720
Kolb E W, Lindley D and Seckel D 1984 *Phys. Rev.* D **30** 1205
Lucchin F and Matarrese S 1985 *Phys. Lett.* B **164** 282
[18] Brandenberger R and Vafa C 1989 *Nucl. Phys.* B **316** 391
Tseytlin A A and Vafa C 1992 *Nucl. Phys.* B **372** 443
[19] Gasperini M, Sánchez N and Veneziano G 1991 *Nucl. Phys.* B **364** 365
[20] Caianiello E R, Gasperini M and Scarpetta G 1991 *Class. Quantum Grav.* **8** 659
Gasperini M 1991 *Proc. First Workshop on 'Advances in Theoretical Physics'
   (Vietri, 1990)* ed E R Caianiello (Singapore: World Scientific) p 77
[21] Gasperini M, Maggiore M and Veneziano G 1997 *Nucl. Phys.* B **494** 315
[22] Rindler W 1956 *Mon. Not. R. Astron. Soc.* **116** 6
[23] Gasperini M 2000 *Phys. Rev.* D **61** 087301
[24] Turner M S and Weinberg E J 1997 *Phys. Rev.* D **56** 4604
[25] Kaloper N, Linde A and Bousso R 1999 *Phys. Rev.* D **59** 043508

[26] Maggiore M and Sturani R 1997 *Phys. Lett.* B **415** 335
[27] Buonanno A, Damour T and Veneziano G 1999 *Nucl. Phys.* B **543** 275
[28] Clancy D, Lidsey J E and Tavakol R 1999 *Phys. Rev.* D **59** 063511
[29] Antoniadis I, Rizos J and Tamvakis K 1994 *Nucl. Phys.* B **415** 497
  Rey S J 1996 *Phys. Rev. Lett.* **77** 1929
  Gasperini M and Veneziano G 1996 *Phys. Lett.* B **387** 715
  Brustein R and Madden R 1997 *Phys. Lett.* B **410** 110
  Brustein R and Madden R 1998 *Phys. Rev.* D **57** 712
  Brandenberger R, Easther R and Maia J 1998 *JHEP* **9808** 007
  Maggiore M and Riotto A 1999 *Nucl. Phys.* B **458** 427
  Foffa S, Maggiore M and Sturani R 1999 *Nucl. Phys.* B **552** 395
[30] Gasperini M and Giovannini M 1993 *Phys. Lett.* B **301** 334
  Gasperini M and Giovannini M 1993 *Class. Quantum Grav.* **10** L133
  Veneziano G 1999 *Phys. Lett.* B **454** 22
[31] Gasperini M and Giovannini M 1992 *Phys. Lett.* B **282** 36
  Gasperini M and Giovannini M 1993 *Phys. Rev.* D **47** 1519
[32] Brustein R, Gasperini M, Giovannini M and Veneziano G 1995 *Phys. Lett.* B **361** 45
[33] Brustein R, Gasperini M and Veneziano G 1997 *Phys. Rev.* D **55** 3882
[34] See, for instance, Fritschel P 1998 *Second Amaldi Conf. on 'Gravitational Waves' (CERN, 1997)* ed E Coccia *et al* (Singapore: World Scientific) p 74
[35] See, for instance, Brillet A 1998 *Second Amaldi Conf. on 'Gravitational Waves' (CERN, 1997)* ed E Coccia *et al* (Singapore: World Scientific) p 86
[36] See, for instance, Hough J 1998 *Second Amaldi Conf. on 'Gravitational Waves' (CERN, 1997)* ed E Coccia *et al* (Singapore: World Scientific) p 97
  See also Bender P these proceedings
[37] Durrer R, Gasperini M, Sakellariadou M and Veneziano G 1998 *Phys. Lett.* B **436** 66
  Durrer R, Gasperini M, Sakellariadou M and Veneziano G 1999 *Phys. Rev.* D **59** 043511
[38] Gasperini M and Veneziano G 1999 *Phys. Rev.* D **59** 043503
[39] Bennet C L *et al* 1994 *Astrophys. J.* **430** 423
[40] Copeland E J, Easther R and Wands D 1997 *Phys. Rev.* D **56** 874
  Copeland E J, Lidsey J and Wands D 1997 *Nucl. Phys.* B **506** 407
  Buonanno A, Meissner K A, Ungarelli C and Veneziano G 1998 *JHEP* **01** 004
[41] Melchiorri A, Vernizzi F, Durrer R and Veneziano G 1999 *Phys. Rev. Lett.* **83** 4464
[42] See, for instance, the web site: http://map.gsfc.nasa.gov
[43] See, for instance, the web site: http://astro.estec.esa.nl/SA-general/Projects/Planck
[44] Gasperini M, Giovannini M and Veneziano G 1995 *Phys. Rev. Lett.* **75** 3796
[45] Turner M S and Widrow L M 1988 *Phys. Rev.* D **37** 2743
[46] Gasperini M 1994 *Phys. Lett.* B **327** 314
  Gasperini M and Veneziano G 1994 *Phys. Rev.* D **50** 2519
[47] Damour T and Polyakov A M 1994 *Nucl. Phys.* B **423** 532
[48] Gasperini M 1997 *Proc. 12th Int. Conf. on 'General Relativity and Gravitational Physics' (Rome)* ed M Bassan *et al* (Singapore: World Scientific) p 85
[49] Mukhanov V F, Feldman H A and Brandenberger R H 1992 *Phys. Rep.* **215** 203
[50] Hwang J 1991 *Ap. J.* **375** 443
[51] Brustein R, Gasperini M, Giovannini M, Mukhanov V and Veneziano G 1995 *Phys.*

Rev. D **51** 6744
[52] Ellis G F R and Bruni S 1989 *Phys. Rev.* D **40** 1804
[53] Abbott R B, Bednarz B and Ellis S D 1986 *Phys. Rev.* D **33** 2147
[54] Gasperini M 1997 *Phys. Rev.* D **56** 4815
[55] Gasperini M, Giovannini M and Veneziano G 1993 *Phys. Rev.* D **48** R439
[56] Grishchuk L P and Sidorov Y V 1989 *Class. Quantum Grav.* **6** L161
Grishchuk L P and Sidorov Y V 1990 *Phys. Rev.* D **42** 3413
Grishchuk L P 1993 *Class. Quantum Grav.* **10** 2449
[57] Brandenberger R, Mukhanov V and Prokopec T 1992 *Phys. Rev. Lett.* **69** 3606
Gasperini M and Giovannini M 1993 *Phys. Lett.* B **301** 334
Gasperini M and Giovannini M 1993 *Class. Quantum Grav.* **10** L133
[58] Grishchuk L P 1975 *Sov. Phys.–JETP* **40** 409
[59] Gasperini M 1998 *String Theory in Curved Spacetimes* ed N Sánchez (Singapore: World Scientific) p 333
[60] Abramowitz M and Stegun I A 1972 *Handbook of Mathematical Functions* (New York: Dover)
[61] Mukhanov V F and Deruelle N 1995 *Phys. Rev.* D **52** 5549
[62] Allen B 1997 *Proc. Les Houches School on 'Astrophysical Sources of Gravitational Waves'* ed J A Mark and J P Lasota (Cambridge: Cambridge University Press) p 373
Allen B and Romano J D 1999 *Phys. Rev.* D **59** 102001
[63] Allen B, Flanagan E E and Papa M A 1999 *Phys. Rev.* D **61** 024024
[64] Grishchuk L P 1998 The detectability of relic (squeezed) gravitational waves by laser interferometers gr-qc/9810055
[65] Gasperini M 1995 *Proc. Second Paris Cosmology Colloquium (Paris, 1994)* ed H J De Vega and N Sanchez (Singapore: World Scientific) p 429
Gasperini M 1996 *String Gravity and Physics at the Planck Energy Scale* ed N Sanchez and A Zichichi (Dordrecht: Kluwer) p 305
Gasperini M 1998 *Second Amaldi Conf. on 'Gravitational Waves' (CERN, 1997)* ed E Coccia *et al* (Singapore: World Scientific) p 62
[66] Maggiore M 1998 High energy physics with gravitational wave experiments *Preprint* gr-qc/9803028
[67] Krauss L M and White M 1992 *Phys. Rev. Lett.* **69** 869
[68] Kaspi V, Taylor J and Ryba M 1994 *Ap. J.* **428** 713
[69] Schwarztmann V F 1969 *JETP Lett.* **9** 184
[70] Copeland E J, Liddle A R, Lidsey J E and Wands D 1998 *Phys. Rev.* D **58** 63508
Copeland E J, Liddle A R, Lidsey J E and Wands D 1998 *Gen. Rel. Grav.* **30** 1711
[71] Galluccio M, Litterio F and Occhionero F 1997 *Phys. Rev. Lett.* **79** 970
[72] Buonanno A, Maggiore M and Ungarelli C 1997 *Phys. Rev.* D **55** 3330
[73] Caldwell R R and Allen B 1992 *Phys. Rev.* D **45** 3447
Battye R A and Shellard E P S 1996 *Phys. Rev.* D **53** 1811
[74] Turner M S and Wilczek F 1990 *Phys. Rev. Lett.* **65** 3080
[75] Khlebnikov S Y and Tkachev I I 1997 *Phys. Rev.* D **56** 653
Bassett B 1997 *Phys. Rev.* D **56** 3439
[76] Peebles P J E and Vilenkin A 1999 *Phys. Rev.* D **59** 063505
Giovannini M 1999 *Phys. Rev.* D **60** 123511
[77] Astone P 1998 *Second Amaldi Conf. on 'Gravitational Waves' (CERN, 1997)* ed E Coccia *et al* (Singapore: World Scientific) p 192

[78] Prodi G 1998 *Talk Given at the Second Int. Conf. 'Around VIRGO' (Tirrenia, Pisa)*
See also Cerdonio M these proceedings

[79] Pegoraro F, Picasso E and Radicati L 1978 *J. Phys. A: Math. Gen.* **11** 1949

[80] Picasso E Private communication

[81] See, for instance, Lobo J A 1998 *Second Amaldi Conf. on 'Gravitational Waves' (CERN, 1997)* ed E Coccia *et al* (Singapore: World Scientific) p 168

[82] Ungarelli C and Vecchio A in preparation

[83] Green M B, Schwartz J and Witten E 1987 *Superstring Theory* (Cambridge: Cambridge University Press)

[84] Metsaev R R and Tseytlin A A 1987 *Nucl. Phys.* B **293** 385

[85] Taylor T R and Veneziano G 1988 *Phys. Lett.* B **213** 459

[86] See, for instance, Antoniadis I and Pioline B 1999 Large dimensions and string physics at a TeV *Preprint* hep-ph/9906480 (CPHT-PC720-0699, March 1999)

[87] Buonanno A, Gasperini M, Maggiore M and Ungarelli C 1997 *Class. Quantum Grav.* **14** L97

[88] Gasperini M and Giovannini M 1992 *Phys. Lett.* B **287** 56

[89] Gibbons G W and Hawking S 1977 *Phys. Rev.* D **15** 2752

Chapter 17

Post-Newtonian computation of binary inspiral waveforms

Luc Blanchet
Département d'Astrophysique Relativiste et de Cosmologie,
Centre National de la Recherche Scientifique (UMR 8629),
Observatoire de Paris, 92195 Meudon Cedex, France

17.1   Introduction

Astrophysical systems known as inspiralling compact binaries are among the most interesting sources to hunt for gravitational radiation in the future network of laser-interferometric detectors, composed of the large-scale interferometers VIRGO and LIGO, and the medium-scale ones GEO and TAMA (see the books [1–3] for reviews, and the contribution of B Schutz in this volume). These systems are composed of two compact objects, i.e. gravitationally-condensed neutron stars or black holes, whose orbit follows an inward spiral, with decreasing orbital radius r and increasing orbital frequency ω. The inspiral is driven by the loss of energy associated with the gravitational-wave emission. Because the dynamics of a binary is essentially aspherical, inspiralling compact binaries are strong emitters of gravitational radiation. Tidal interactions between the compact objects are expected to play a little role during most of the inspiral phase; the mass transfer (in the case of neutron stars) does not occur until very late, near the final coalescence. Inspiralling compact binaries are very clean systems, essentially dominated by gravitational forces. Therefore, the relevant model for describing the inspiral phase consists of two point-masses moving under their mutual gravitational attraction. As a simplification for the theoretical analysis, the orbit of inspiralling binaries can be considered to be circular, apart from the gradual inspiral, with a good approximation. At some point in the evolution, there will be a transition from the adiabatic inspiral to the plunge of the two objects followed by the collision and final merger. Evidently the model of point-masses breaks down at this point, and is to be replaced by a fully relativistic numerical computation of the plunge and merger (see the contribution of E Seidel in this volume).

Currently the theoretical prediction from general relativity for the gravitational waves emitted during the inspiral phase is determined using the post-Newtonian approximation (see [4, 5] for reviews). This is possible because the dynamics of inspiralling compact binaries, though very relativistic, is not *fully* relativistic: the orbital velocity v is always less than one third of c (say). However, because $1/3$ is far from negligible as compared to 1, the gravitational-radiation waveform should be predicted up to a high post-Newtonian order. In particular, the radiation *reaction* onto the orbit, which triggers the inspiral, is to be determined with the maximal precision, corresponding to at least the second and maybe the third post-Newtonian (3PN, or $1/c^6$) order [6,7]. Notice that the zeroth order in this post-Newtonian counting corresponds to the dominant radiation reaction force (already of the order of 2.5PN relative to the Newtonian force), which is due to the change in the quadrupole moment of the source. Actually, the method is not to compute directly the radiation reaction force but to determine the inspiral rate from the energy balance equation relating the mechanical loss of energy in the binary's centre of mass to the total emitted flux at infinity.

The implemented strategy is to develop a formalism for the emission and propagation of gravitational waves from a general isolated system, and only then, once some general formulae valid to some prescribed post-Newtonian order are in our hands, to apply the formalism to compact binaries. Hence, we consider in this paper a particular formalism applicable to a general description of matter, under the tenet of validity of the post-Newtonian expansion, namely that the matter should be slowly moving, weakly stressed and self-gravitating. Within this formalism we compute the retarded far field of the source by means of a formal post-Minkowskian expansion, valid in the exterior of the source, and parametrized by some appropriately defined multipole moments describing the source. From the post-Minkowskian expansion we obtain a relation (correct up to the prescribed post-Newtonian order) between the *radiative* multipole moments parametrizing the metric field at infinity, and the source multipole moments. On the other hand, the source multipole moments are obtained as some specific integrals extending over the distribution of matter fields in the source and the contribution of the gravitational field itself. The source moments are computed separately up to the same post-Newtonian order. The latter formalism has been developed by Blanchet, Damour and Iyer [8–14]. More recently, a different formalism has been proposed and implemented by Will and Wiseman [15] (see also [16, 17]). The two formalisms are equivalent at the most general level, but the details of the computations are quite far apart. In the second stage, one applies the formalism to a system of point-particles (modelling compact objects) by substituting for the matter stress–energy tensor that expression, involving delta-functions, which is appropriate for point-particles. This entails some divergencies due to the infinite self-field of point-particles. Our present method is to cure them systematically by means of a variant of the Hadamard regularization (based on the concept of 'partie finie') [18, 19].

In this paper, we first analyse the binary inspiral gravitational waveform at the simplest Newtonian approximation. Notably, we spend some time describing the relative orientation of the binary with respect to the detector. Then we compute, still at the 'Newtonian order' (corresponding, in fact, to the quadrupole approximation), the evolution in the course of time of the orbital phase of the binary, which is a crucial quantity to predict. Next, we review the main steps of our general wave-generation formalism, with emphasis on the definition of the various types of multipole moments which are involved. At last, we present the result for the binary inspiral waveform whose current post-Newtonian precision is 2PN in the wave amplitude and 2.5PN in the orbital phase (that is $1/c^5$ beyond the quadrupole radiation reaction). However, since our ultimate aim is to construct accurate templates to be used in the data analysis of detectors, it is appropriate to warm up with a short review of the optimal filtering technique which will be used for hunting the inspiral binary waveform (see [20] for an extended review).

17.2   Summary of optimal signal filtering

Let $o(t)$ be the raw output of the detector, which is made of the superposition of the useful gravitational-wave signal $h(t)$ and of noise $n(t)$:

$$o(t) = h(t) + n(t). \tag{17.1}$$

The noise is assumed to be a stationary Gaussian random variable, with zero expectation value,

$$\overline{n(t)} = 0, \tag{17.2}$$

and with (supposedly known) frequency-dependent power spectral density $S_n(\omega)$ satisfying

$$\overline{\tilde{n}(\omega)\tilde{n}^*(\omega')} = 2\pi\delta(\omega - \omega')S_n(\omega), \tag{17.3}$$

where $\tilde{n}(\omega)$ is the Fourier transform of $n(t)$. In (17.2) and (17.3), we denote by an upper bar the average over many realizations of noise in a large ensemble of detectors. From (17.3), we have $S_n(\omega) = S_n^*(\omega) = S_n(-\omega) > 0$.

Looking for the signal $h(t)$ in the output of the detector $o(t)$, the experimenters construct the correlation $c(t)$ between $o(t)$ and a filter $q(t)$, i.e.

$$c(t) = \int_{-\infty}^{+\infty} dt' \, o(t')q(t + t'), \tag{17.4}$$

and divide $c(t)$ by the square root of its variance (or correlation noise). Thus, the experimenters consider the ratio

$$\sigma[q](t) = \frac{c(t)}{(\overline{c^2(t)} - \overline{c(t)}^2)^{1/2}} = \frac{\int_{-\infty}^{+\infty} \frac{d\omega}{2\pi} \tilde{o}(\omega)\tilde{q}^*(\omega)e^{i\omega t}}{\left(\int_{-\infty}^{+\infty} \frac{d\omega}{2\pi} S_n(\omega)|\tilde{q}(\omega)|^2\right)^{1/2}}, \tag{17.5}$$

where $\tilde{o}(\omega)$ and $\tilde{q}(\omega)$ are the Fourier transforms of $o(t)$ and $q(t)$. The expectation value (or ensemble average) of this ratio defines the filtered signal-to-noise ratio

$$\rho[q](t) = \overline{\sigma[q](t)} = \frac{\int_{-\infty}^{+\infty} \frac{d\omega}{2\pi} \tilde{h}(\omega)\tilde{q}^*(\omega)e^{i\omega t}}{\left(\int_{-\infty}^{+\infty} \frac{d\omega}{2\pi} S_n(\omega)|\tilde{q}(\omega)|^2\right)^{1/2}}. \qquad (17.6)$$

The optimal filter (or Wiener filter) which maximizes the signal-to-noise (17.6) at a particular instant $t = 0$ (say), is given by the matched filtering theorem as

$$\tilde{q}(\omega) = \gamma \frac{\tilde{h}(\omega)}{S_n(\omega)}, \qquad (17.7)$$

where γ is an arbitrary real constant. The optimal filter (17.7) is matched on the expected signal $\tilde{h}(\omega)$ itself, and weighted by the inverse of the power spectral density of the noise. The maximum signal to noise, corresponding to the optimal filter (17.7), is given by

$$\rho = \left(\int_{-\infty}^{+\infty} \frac{d\omega}{2\pi} \frac{|\tilde{h}(\omega)|^2}{S_n(\omega)}\right)^{1/2} = \langle h, h \rangle^{1/2}. \qquad (17.8)$$

This is the best achievable signal-to-noise ratio with a linear filter. In (17.8), we have used, for any two real functions $f(t)$ and $g(t)$, the notation

$$\langle f, g \rangle = \int_{-\infty}^{+\infty} \frac{d\omega}{2\pi} \frac{\tilde{f}(\omega)\tilde{g}^*(\omega)}{S_n(\omega)} \qquad (17.9)$$

for an inner scalar product satisfying $\langle f, g \rangle = \langle f, g \rangle^* = \langle g, f \rangle$.

   In practice, the signal $h(t)$ or $\tilde{h}(\omega)$ is of known form (given, for instance, by (17.27)–(17.32) later) but depends on an unknown set of parameters which describe the source of radiation, and are to be measured. The experimenters must therefore use a whole family of filters analogous to (17.7) but in which the signal is parametrized by a whole family of 'test' parameters which are *a priori* different from the actual source parameters. Thus, one will have to define and use a lattice of filters in the parameter space. The set of parameters maximizing the signal to noise (17.6) is equal, by the matched filtering theorem, to the set of source parameters. However, in a single detector, the experimenters maximize the ratio (17.5) rather than the signal to noise (17.6), and therefore make errors on the determination of the parameters, depending on a particular realization of noise in the detector. If the signal-to-noise ratio is high enough, the measured values of the parameters are Gaussian distributed around the source parameters, with variances and correlation coefficients given by the covariance matrix, the computation of which we now recall. Since the optimal filter (17.7) is defined up to an arbitrary multiplicative constant, it is convenient to treat separately a constant amplitude parameter in front of the signal (involving, in general, the distance of the source). We shall thus write the signal in the form

$$\tilde{h}(\omega; A, \lambda_a) = A \tilde{k}(\omega; \lambda_a), \qquad (17.10)$$

where A denotes some amplitude parameter. The function $\tilde{k}$ depends only on the other parameters, collectively denoted by λ_a where the label a ranges on the values $1, \ldots, N$. The family of matched filters (or 'templates') we consider is defined by

$$\tilde{q}(\omega; {}_t\lambda_a) = \gamma' \frac{\tilde{k}(\omega; {}_t\lambda_a)}{S_n(\omega)}, \tag{17.11}$$

where ${}_t\lambda_a$ is a set of test parameters, assumed to be all independent, and γ' is arbitrary. By substituting (17.11) into (17.5) and choosing $t = 0$, we get, with the notation of (17.9),

$$\sigma({}_t\lambda) = \frac{\langle o, k({}_t\lambda) \rangle}{\langle k({}_t\lambda), k({}_t\lambda) \rangle^{1/2}}. \tag{17.12}$$

(Note that σ is in fact a function of both the parameters λ_a and ${}_t\lambda_a$.) Now the experimenters choose as their best estimate of the source parameters λ_a the *measured* parameters ${}_m\lambda_a$ which among all the test parameters ${}_t\lambda_a$ (independently) maximize (17.12), i.e. which satisfy

$$\frac{\partial \sigma}{\partial {}_t\lambda_a}({}_m\lambda) = 0, \quad a = 1, \ldots, N. \tag{17.13}$$

Assuming that the signal to noise is high enough, we can work out (17.13) up to the first order in the difference between the actual source parameters and the measured ones,

$$\delta\lambda_a = \lambda_a - {}_m\lambda_a. \tag{17.14}$$

As a result, we obtain

$$\delta\lambda_a = C_{ab} \left\{ -\langle n, \frac{\partial h}{\partial \lambda_b} \rangle + \frac{\langle n, h \rangle}{\langle h, h \rangle} \langle h, \frac{\partial h}{\partial \lambda_b} \rangle \right\}, \tag{17.15}$$

where a summation is understood on the dummy label b, and where the matrix C_{ab} (with $a, b = 1, \ldots, N$) is the inverse of the Fisher information matrix

$$\mathcal{D}_{ab} = \left\langle \frac{\partial h}{\partial \lambda_a}, \frac{\partial h}{\partial \lambda_b} \right\rangle - \frac{1}{\langle h, h \rangle} \left\langle h, \frac{\partial h}{\partial \lambda_a} \right\rangle \left\langle h, \frac{\partial h}{\partial \lambda_b} \right\rangle \tag{17.16}$$

(we have $C_{ab}\mathcal{D}_{bc} = \delta_{ac}$). On the right-hand sides of (17.15) and (17.16), the signal is equally (with this approximation) parametrized by the measured or actual parameters. Since the noise is Gaussian, so are, by (17.15), the variables $\delta\lambda_a$ (indeed, $\delta\lambda_a$ result from a linear operation on the noise variable). The expectation value and quadratic moments of the distribution of these variables are readily obtained from the facts that $\overline{\langle n, f \rangle} = 0$ and $\overline{\langle n, f \rangle \langle n, g \rangle} = \langle f, g \rangle$ for any deterministic functions f and g (see (17.2) and (17.3)). We then obtain

$$\overline{\delta\lambda_a} = 0,$$

$$\overline{\delta\lambda_a \delta\lambda_b} = C_{ab}. \tag{17.17}$$

Thus, the matrix $\mathcal{C}_{ab}$ (the inverse of (17.16)) is the matrix of variances and correlation coefficients, or covariance matrix, of the variables $\delta\lambda_a$. The probability distribution of $\delta\lambda_a$ reads as

$$P(\delta\lambda_a) = \frac{1}{\sqrt{(2\pi)^{N+1}\det\mathcal{C}}} \exp\left\{-\frac{1}{2}\mathcal{D}_{ab}\delta\lambda_a\delta\lambda_b\right\}, \qquad (17.18)$$

where $\det\mathcal{C}$ is the determinant of $\mathcal{C}_{ab}$. A similar analysis can be done for the measurement of the amplitude parameter A of the signal.

17.3 Newtonian binary polarization waveforms

The source of gravitational waves is a binary system made of two point-masses moving on a circular orbit. We assume that the masses do not possess any intrinsic spins, so that the motion of the binary takes place in a plane. To simplify the presentation we suppose that the centre of mass of the binary is at rest with respect to the detector. The detector is a large-scale laser-interferometric detector like VIRGO or LIGO, with two perpendicular arms (with length 3 km in the case of VIRGO). The two laser beams inside the arms are separated by the beam-splitter which defines the central point of the interferometer. We introduce an orthonormal right-handed triad $(\boldsymbol{X}, \boldsymbol{Y}, \boldsymbol{Z})$ linked with the detector, with $\boldsymbol{X}$ and $\boldsymbol{Y}$ pointing along the two arms of the interferometer, and $\boldsymbol{Z}$ pointing toward the zenithal direction. We denote by $\boldsymbol{n}$ the direction of the detector as seen from the source, that is, $-\boldsymbol{n}$ is defined as the unit vector pointing from the centre of the interferometer to the binary's centre of mass. We introduce some spherical angles α and β such that

$$-\boldsymbol{n} = \boldsymbol{X}\sin\alpha\cos\beta + \boldsymbol{Y}\sin\alpha\sin\beta + \boldsymbol{Z}\cos\alpha. \qquad (17.19)$$

Thus, the plane $\beta = $ constant defines the plane which is vertical, as seen from the detector, and which contains the source. Next, we introduce an orthonormal right-handed triad $(\boldsymbol{x}, \boldsymbol{y}, \boldsymbol{z})$ which is linked to the binary's orbit, with $\boldsymbol{x}$ and $\boldsymbol{y}$ located in the orbital plane, and $\boldsymbol{z}$ along the normal to the orbital plane. The vector $\boldsymbol{x}$ is chosen to be perpendicular to $\boldsymbol{n}$; thus, $\boldsymbol{n}$ is within the plane formed by $\boldsymbol{y}$ and $\boldsymbol{z}$. The orientation of this triad is 'right-hand' with respect to the sense of motion. We denote by i the inclination angle, namely the angle between the direction of the source or line-of-sight $\boldsymbol{n}$ and the normal $\boldsymbol{z}$ to the orbital plane. Since $\boldsymbol{z}$ is right-handed with respect to the sense of motion we have $0 \le i \le \pi$. Furthermore, we define two unit vectors $\boldsymbol{p}$ and $\boldsymbol{q}$, called the polarization vectors, in the plane orthogonal to $\boldsymbol{n}$ (or plane of the sky). We choose $\boldsymbol{p} = \boldsymbol{x}$ and define $\boldsymbol{q}$ in such a way that the triad $(\boldsymbol{n}, \boldsymbol{p}, \boldsymbol{q})$ is right-handed; thus

$$\boldsymbol{n} = \boldsymbol{y}\sin i + \boldsymbol{z}\cos i, \qquad (17.20)$$

$$\boldsymbol{p} = \boldsymbol{x}, \qquad (17.21)$$

$$\boldsymbol{q} = \boldsymbol{y}\cos i - \boldsymbol{z}\sin i. \qquad (17.22)$$

Notice that the direction $\boldsymbol{p} \equiv \boldsymbol{x}$ is one of the 'ascending node' N of the binary, namely the point at which the bodies cross the plane of the sky moving toward the detector. Thus, the polarization vectors $\boldsymbol{p}$ and $\boldsymbol{q}$ lie, respectively, along the major and minor axis of the projection onto the plane of the sky of the (circular) orbit, with $\boldsymbol{p}$ pointing toward N using the standard practice of celestial mechanics. Finally, let us denote by ξ the polarization angle between $\boldsymbol{p}$ and the vertical plane $\beta = $ constant; that is, ξ is the angle between the vertical and the direction of the node N. We have

$$\boldsymbol{n} = -\boldsymbol{X}\sin\alpha\cos\beta - \boldsymbol{Y}\sin\alpha\sin\beta - \boldsymbol{Z}\cos\alpha, \tag{17.23}$$

$$\boldsymbol{p} = \boldsymbol{X}(\cos\xi\cos\alpha\cos\beta + \sin\xi\sin\beta)$$
$$+ \boldsymbol{Y}(\cos\xi\cos\alpha\sin\beta - \sin\xi\cos\beta) - \boldsymbol{Z}\cos\xi\sin\alpha, \tag{17.24}$$

$$\boldsymbol{q} = \boldsymbol{X}(-\sin\xi\cos\alpha\cos\beta + \cos\xi\sin\beta)$$
$$+ \boldsymbol{Y}(-\sin\xi\cos\alpha\sin\beta - \cos\xi\cos\beta) + \boldsymbol{Z}\sin\xi\sin\alpha. \tag{17.25}$$

Defining all these angles, the relative orientation of the binary with respect to the interferometric detector is entirely determined. Indeed using (17.22) and (17.25) one relates the triad $(\boldsymbol{x}, \boldsymbol{y}, \boldsymbol{z})$ associated with the source to the triad $(\boldsymbol{X}, \boldsymbol{Y}, \boldsymbol{Z})$ linked with the detector.

The gravitational wave as it propagates through the detector in the wave zone of the source is described by the so-called transverse and traceless (TT) asymptotic waveform $h_{ij}^{\text{TT}} = (g_{ij} - \delta_{ij})^{\text{TT}}$, where g_{ij} denotes the spatial covariant metric in a coordinate system adapted to the wave zone, and δ_{ij} is the Kronecker metric. Neglecting terms dying out like $1/R^2$ in the distance to the source, the two polarization states of the wave, customarily denoted h_+ and $h_\times$, are given by

$$h_+ = \tfrac{1}{2}(p_i p_j - q_i q_j)h_{ij}^{\text{TT}}, \tag{17.26}$$

$$h_\times = \tfrac{1}{2}(p_i q_j + p_j q_i)h_{ij}^{\text{TT}}, \tag{17.27}$$

where p_i and q_i are the components of the polarization vectors. The detector is directly sensitive to a linear combination of the polarization waveforms h_+ and $h_\times$ given by

$$h(t) = F_+ h_+(t) + F_\times h_\times(t), \tag{17.28}$$

where F_+ and $F_\times$ are the so-called beam-pattern functions of the detector, which are some given functions (for a given type of detector) of the direction of the source α, β and of the polarization angle ξ. This $h(t)$ is the gravitational-wave signal looked for in the data analysis of section 17.2, and used to construct the optimal filter (17.10). In the case of the laser-interferometric detector we have

$$F_+ = \tfrac{1}{2}(1 + \cos^2\alpha)\cos 2\beta \cos 2\xi + \cos\alpha \sin 2\beta \sin 2\xi, \tag{17.29}$$

$$F_\times = -\tfrac{1}{2}(1 + \cos^2\alpha)\cos 2\beta \sin 2\xi + \cos\alpha \sin 2\beta \cos 2\xi. \tag{17.30}$$

The orbital plane and the direction of the node N are fixed so the polarization angle ξ is constant (in the case of spinning particles, the orbital plane precesses around the direction of the total angular momentum, and angle ξ varies). Thus, the gravitational wave $h(t)$ depends on time only through the two polarization waveforms $h_+(t)$ and $h_\times(t)$. In turn, these waveforms depend on time through the binary's orbital phase $\phi(t)$ and the orbital frequency $\omega(t) = d\phi(t)/dt$. The orbital phase is defined as the angle, oriented in the sense of motion, between the ascending node N and the direction of one of the particles, conventionally particle 1 (thus $\phi = 0$ modulo 2π when the two particles lie along $\boldsymbol{p}$, with particle 1 at the ascending node). In the absence of any radiation reaction, the orbital frequency would be constant, and so the phase would evolve linearly with time. Because of the radiation reaction forces, the actual variation of $\phi(t)$ is nonlinear, and the orbit spirals in and shrinks to zero-size to account, via the Kepler third law, for the gravitational-radiation energy loss. The main problem of the construction of accurate templates for the detection of inspiralling compact binaries is the prediction of the time variation of the phase $\phi(t)$. Indeed, because of the accumulation of cycles, most of the accessible information allowing accurate measurements of the binary's intrinsic parameters (such as the two masses) is contained within the phase, and rather less accurate information is available in the wave amplitude itself. For instance, the relative precision in the determination of the distance R to the source, which affects the wave amplitude, is less than for the masses, which strongly affect the phase evolution [6, 7]. Hence, we can often neglect the higher-order contributions to the amplitude, which means retaining only the dominant harmonics in the waveform, which corresponds to a frequency at twice the orbital frequency.

Once the functions $\phi(t)$ and $\omega(t)$ are known they must be inserted into the polarization waveforms computed by means of some wave-generation formalism. For instance, using the quadrupole formalism, which neglects all the harmonics but the dominant one, we find

$$h_+ = -\frac{2G\mu}{c^2 R} \left(\frac{Gm\omega}{c^3}\right)^{2/3} (1 + \cos^2 i) \cos 2\phi, \qquad (17.31)$$

$$h_\times = -\frac{2G\mu}{c^2 R} \left(\frac{Gm\omega}{c^3}\right)^{2/3} (2\cos i) \sin 2\phi \qquad (17.32)$$

where R denotes the absolute luminosity distance of the binary's centre of mass; the mass parameters are given by

$$m = m_1 + m_2; \quad \mu = \frac{m_1 m_2}{m}; \quad \nu = \frac{\mu}{m}. \qquad (17.33)$$

This last parameter ν, introduced for later convenience, is the ratio between the reduced mass and the total mass, and is such that $0 < \nu \leq 1/4$ with $\nu \to 0$ in the test-mass limit and $\nu = 1/4$ in the case of two equal masses.

17.4   Newtonian orbital phase evolution

Let $\mathbf{y}_1(t)$ and $\mathbf{y}_2(t)$ be the the two trajectories of the masses m_1 and m_2, and $\mathbf{y} = \mathbf{y}_1 - \mathbf{y}_2$ be their relative position, and denote $r = |\mathbf{y}|$. The velocities are $\mathbf{v}_1(t) = \mathrm{d}\mathbf{y}_1/\mathrm{d}t$, $\mathbf{v}_2(t) = \mathrm{d}\mathbf{y}_2/\mathrm{d}t$ and $\mathbf{v}(t) = \mathrm{d}\mathbf{y}/\mathrm{d}t$. The Newtonian equations of motion read as

$$\frac{\mathrm{d}\mathbf{v}_1}{\mathrm{d}t} = -\frac{Gm_2}{r^3}\mathbf{y}; \quad \frac{\mathrm{d}\mathbf{v}_2}{\mathrm{d}t} = \frac{Gm_1}{r^3}\mathbf{y}. \tag{17.34}$$

The difference between these two equations yields the relative acceleration,

$$\frac{\mathrm{d}\mathbf{v}}{\mathrm{d}t} = -\frac{Gm}{r^3}\mathbf{y}. \tag{17.35}$$

We place ourselves into the Newtonian centre-of-mass frame defined by

$$m_1\mathbf{y}_1 + m_2\mathbf{y}_2 = \mathbf{0}, \tag{17.36}$$

in which frame the individual trajectories $\mathbf{y}_1$ and $\mathbf{y}_2$ are related to the relative one $\mathbf{y}$ by

$$\mathbf{y}_1 = \frac{m_2}{m}\mathbf{y}; \quad \mathbf{y}_2 = -\frac{m_1}{m}\mathbf{y}. \tag{17.37}$$

The velocities are given similarly by

$$\mathbf{v}_1 = \frac{m_2}{m}\mathbf{v}; \quad \mathbf{v}_2 = -\frac{m_1}{m}\mathbf{v}. \tag{17.38}$$

In principle, the binary's phase evolution $\phi(t)$ should be determined from a knowledge of the radiation reaction forces acting locally on the orbit. At the Newtonian order, this means considering the 'Newtonian' radiation reaction force, which is known to contribute to the total acceleration only at the 2.5PN level, i.e. $1/c^5$ smaller than the Newtonian acceleration (where $5 = 2s + 1$, with $s = 2$ the helicity of the graviton). A simpler computation of the phase is to deduce it from the energy balance equation between the loss of centre-of-mass energy and the total flux emitted at infinity in the form of waves. In the case of circular orbits one needs only to find the decrease of the orbital separation r and for that purpose the balance of energy is sufficient. Relying on an energy balance equation is the method we follow for computing the phase of inspiralling binaries in higher post-Newtonian approximations (see section 17.6). Thus, we write

$$\frac{\mathrm{d}E}{\mathrm{d}t} = -\mathcal{L}, \tag{17.39}$$

where E is the centre-of-mass energy, given at the Newtonian order by

$$E = -\frac{Gm_1m_2}{2r}, \tag{17.40}$$

and where $\mathcal{L}$ denotes the total energy flux (or gravitational 'luminosity'), deduced to the Newtonian order from the quadrupole formula of Einstein:

$$\mathcal{L} = \frac{G}{5c^5} \frac{d^3 Q_{ij}}{dt^3} \frac{d^3 Q_{ij}}{dt^3}. \tag{17.41}$$

The quadrupole moment is merely the Newtonian (trace-free) quadrupole of the source, which reads in the case of the point-particle binary as

$$Q_{ij} = m_1 (y_1^i y_1^j - \tfrac{1}{3}\delta^{ij} y_1^2) + 1 \leftrightarrow 2. \tag{17.42}$$

In the mass-centred frame (17.36) we get

$$Q_{ij} = \mu (y^i y^j - \tfrac{1}{3}\delta^{ij} r^2). \tag{17.43}$$

The third time derivative of Q_{ij} needed in the quadrupole formula (17.41) is easily obtained. When an acceleration is generated we replace it by the Newtonian equation of motion (17.35). In the case of a circular orbit we get

$$\frac{d^3 Q_{ij}}{dt^3} = -4 \frac{Gm\mu}{r^3} (y^i v^j + y^j v^i) \tag{17.44}$$

(this is automatically trace-free because $\mathbf{y} \cdot \mathbf{v} = 0$). Replacing (17.44) into (17.41) leads to the 'Newtonian' flux

$$\mathcal{L} = \frac{32}{5} \frac{G^3 m^2 \mu^2}{c^5 r^4} v^2. \tag{17.45}$$

A better way to express the flux is in terms of some dimensionless quantities, namely the mass ratio v given in (17.33), and a very convenient post-Newtonian parameter defined from the orbital frequency ω by

$$x = \left(\frac{Gm\omega}{c^3}\right)^{2/3}. \tag{17.46}$$

Notice that x is of formal order $O(1/c^2)$ in the post-Newtonian expansion. Thanks to the Kepler law $Gm = r^3 \omega^2$ we transform (17.45) and arrive at

$$\mathcal{L} = \frac{32}{5} \frac{c^5}{G} v^2 x^5. \tag{17.47}$$

In this form the only factor having a dimension is

$$\frac{c^5}{G} \approx 3.63 \times 10^{52} \text{ W}, \tag{17.48}$$

which is the Planck unit of a power, which turns out to be independent of the Planck constant. (Notice that instead of c^5/G the inverse ratio G/c^5 appears as a factor in the quadrupole formula (17.41).) On the other hand, we find that E reads simply

$$E = -\tfrac{1}{2}\mu c^2 x. \tag{17.49}$$

Next we replace (17.47) and (17.49) into the balance equation (17.39), and find in this way an ordinary differential equation which is easily integrated for the unknown x. We introduce for later convenience the dimensionless time variable

$$\tau = \frac{c^3 \nu}{5Gm}(t_c - t), \tag{17.50}$$

where t_c is a constant of integration. Then the solution reads

$$x(t) = \tfrac{1}{4}\tau^{-1/4}. \tag{17.51}$$

It is clear that t_c represents the instant of coalescence, at which (by definition) the orbital frequency diverges to infinity. Then a further integration yields $\phi = \int \omega \, dt = -\tfrac{5}{\nu}\int x^{2/3}\, d\tau$, and we get the looked for result

$$\phi_c - \phi(t) = \frac{1}{\nu}\tau^{5/8}, \tag{17.52}$$

where ϕ_c denotes the constant phase at the instant of coalescence. It is often useful to consider the number $\mathcal{N}$ of gravitational-wave cycles which are left until the final coalescence starting from some frequency ω:

$$\mathcal{N} = \frac{\phi_c - \phi}{\pi} = \frac{1}{32\pi\nu}x^{-5/2}. \tag{17.53}$$

As we see the post-Newtonian order of magnitude of $\mathcal{N}$ is c^{+5}, that is the inverse of the order c^{-5} of radiation reaction effects. As a matter of fact, $\mathcal{N}$ is a large number, approximately equal to 1.6×10^4 in the case of two neutron stars between 10 and 1000 Hz (roughly the frequency bandwidth of the detector VIRGO). Data analysts of detectors have estimated that, in order not to suffer a too severe reduction of signal to noise, one should monitor the phase evolution with an accuracy comparable to one gravitational-wave cycle (i.e. $\delta\mathcal{N} \sim 1$) or better. Now it is clear, from a post-Newtonian point of view, that since the 'Newtonian' number of cycles given by (17.53) is formally of order c^{+5}, any post-Newtonian correction therein which is larger than order c^{-5} is expected to contribute to the phase evolution more than that allowed by the previous estimate. Therefore, one expects that in order to construct accurate templates it will be necessary to include into the phase the post-Newtonian corrections up to at least the 2.5PN or $1/c^5$ order. This expectation has been confirmed by various studies [21–24] which showed that in advanced detectors the 2.5PN or, better, the 3PN approximation is required in the case of inspiralling neutron star binaries. Notice that 3PN here means 3PN in the centre-of-mass energy E, which is deduced from the 3PN equations of motion, as well as in the total flux $\mathcal{L}$, which is computed from a 3PN wave-generation formalism. For the moment the phase has been completed to the 2.5PN order [15, 25–27]; the 3PN order is still incomplete (but, see [13, 28, 29]).

17.5   Post-Newtonian wave generation

17.5.1   Field equations

We consider a general compact-support stress–energy tensor $T^{\mu\nu}$ describing the isolated source, and we look for the solutions, in the form of a (formal) post-Newtonian expansion, of the Einstein field equations,

$$R^{\mu\nu} - \frac{1}{2} g^{\mu\nu} R = \frac{8\pi G}{c^4} T^{\mu\nu}, \tag{17.54}$$

and thus also of their consequence, the equations of motion $\nabla_\nu T^{\mu\nu} = 0$ of the source. We impose the condition of harmonic coordinates, i.e. the gauge condition

$$\partial_\nu h^{\mu\nu} = 0; \quad h^{\mu\nu} = \sqrt{-g} g^{\mu\nu} - \eta^{\mu\nu}, \tag{17.55}$$

where g and $g^{\mu\nu}$ denote the determinant and inverse of the covariant metric $g_{\mu\nu}$, and where $\eta^{\mu\nu}$ is a Minkowski metric: $\eta^{\mu\nu} = \mathrm{diag}(-1, 1, 1, 1)$. Then the Einstein field equations (17.54) can be replaced by the so-called *relaxed* equations, which take the form of simple wave equations,

$$\Box h^{\mu\nu} = \frac{16\pi G}{c^4} \tau^{\mu\nu}, \tag{17.56}$$

where the box operator is the flat d'Alembertian $\Box = \eta^{\mu\nu} \partial_\mu \partial_\nu$, and where the source term $\tau^{\mu\nu}$ can be viewed as the stress–energy pseudotensor of the matter and gravitational fields in harmonic coordinates. It is given by

$$\tau^{\mu\nu} = |g| T^{\mu\nu} + \frac{c^4}{16\pi G} \Lambda^{\mu\nu}. \tag{17.57}$$

$\tau^{\mu\nu}$ is not a generally-covariant tensor, but only a Lorentz tensor relative to the Minkowski metric $\eta_{\mu\nu}$. As a consequence of the gauge condition (17.55), $\tau^{\mu\nu}$ is conserved in the usual sense,

$$\partial_\nu \tau^{\mu\nu} = 0 \tag{17.58}$$

(this is equivalent to $\nabla_\nu T^{\mu\nu} = 0$). The gravitational source term $\Lambda^{\mu\nu}$ is a quite complicated, highly nonlinear (quadratic at least) functional of $h^{\mu\nu}$ and its first- and second-spacetime derivatives.

We supplement the resolution of the field equations (17.55) and (17.56) by the requirement that the source does not receive any radiation from other sources located very far away. Such a requirement of 'no-incoming radiation' is to be imposed at Minkowskian past null infinity (taking advantage of the presence of the Minkowski metric $\eta_{\mu\nu}$); this corresponds to the limit $r = |\mathbf{x}| \to +\infty$ with $t + r/c = $ constant. (Please do not confuse this r with the same r denoting the separation between the two bodies in section 17.4.) The precise formulation of the no-incoming radiation condition is

$$\lim_{\substack{r \to +\infty \\ t+\frac{r}{c}=\text{constant}}} \left[\frac{\partial}{\partial r}(rh^{\mu\nu}) + \frac{\partial}{c\partial t}(rh^{\mu\nu}) \right] (\mathbf{x}, t) = 0. \tag{17.59}$$

In addition, $r \partial_\lambda h^{\mu\nu}$ should be bounded in the same limit. Actually we often adopt, for technical reasons, the more restrictive condition that the field is stationary before some finite instant $-\mathcal{T}$ in the past (refer to [8] for details). With the no-incoming radiation condition (17.58) or (17.59) we transform the differential Einstein equation (17.56) into the equivalent integro-differential system

$$h^{\mu\nu} = \frac{16\pi G}{c^4} \Box_R^{-1} \tau^{\mu\nu}, \tag{17.60}$$

where $\Box_R^{-1}$ denotes the standard retarded inverse d'Alembertian given by

$$(\Box_R^{-1}\tau)(\boldsymbol{x}, t) = -\frac{1}{4\pi} \int \frac{d^3\boldsymbol{x}'}{|\boldsymbol{x} - \boldsymbol{x}'|} \tau(\boldsymbol{x}', t - |\boldsymbol{x} - \boldsymbol{x}'|/c). \tag{17.61}$$

17.5.2   Source moments

In this section we shall solve the field equations (17.55) and (17.56) in the exterior of the isolated source by means of a multipole expansion, parametrized by some appropriate source multipole moments. The particularity of the moments we shall obtain, is that they are defined from the *formal* post-Newtonian expansion of the pseudotensor $\tau^{\mu\nu}$, supposing that the latter expansion can be iterated to any order. Therefore, these source multipole moments are physically valid only in the case of a slowly-moving source (slow internal velocities; weak stresses). The general structure of the post-Newtonian expansion involves besides the usual powers of $1/c$ some arbitrary powers of the logarithm of c, say

$$\overline{\tau}^{\mu\nu}(t, \boldsymbol{x}, c) = \sum_{p,q} \frac{(\ln c)^q}{c^p} \tau_{pq}^{\mu\nu}(t, \boldsymbol{x}), \tag{17.62}$$

where the overbar denotes the formal post-Newtonian expansion, and where $\tau_{pq}^{\mu\nu}$ are the functional coefficients of the expansion (p, q are integers, including zero). Now, the general multipole expansion of the metric field $h^{\mu\nu}$, denoted by $\mathcal{M}(h^{\mu\nu})$, is found by requiring that when re-developed into the near-zone, i.e. in the limit where $r/c \to 0$ (this is equivalent with the formal re-expansion when $c \to \infty$), it *matches* with the multipole expansion of the post-Newtonian expansion $\overline{h}^{\mu\nu}$ (whose structure is similar to (17.62)) in the sense of the mathematical technics of matched asymptotic expansions. We find [11, 14] that the multipole expansion $\mathcal{M}(h^{\mu\nu})$ satisfying the matching is uniquely determined, and is composed of the sum of two terms,

$$\mathcal{M}(h^{\mu\nu}) = \text{finite part} \, \Box_R^{-1}[\mathcal{M}(\Lambda^{\mu\nu})] - \frac{4G}{c^4} \sum_{l=0}^{+\infty} \frac{(-)^l}{l!} \partial_L \left\{ \frac{1}{r} \mathcal{F}_L^{\mu\nu}(t - r/c) \right\}. \tag{17.63}$$

The first term, in which $\Box_R^{-1}$ is the flat retarded operator (17.61), is a particular solution of the Einstein field equations in vacuum (outside the source), i.e. it satisfies $\Box h_{\text{part}}^{\mu\nu} = \mathcal{M}(\Lambda^{\mu\nu})$. The second term is a retarded solution of the source-free homogeneous wave equation, i.e. $\Box h_{\text{hom}}^{\mu\nu} = 0$. We denote $\partial_L = \partial_{i_1} \ldots \partial_{i_l}$ where $L = i_1 \ldots i_l$ is a multi-index composed of l indices; the l summations over the indices $i_1 \ldots i_l$ are not indicated in (17.63). The 'multipole moments' parametrizing this homogeneous solution are given explicitly by (with $u = t - r/c$)

$$\mathcal{F}_L^{\mu\nu}(u) = \text{finite part} \int d^3x \, \hat{x}_L \int_{-1}^1 dz \, \delta_l(z) \bar{\tau}^{\mu\nu}(\boldsymbol{x}, u + z|\boldsymbol{x}|/c), \qquad (17.64)$$

where the integrand contains the *post-Newtonian* expansion of the pseudostress–energy tensor $\bar{\tau}^{\mu\nu}$, whose structure reads like (17.62). In (17.64), we denote the symmetric-trace-free (STF) projection of the product of l vectors x^i with a hat, so that $\hat{x}_L = \text{STF}(x^L)$, with $x^L = x^{i_1} \ldots x^{i_l}$ and $L = i_1 \ldots i_l$; for instance, $\hat{x}_{ij} = x_i x_j - \frac{1}{3}\delta_{ij}\boldsymbol{x}^2$. The function $\delta_l(z)$ is given by

$$\delta_l(z) = \frac{(2l+1)!!}{2^{l+1}l!}(1 - z^2)^l, \qquad (17.65)$$

and satisfies the properties

$$\int_{-1}^1 dz \delta_l(z) = 1; \qquad \lim_{l \to +\infty} \delta_l(z) = \delta(z) \qquad (17.66)$$

(where $\delta(z)$ is the Dirac measure). Both terms in (17.63) involve an operation of taking a finite part. This finite part can be defined precisely by means of an analytic continuation (see [14] for details), but it is in fact basically equivalent to taking the finite part of a divergent integral in the sense of Hadamard [18]. Notice, in particular, that the finite part in the expression of the multipole moments (17.64) deals with the behaviour of the integral *at infinity*: $r \to \infty$ (without the finite part the integral would be divergent because of the factor $x_L = r^l n_L$ in the integrand and the fact that the pseudotensor $\bar{\tau}^{\mu\nu}$ is not of compact support).

The result (17.63)–(17.64) permits us to define a very convenient notion of the *source* multipole moments (by opposition to the *radiative* moments defined below). Quite naturally, the source moments are constructed from the ten components of the tensorial function $\mathcal{F}_L^{\mu\nu}(u)$. Among these components four can be eliminated using the harmonic gauge condition (17.55), so in the end we find only six independent source multipole moments. Furthermore, it can be shown that by changing the harmonic gauge in the exterior zone one can further reduce the number of independent moments to only two. Here we shall report the result for the 'main' multipole moments of the source, which are the mass-type moment I_L and current-type J_L (the other moments play a small role starting only at highorder in the post-Newtonian expansion). We have [14]

$$I_L(u) = \text{finite part} \int d^3x \int_{-1}^{1} dz \left\{ \delta_l \hat{x}_L \Sigma - \frac{4(2l+1)}{c^2(l+1)(2l+3)} \delta_{l+1} \hat{x}_{iL} \partial_t \Sigma_i \right.$$

$$\left. + \frac{2(2l+1)}{c^4(l+1)(l+2)(2l+5)} \delta_{l+2} \hat{x}_{ijL} \partial_t^2 \Sigma_{ij} \right\}, \tag{17.67}$$

$$J_L(u) = \text{finite part} \int d^3x \int_{-1}^{1} dz\, \varepsilon_{ab<i_l} \left\{ \delta_l \hat{x}_{L-1>a} \Sigma_b \right.$$

$$\left. - \frac{2l+1}{c^2(l+2)(2l+3)} \delta_{l+1} \hat{x}_{L-1>ac} \partial_t \Sigma_{bc} \right\}. \tag{17.68}$$

Here the integrand is evaluated at the instant $u + z|\mathbf{x}|/c$, ε_{abc} is the Levi-Civita symbol, $\langle L \rangle$ is the STF projection, and we employ the notation

$$\Sigma = \frac{\overline{\tau}^{00} + \overline{\tau}^{ii}}{c^2}; \quad \Sigma_i = \frac{\overline{\tau}^{0i}}{c}; \quad \Sigma_{ij} = \overline{\tau}^{ij} \tag{17.69}$$

(with $\overline{\tau}^{ii} = \delta_{ij} \overline{\tau}^{ij}$). The multipole moments I_L, J_L are valid formally up to any post-Newtonian order, and constitute a generalization in the nonlinear theory of the usual mass and current Newtonian moments (see, [14] for details). It can be checked that, when considered at the 1PN order, these moments agree with the different expressions obtained in [9] (case of mass moments) and in [10] (current moments).

17.5.3   Radiative moments

In linearized theory, where we can neglect the gravitational source term $\Lambda^{\mu\nu}$ in (17.57), as well as the first term in (17.63), the source multipole moments coincide with the so-called radiative multipole moments, defined as the coefficients of the multipole expansion of the $1/r$ term in the distance to the source at retarded times $t - r/c = $ constant. However, in full nonlinear theory, the first term in (17.63) will bring another contribution to the $1/r$ term at future null infinity. Therefore, the source multipole moments are not the 'measured' ones at infinity, and so they must be related to the real observables of the field at infinity which are constituted by the radiative moments. It has been known for a long time that the harmonic coordinates do not belong to the class of Bondi coordinate systems at infinity, because the expansion of the harmonic metric when $r \to \infty$ with $t - r/c = $ constant involves, in addition to the normal powers of $1/r$, some powers of the *logarithm* of r. Let us change the coordinates from harmonic to some Bondi-type or 'radiative' coordinates $(\mathbf{X}, T)$ such that the metric admits a power-like expansion without logarithms when $R \to \infty$ with $T - R/c = $ constant and $R = |\mathbf{X}|$ (it can be shown that the condition to be satisfied by the radiative coordinate system is that the retarded time $T - R/c$ becomes asymptotically null at infinity). For the purpose of deriving the formula (17.73) below it is sufficient to transform the coordinates according to

$$T - \frac{R}{c} = t - \frac{r}{c} - \frac{2GM}{c^3} \ln\left(\frac{r}{r_0}\right), \tag{17.70}$$

where M denotes the ADM mass of the source and r_0 is a gauge constant. In radiative coordinates it is easy to decompose the $1/R$ term of the metric into multipoles and to define in that way the radiative multipole moments U_L (mass-type; where $L = i_1 \ldots i_l$ with $l \geq 2$) and V_L (current-type; with $l \geq 2$). (Actually, it is often simpler to bypass the need for transforming the coordinates from harmonic to radiative by considering directly the TT projection of the spatial components of the harmonic metric at infinity.) The formula for the definition of the radiative moments is

$$h_{ij}^{TT} = -\frac{4G}{c^2 R} \mathcal{P}_{ijab}(\mathbf{N}) \sum_{l=2}^{+\infty} \frac{1}{c^l l!} \left\{ N_{L-2} U_{abL-2}(T - R/c) \right.$$
$$\left. - \frac{2l}{c(l+1)} N_{cL-2} \varepsilon_{cd(a} V_{b)dL-2}(T - R/c) \right\} + O\left(\frac{1}{R^2}\right) \tag{17.71}$$

where $\mathbf{N}$ is the vector $N_i = N^i = X^i/R$ (for instance $N_{L-2} = N_{i_1} \ldots N_{i_{l-2}}$), and $\mathcal{P}_{ijab}$ denotes the TT projector

$$\mathcal{P}_{ijab} = (\delta_{ia} - N_i N_a)(\delta_{jb} - N_j N_b) - \tfrac{1}{2}(\delta_{ij} - N_i N_j)(\delta_{ab} - N_a N_b). \tag{17.72}$$

In the limit of linearized gravity the radiative multipole moments U_L, V_L agree with the lth time derivatives of the source moments I_L, J_L. Let us give, without proof, the result for the expression of the radiative mass-quadrupole moment U_{ij} including relativistic corrections up to the 3PN or $1/c^6$ order inclusively [12, 13]. The calculation involves implementing explicitly a post-Minkowskian algorithm defined in [8] for the computation of the nonlinearities due to the first term of (17.63). We find ($U \equiv T - R/c$)

$$U_{ij}(U) = M_{ij}^{(2)}(U) + 2\frac{GM}{c^3} \int_0^{+\infty} dv\, M_{ij}^{(4)}(U - v) \left[\ln\left(\frac{cv}{2r_0}\right) + \frac{11}{12}\right]$$
$$+ \frac{G}{c^5}\left\{ -\frac{2}{7} \int_0^{+\infty} dv\, [M_{a<i}^{(3)} M_{j>a}^{(3)}](U - v) - \frac{2}{7} M_{a<i}^{(3)} M_{j>a}^{(2)}(U) \right.$$
$$\left. - \frac{5}{7} M_{a<i}^{(4)} M_{j>a}^{(1)}(U) + \frac{1}{7} M_{a<i}^{(5)} M_{j>a}(U) + \frac{1}{3}\varepsilon_{ab<i} M_{j>a}^{(4)} J_b(U) \right\}$$
$$+ 2\left(\frac{GM}{c^3}\right)^2 \int_0^{+\infty} dv\, M_{ij}^{(5)}(U - v)$$
$$\times \left[\ln^2\left(\frac{cv}{2r_0}\right) + \frac{57}{70} \ln\left(\frac{cv}{2r_0}\right) + \frac{124\,627}{44\,100}\right]$$
$$+ O\left(\frac{1}{c^7}\right). \tag{17.73}$$

The superscript (n) denotes n time derivations. The quadrupole moment M_{ij} entering this formula is closely related to the source quadrupole I_{ij},

$$M_{ij} = I_{ij} + \frac{2G}{3c^5}\{K^{(3)}I_{ij} - K^{(2)}I_{ij}^{(1)}\} + O\left(\frac{1}{c^7}\right), \qquad (17.74)$$

where K is the Newtonian moment of inertia (see equation (4.24) in [27]; we are using here a mass-centred frame so that the mass-dipole moment I_i is zero). The Newtonian term in (17.73) corresponds to the quadrupole formalism. Next, there is a quadratic nonlinear correction term with multipole interaction $M \times M_{ij}$ which represents the effect of tails of gravitational waves (scattering of linear waves off the spacetime curvature generated by the mass M). This correction is of order $1/c^3$ or 1.5PN and takes the form of a non-local integral with logarithmic kernel [30]. It is responsible notably for the term proportional to $\pi\tau^{1/4}$ in the formula for the phase (17.87) below. The next correction, of order $1/c^5$ or 2.5PN, is constituted by quadratic interactions between two mass-quadrupoles, and between a mass-quadrupole and the constant current dipole [12]. This term contains also a non-local integral, which is due to the radiation of gravitational waves by the distribution of the stress–energy of linear waves [12,30–32]. Finally, at the 3PN order in (17.73) the first cubic nonlinear interaction appears, which is of the type $(M \times M \times M_{ij})$ and corresponds to the tails generated by the tails themselves [13].

17.6   Inspiral binary waveform

To conclude, let us give (without proof) the result for the two polarization waveforms $h_+(t)$ and $h_\times(t)$ of the inspiralling compact binary developed to 2PN order in the amplitude and to 2.5PN order in the phase. The calculation was done by Blanchet, Damour, Iyer, Will and Wiseman [15,25–27,33], based on the formalism reviewed in section 17.5 and, independently, on that defined in [15]. Following [33] we present the polarization waveforms in a form which is ready for use in the data analysis of binary inspirals in the detectors VIRGO and LIGO (the analysis will be based on the optimal filtering technique reviewed in section 17.2). We find, extending the Newtonian formulae in section 17.3,

$$h_{+,\times} = \frac{2G\mu}{c^2 R}\left(\frac{Gm\omega}{c^3}\right)^{2/3}$$
$$\times \{H_{+,\times}^{(0)} + x^{1/2}H_{+,\times}^{(1/2)} + xH_{+,\times}^{(1)} + x^{3/2}H_{+,\times}^{(3/2)} + x^2 H_{+,\times}^{(2)}\}, \quad (17.75)$$

where the various post-Newtonian terms, ordered by x, are given for the plus polarization by

$$H_+^{(0)} = -(1 + c_i^2)\cos 2\psi, \qquad (17.76)$$

$$H_+^{(1/2)} = -\frac{s_i}{8}\frac{\delta m}{m}[(5 + c_i^2)\cos\psi - 9(1 + c_i^2)\cos 3\psi], \qquad (17.77)$$

$$H_+^{(1)} = \tfrac{1}{6}[(19 + 9c_i^2 - 2c_i^4) - v(19 - 11c_i^2 - 6c_i^4)]\cos 2\psi$$
$$- \tfrac{4}{3}s_i^2(1 + c_i^2)(1 - 3v)\cos 4\psi, \tag{17.78}$$

$$H_+^{(3/2)} = \frac{s_i}{192}\frac{\delta m}{m}\{[(57 + 60c_i^2 - c_i^4) - 2v(49 - 12c_i^2 - c_i^4)]\cos\psi$$
$$- \tfrac{27}{2}[(73 + 40c_i^2 - 9c_i^4) - 2v(25 - 8c_i^2 - 9c_i^4)]\cos 3\psi$$
$$+ \tfrac{625}{2}(1 - 2v)s_i^2(1 + c_i^2)\cos 5\psi\} - 2\pi(1 + c_i^2)\cos 2\psi, \tag{17.79}$$

$$H_+^{(2)} = \tfrac{1}{120}[(22 + 396c_i^2 + 145c_i^4 - 5c_i^6) + \tfrac{5}{3}v(706 - 216c_i^2 - 251c_i^4 + 15c_i^6)$$
$$- 5v^2(98 - 108c_i^2 + 7c_i^4 + 5c_i^6)]\cos 2\psi$$
$$+ \tfrac{2}{15}s_i^2[(59 + 35c_i^2 - 8c_i^4) - \tfrac{5}{3}v(131 + 59c_i^2 - 24c_i^4)$$
$$+ 5v^2(21 - 3c_i^2 - 8c_i^4)]\cos 4\psi$$
$$- \tfrac{81}{40}(1 - 5v + 5v^2)s_i^4(1 + c_i^2)\cos 6\psi$$
$$+ \frac{s_i}{40}\frac{\delta m}{m}\{[11 + 7c_i^2 + 10(5 + c_i^2)\ln 2]\sin\psi - 5\pi(5 + c_i^2)\cos\psi$$
$$- 27[7 - 10\ln(3/2)](1 + c_i^2)\sin 3\psi + 135\pi(1 + c_i^2)\cos 3\psi\}, \tag{17.80}$$

and for the cross-polarization by

$$H_\times^{(0)} = -2c_i\sin 2\psi, \tag{17.81}$$

$$H_\times^{(1/2)} = -\frac{3}{4}s_i c_i\frac{\delta m}{m}[\sin\psi - 3\sin 3\psi], \tag{17.82}$$

$$H_\times^{(1)} = \frac{c_i}{3}[(17 - 4c_i^2) - v(13 - 12c_i^2)]\sin 2\psi - \tfrac{8}{3}(1 - 3v)c_i s_i^2\sin 4\psi, \tag{17.83}$$

$$H_\times^{(3/2)} = \frac{s_i c_i}{96}\frac{\delta m}{m}\{[(63 - 5c_i^2) - 2v(23 - 5c_i^2)]\sin\psi$$
$$- \tfrac{27}{2}[(67 - 15c_i^2) - 2v(19 - 15c_i^2)]\sin 3\psi$$
$$+ \tfrac{625}{2}(1 - 2v)s_i^2\sin 5\psi\} - 4\pi c_i\sin 2\psi, \tag{17.84}$$

$$H_\times^{(2)} = \frac{c_i}{60}[(68 + 226c_i^2 - 15c_i^4) + \tfrac{5}{3}v(572 - 490c_i^2 + 45c_i^4)$$
$$- 5v^2(56 - 70c_i^2 + 15c_i^4)]\sin 2\psi$$
$$+ \tfrac{4}{15}c_i s_i^2[(55 - 12c_i^2) - \tfrac{5}{3}v(119 - 36c_i^2) + 5v^2(17 - 12c_i^2)]\sin 4\psi$$
$$- \tfrac{81}{20}(1 - 5v + 5v^2)c_i s_i^4\sin 6\psi$$
$$- \frac{3}{20}s_i c_i\frac{\delta m}{m}\{[3 + 10\ln 2]\cos\psi + 5\pi\sin\psi$$
$$- 9[7 - 10\ln(3/2)]\cos 3\psi - 45\pi\sin 3\psi\}. \tag{17.85}$$

The notation is consistent with sections 17.3 and 17.4. In particular, the post-Newtonian parameter x is defined by (17.46). We use the shorthands $c_i = \cos i$

and $s_i = \sin i$ where i is the inclination angle. The basic phase variable ψ entering the waveforms is defined by

$$\psi = \phi - \frac{2Gm\omega}{c^3} \ln\left(\frac{\omega}{\omega_0}\right), \tag{17.86}$$

where ϕ is the actual orbital phase of the binary, and where ω_0 can be chosen as the seismic cut-off of the detector (see [33] for details). As for the phase evolution $\phi(t)$, it is given up to 2.5PN order, generalizing the Newtonian formula (17.52), by

$$\phi(t) = \phi_0 - \frac{1}{\nu}\left\{\tau^{5/8} + \left(\frac{3715}{8064} + \frac{55}{96}\nu\right)\tau^{3/8} - \frac{3}{4}\pi\tau^{1/4}\right.$$
$$+ \left(\frac{9\,275\,495}{14\,450\,688} + \frac{284\,875}{258\,048}\nu + \frac{1855}{2048}\nu^2\right)\tau^{1/8}$$
$$\left. + \left(-\frac{38\,645}{172\,032} - \frac{15}{2048}\nu\right)\pi\ln\tau\right\}, \tag{17.87}$$

where ϕ_0 is a constant and where we recall that the dimensionless time variable τ was given by (17.50). The frequency is equal to the time derivative of (17.87), hence

$$\omega(t) = \frac{c^3}{8Gm}\left\{\tau^{-3/8} + \left(\frac{743}{2688} + \frac{11}{32}\nu\right)\tau^{-5/8} - \frac{3}{10}\pi\tau^{-3/4}\right.$$
$$+ \left(\frac{1\,855\,099}{14\,450\,688} + \frac{56\,975}{258\,048}\nu + \frac{371}{2048}\nu^2\right)\tau^{-7/8}$$
$$\left. + \left(-\frac{7729}{21\,504} - \frac{3}{256}\nu\right)\pi\tau^{-1}\right\}. \tag{17.88}$$

We have checked that both waveforms (17.76)–(17.81) and phase/frequency (17.87)–(17.88) agree in the test mass limit $\nu \to 0$ with the results of linear black hole perturbations as given by Tagoshi and Sasaki [34].

References

[1] Ciufolini I and Fidecaro F (ed) 1996 *Gravitational Waves, Sources and Detectors* (Singapore: World Scientific)
[2] Marck J-A and Lasota J-P (ed) 1997 *Relativistic Gravitation and Gravitational Radiation* (Cambridge: Cambridge University Press)
[3] Davier M and Hello P (ed) 1997 *Gravitational Wave Data Analysis* (Gif-sur-Yvette: Edition Frontières)
[4] Will C M 1994 *Proc. 8th Nishinomiya–Yukawa Symposium on Relativistic Cosmology* ed M Sasaki (Universal Academic)
[5] Blanchet L 1997 *Relativistic Gravitation and Gravitational Radiation* ed J-A Marck and J-P Lasota (Cambridge: Cambridge Univerity Press)

[6] Cutler C *et al* 1993 *Phys. Rev. Lett.* **70** 2984
[7] Cutler C and Flanagan E 1994 *Phys. Rev.* D **49** 2658
[8] Blanchet L and Damour T 1986 *Phil. Trans. R. Soc.* A **320** 379
[9] Blanchet L and Damour T 1989 *Ann. Inst. H. Poincaré (Phys. Théorique)* **50** 377
[10] Damour T and Iyer B R 1991 *Ann. Inst. H. Poincaré (Phys. Théorique)* **54** 115
[11] Blanchet L 1995 *Phys. Rev.* D **51** 2559
[12] Blanchet L 1998 *Class. Quantum Grav.* **15** 89
[13] Blanchet L 1998 *Class. Quantum Grav.* **15** 113
[14] Blanchet L 1998 *Class. Quantum Grav.* **15** 1971
[15] Will C M and Wiseman A G 1996 *Phys. Rev.* D **54** 4813
[16] Epstein R and Wagoner R V 1975 *Astrophys. J.* **197** 717
[17] Thorne K S 1980 *Rev. Mod. Phys.* **52** 299
[18] Hadamard J 1932 *Le problème de Cauchy et les Équations aux Dérivées Partielles Linéaires Hyperboliques* (Paris: Hermann)
[19] Blanchet L and Faye G 2000 *J. Math. Phys.* **41** 7675
Blanchet L and Faye G *Preprint* gr-qc 0004008
[20] Wainstein L A and Zubakov L D 1962 *Extraction of Signals from Noise* (Englewood Cliffs, NJ: Prentice-Hall)
[21] Cutler C, Finn L S, Poisson E and Sussman G J 1993 *Phys. Rev.* D **47** 1511
[22] Tagoshi H and Nakamura T 1994 *Phys. Rev.* D **49** 4016
[23] Poisson E 1995 *Phys. Rev.* D **52** 5719
Poisson E 1997 *Phys. Rev.* D **55** 7980
[24] Damour T, Iyer B R and Sathyaprakash B S 1998 *Phys. Rev.* D **57** 885
[25] Blanchet L, Damour T, Iyer B R, Will C M and Wiseman A G 1995 *Phys. Rev. Lett.* **74** 3515
[26] Blanchet L, Damour T and Iyer B R 1995 *Phys. Rev.* D **51** 5360
[27] Blanchet L 1996 *Phys. Rev.* D **54** 1417
[28] Jaranowski P and Schäfer G 1998 *Phys. Rev.* D **57** 7274
[29] Blanchet L and Faye G 2000 *Phys. Lett.* A to appear
Blanchet L and Faye G *Preprint* gr-qc 0004009
[30] Blanchet L and Damour T 1992 *Phys. Rev.* D **46** 4304
[31] Wiseman A G and Will C M 1991 *Phys. Rev.* D **44** R2945
[32] Thorne K S 1992 *Phys. Rev.* D **45** 520
[33] Blanchet L, Iyer B R, Will C M and Wiseman A G 1996 *Class. Quantum Grav.* **13** 575
[34] Tagoshi H and Sasaki M 1994 *Prog. Theor. Phys.* **92** 745

PART 5

NUMERICAL RELATIVITY

Edward Seidel

*Max-Planck-Institut für Gravitationsphysik,
Albert-Einstein-Institut, Am Mühlenberg 5, D-14473, Potsdam,
Germany and University of Illinois, NCSA and Departments of
Physics and Astronomy, Champaign, IL 61820, USA*

Chapter 18

Numerical relativity

General relativity is the fundamental theory of gravity, which is governed by an extremely complex set of coupled, nonlinear, hyperbolic-elliptic partial differential equations. General solutions to these equations, needed to fully understand their implications as a fundamental theory of physics, are elusive. Additionally, the astrophysics of compact objects, which requires Einstein's theory of general relativity for understanding phenomena such as black holes and neutron stars, is attracting increasing attention. The largest parallel supercomputers are finally approaching the speed and memory required to solve the complete set of Einstein's equations for the first time since they were written over 80 years ago, allowing one to attempt full 3D simulations of such exciting events as colliding black holes and neutron stars. In this paper we review the computational effort in this direction, and discuss a new 3D multipurpose parallel code called 'Cactus' for general relativistic astrophysics. Directions for further work are indicated where appropriate.

18.1 Overview

This article is intended to provide an introduction and overview to numerical relativity, following a set of lectures given at the Como SIGRAV School on gravitational waves in Spring, 1999. We have based them heavily on several previous articles, especially [1, 2], but we have tried to extend and update them where significant new work has been done.

The Einstein equations for the structure of spacetime have remained essentially unchanged since their discovery nearly a century ago, providing the underpinnings of modern theories of gravity, astrophysics and cosmology. The theory is essential in describing phenomena such as black holes, compact objects, supernovae, and the formation of structure in the universe. Unfortunately, the equations are a set of highly complex, coupled, nonlinear partial differential equations involving ten functions of four independent variables. They are among the most complicated equations in mathematical physics. For this reason, in spite of more than 80 years of intense study, the solution space to the full set of equations is essentially unknown. Most of what we know about this fundamental theory of physics has been gleaned from linearized solutions, highly idealized solutions possessing a high degree of symmetry (e.g., static, or spherically or axially symmetric), or from perturbations of these solutions.

Over the last several decades a growing research area, called numerical relativity, has developed, where computers are employed to construct numerical solutions to these equations. Although much has been learned through this approach, progress has been slow due to the complexity of the equations and inadequate computer power. For example, an important astrophysical application is the 3D spiralling coalescence of two black holes (BH) or neutron stars (NS), which will generate strong sources of gravitational waves. As has been emphasized by Flanagan and Hughes, one of the best candidates for early detection by the laser interferometer network is increasingly considered to be BH mergers [3,4]. The imminent arrival of data from the long awaited gravitational-wave interferometers (see, e.g., [3] and references therein) has provided a sense of urgency in understanding these strong sources of gravitational waves. Such understanding can be obtained only through large-scale computer simulations using the full machinery of numerical relativity.

Furthermore, the gravitational-wave signals are likely to be so weak by the time they reach the detectors that reliable detection may be difficult without prior knowledge of the merger waveform. These signals can be properly interpreted, or perhaps even detected, only with a detailed comparison between the observational data and a set of theoretically determined 'waveform templates'. In most cases, these waveform templates needed for gravitational-wave data analysis have to be generated by large-scale computer simulations, adding to the urgency of developing numerical relativity. However, a realistic 3D simulation based on the full Einstein equations is a highly non-trivial task—one can estimate the time required for a reasonably accurate 3D simulation of, say, the coalescence of a compact object binary, to be at least 100 000 Cray Y-MP hours!

However, there is good reason for optimism that such problems can be solved within the next decade. Scalable parallel computers, and efficient algorithms that exploit them, are quickly revolutionizing computational science, and numerical relativity is a great beneficiary of these developments. Over the last years the community has developed 3D codes designed to solve the complete set of Einstein equations that run very efficiently on large-scale parallel computers. We will describe below one such code, called 'Cactus', that has achieved 142 GFlops on a 1024 node Cray T3E-1200, which is more than 2000 times faster than 2D codes of a few years ago running on a Cray Y-MP (which also had only about 0.5% the memory capacity of the large T3E). Such machines are expected to scale up rapidly as faster processors are connected together in even higher numbers, achieving Teraflop performance on real applications in a few years.

Numerical relativity requires not only large computers and efficient codes, but also a wide variety of numerical algorithms for evolving and analysing the solution.   Because of this richness and complexity of the equations, and the interesting applications to problems such as black holes and neutron stars, natural collaborations have developed between applied mathematicians, physicists, astrophysicists and computational scientists in the development of a single code to attack these problems. There are various large-scale collaborative efforts in recent years in this direction, including the NSF Black Hole Grand Challenge Project (recently concluded), the NASA Neutron Star Grand Challenge Project, the NCSA/Potsdam/Wash U numerical relativity collaboration, and most recently a large European collaboration of ten institutions funded by the EU [5].

We will describe the Cactus Computational Toolkit, along with some of its algorithms and capabilities of this code, and a number of its applications to problems of black holes, gravitational waves, and neutron stars. In the next sections we will first give a brief description of the numerical formulation of the theory of general relativity, and discuss particular difficulties associated with numerical relativity. The discussion will necessarily be brief. Examples are mostly drawn from work carried out by the NCSA/Potsdam/Wash U numerical relativity collaboration. We also provide URL addresses for web pages containing graphics and movies of some of our results.

To conclude this brief introduction, a statement of where we stand in terms of simulating general relativistic compact objects is in order. The NSF black hole grand challenge project and related work achieved long term stable evolution of single black hole spacetimes under certain conditions [6–8], but there is still a long way to go before the spiralling coalescence can be computed. The presently on-going NASA neutron star grand challenge project recently succeeded in evolving grazing collision of two neutron stars using the full Einstein-relativistic hydrodynamic system of equations, with a simple equation of state, and the Japanese groups also report preliminary success in evolving several orbits with a fully relativistic GR-hydro code [9]. The recently funded EU Network [5] will continue on the momentum of these projects. However, the final goal of a full solution of the problem including radiation transport and magnetohydrodynamics for comparison between numerical simulations and observations in gravitational-wave astronomy (waveform templates) and high-energy astronomy (γ-ray bursts) will take many more years, hopefully building on the effort described in this paper. So although much work has been done, much more remains, and new community tools, including Cactus, are being made available to all groups. We are very optimistic about the future, and hope to see more involvement in this effort across the relativity and astrophysics communities.

18.2   Einstein equations for relativity

The generality and complexity of the Einstein equations make them an excellent and fertile testing ground for a variety of broadly significant computing issues. They form a system of dozens of coupled, nonlinear equations, with thousands of terms, of mixed hyperbolic-elliptic type, and even undefined types, depending on coordinate conditions. This rich and general structure of the equations implies that the techniques developed to solve our problems will be immediately applicable to a large family of diverse scientific applications.

The system of equations breaks up naturally into a set of constraint equations, which are elliptic in nature, evolution equations, which are 'hyperbolic' in nature (more on this below), and gauge equations, which can be chosen arbitrarily (often leading to more elliptic equations). The evolution equations guarantee (mathematically) that the elliptic constraints are satisfied at all times provided the initial data satisfied them. This implies that the initial data are not freely specifiable. Moreover, although the constraints are satisfied mathematically during evolution, it will not be so numerically. These problems are discussed in turn below. First, however, we point out that a much simpler theory, familiar to many, has all of these same features. Maxwell's equations describing electromagnetic radiation have: (a) elliptic constraint equations, demanding that in vacuum the divergence of the electric and magnetic fields vanish at all times; (b) evolution equations, determining the time development of these fields, given suitable initial data satisfying the elliptic constraint equations; and (c) gauge conditions that can be applied freely to certain variables in the theory, such as some components of the vector potential. Some choices of vector potential lead to hyperbolic evolution equations for the system, and some do not. We will find all of these features present in the much more complicated Einstein equations, so it is useful to keep Maxwell's equations in mind when reading the next sections.

In the standard $3 + 1$ ADM approach to general relativity [10], the basic building block of the theory—the spacetime metric—is written in the form

$$\mathrm{d}s^2 = -(\alpha^2 - \beta^a \beta_a)\,\mathrm{d}t^2 + 2\beta_a\,\mathrm{d}x^a\,\mathrm{d}t + \gamma_{ab}\,\mathrm{d}x^a\,\mathrm{d}x^b, \qquad (18.1)$$

using geometrized units such that the gravitational constant G and the speed of light c are both equal to unity. Throughout this paper, we use Latin indices to label spatial coordinates, running from 1 to 3. The ten functions $(\alpha, \beta^a, \gamma_{ab})$ are functions of the spatial coordinates x^a and time t. Indices are raised and lowered by the 'spatial 3-metric' γ_{ab}. Notice that the geometry on a 3D spacelike hypersurface of constant time (i.e. $\mathrm{d}t = 0$) is determined by γ_{ab}. As we will see below, the Einstein equations control the evolution in time of this 3D geometry described by γ_{ab}, given appropriate initial conditions. The lapse function α and the shift vector β^a determine how the slices are threaded by the spatial coordinates. Together, α and β^a represent the coordinate degrees of freedom inherent in the covariant formulation of Einstein's equations, and can therefore be chosen, in some sense, 'freely', as discussed below.

This formulation of the equations assumes that one begins with an everywhere spacelike slice of spacetime, that should be evolved forward in time. Due to limited space, we will not discuss promising alternate treatments, based on either characteristic, or null foliations of spacetime [11], or on asymptotically null slices of spacetime [12–15].

18.2.1   Constraint equations

The constraints can be considered as the relativistic generalization of the Poisson equation of Newtonian gravity, but instead of a single linear elliptic equation there are now four, coupled, highly nonlinear elliptic equations, known as the Hamiltonian and momentum constraints. Under certain conditions, the equations decouple and can be solved independently and more easily, and this is how they have usually been treated. Recently, techniques have been developed that allow one to solve the constraints in a more general setting, without making restrictive assumptions that lead to decoupling [16–19]. In such a system the four constraint equations are solved simultaneously. This may prove useful in generating new classes of initial data. However, at present there is no satisfactory algorithm for controlling the physics content of the data generated. The major remaining work in this direction is to develop a scheme that is capable of constructing the initial data that describe a *given* physical system. That is, although we have schemes available to solve many variations on the initial value problem, it is difficult to specify in advance, for example, what are the precise spins and momenta of two black holes in orbit, or even if the holes *are* in orbit. This can generally only be determined after the equations have been solved and analysed.

The elliptic operators for these equations are usually symmetric, but they are otherwise the most general type, with all first and mixed second derivative terms present. The boundary conditions, which can break the symmetry, are usually linear conditions that involve derivatives of the fields being solved. In any case, once the initial value equations have been solved, initial data for the evolution problem result.

We illustrate the central idea of constructing initial data with vacuum spacetimes for simplicity. The application of the algorithm presented here to a general spacetime with matter source is currently routine in numerical relativity. The full 4D Einstein equations can be decomposed into six evolution equations and four constraint equations. The constraints may be subdivided, in turn, into one Hamiltonian (or energy) constraint equation,

$$R + (\operatorname{tr} K)^2 - K^{ab} K_{ab} = 0, \tag{18.2}$$

and three momentum constraint equations (or one vector equation),

$$D_b(K^{ab} - \gamma^{ab} \operatorname{tr} K) = 0. \tag{18.3}$$

In these equations K_{ab} is the extrinsic curvature of the slice, related to the time derivative of γ_{ab} by

$$K_{ab} = -\frac{1}{2\alpha}(\partial_t \gamma_{ab} - D_a \beta_b - D_b \beta_a). \tag{18.4}$$

Here we have introduced the 3D spatial covariant derivative operator D_a associated with the 3-metric γ_{ab} (i.e. $D_a \gamma_{bc} = 0$), and the 3D scalar curvature R computed from γ_{ab}. These four constraint equations can be used to determine initial data for γ_{ab} and K_{ab}, which are to be evolved with the evolution equations discussed below. These equations (18.2) and (18.3) are referred to as constraints because, as in the case of electrodynamics, they contain no time derivatives of the fundamental fields γ_{ab} and K_{ab}, and hence do not propagate the solution in time.

Next, we will sketch the standard method for obtaining a solution to these constraint equations. We follow York and coworkers (e.g., [20]) by writing the 3-metric and extrinsic curvature in 'conformal form', and also make use of the simplifying assumption tr $K = 0$ which causes the Hamiltonian and momentum constraints to completely decouple (note that actually the equations decoupled with tr $K = $ constant but we will discuss only the simplest case here). We write

$$\gamma_{ab} = \Psi^4 \hat{\gamma}_{ab}, \quad K_{ab} = \Psi^{-2} \hat{K}_{ab}, \tag{18.5}$$

where $\hat{\gamma}_{ab}$ and the transverse-tracefree part of $\hat{K}_{ab}$ is regarded as given, i.e. chosen to represent the physical system that we want to study. Under the conformal transformation, with tr $K = 0$ we find that the momentum constraint becomes

$$\hat{D}_b \hat{K}^{ab} = 0, \tag{18.6}$$

where $\hat{D}_a$ is the 3D covariant derivative associated with $\hat{\gamma}_{ab}$ (i.e. $\hat{D}_a \hat{\gamma}_{ab} = 0$). In vacuum, black hole spacetimes $\hat{K}_{ab}$ can often be solved analytically. For more details on how to solve the momentum constraints in complicated situations, see [10, 21, 22].

The remaining unknown function Ψ, must satisfy the Hamiltonian constraint. The conformal transformation of the scalar curvature is

$$R = \Psi^{-4} \hat{R} - 8 \Psi^{-5} \hat{\Delta} \Psi, \tag{18.7}$$

where $\hat{\Delta} = \hat{\gamma}^{ab} \hat{D}_a \hat{D}_b$ and $\hat{R}$ is the scalar curvature of the known metric $\hat{\gamma}_{ab}$. Plugging this back into the Hamiltonian constraint and dividing through by $-8 \Psi^{-5}$, we obtain

$$\hat{\Delta} \Psi - \tfrac{1}{8} \Psi \hat{R} + \tfrac{1}{8} \Psi^{-7} (\hat{K}_{ab} \hat{K}^{ab}) = 0, \tag{18.8}$$

an elliptic equation for the conformal factor Ψ.

To summarize, one first specifies $\hat{\gamma}_{ab}$ and the transverse-tracefree part of $\hat{K}_{ab}$ 'at will', choosing them to be something 'closest' to the spacetime one wants to study. Then one solves (18.6) for the conformal extrinsic curvature $\hat{K}_{ab}$. Finally, (18.8) is solved for the conformal factor Ψ, so the full solution γ_{ab} and K_{ab} can be reconstructed. In this process the elliptic equations are solved by standard techniques, for example, the conjugate gradient [23] or multigrid methods [24]. In situations where there is a black hole singularity, there could be added complications in solving the elliptic equations, and special treatments would have to be introduced, for example, the 'puncture' treatment of [25], or employing an 'isometry' operation to provide boundary conditions on black hole throats, ensuring identical spatial geometries inside and outside the throat (see, e.g., [21, 26], or [27] for more details).

While this is a well-established process for generating an initial data set for numerical study, there is a fundamental difficulty in using this approach to generate initial data corresponding to a physical system one wants to evolve, for example, a coalescing binary system. It is not clear how to choose the 'closest' $\hat{\gamma}_{ab}$, and the corresponding free components in $\hat{K}_{ab}$, so that the resulting γ_{ab} and K_{ab} represent the inspiralling system at its late stage of inspiral. This late stage is the so-called 'intermediate challenge problem' of binary black holes [28], an area of much current interest.

18.2.2 Evolution equations

18.2.2.1 The standard evolution system

With the initial data γ_{ab} and K_{ab} specified, we now consider their evolution in time. There are six evolution equations for the 3-metric γ_{ab} that are second order in time, resulting from projections of the full 4D Einstein equations onto the 3D spacelike slice [10]. These are most often written as a first-order in-time system of twelve evolution equations, usually referred to as the 'ADM' evolution system [10, 29]:

$$\partial_t \gamma_{ab} = -2\alpha K_{ab} + D_a \beta_b + D_b \beta_a \tag{18.9}$$

$$\partial_t K_{ab} = -D_a D_b \alpha + \alpha [R_{ab} + (\text{tr } K) K_{ab} - 2K_{ac} K^c{}_b]$$
$$+ \beta^c D_c K_{ab} + K_{ac} D_b \beta^c + K_{cb} D_a \beta^c. \tag{18.10}$$

Here R_{ab} is the Ricci tensor of the 3D spacelike slice labelled by a constant value of t. Note that these are quantities defined only on a $t =$ constant hypersurface, and require only the 3-metric γ_{ab} in their construction. Do not confuse them with the conventional 4D objects! The complete set of Einstein equations are contained in constraint equations (18.2), (18.3) and the evolution equations (18.10), (18.9). Note that (18.9) is simply the definition of the extrinsic curvature K_{ab} (18.4). These equations are analogous to the evolution equations for the electric and magnetic fields of electrodynamics. Given the 'lapse' α and 'shift' β^a, discussed below, they allow one to advance the system forward in time.

18.2.2.2 Hyperbolic evolution systems

The evolution equations (18.10) and (18.9) have been presented in the 'standard ADM form', which has served numerical relativity well over the last few decades. However, the equations are enormously complicated; the complication is hidden in the definition of the curvature tensor R_{ab} and the covariant differentiation operator D_a. In particular, although they describe physical information propagating with a finite speed, the system does not form a hyperbolic system, and is not necessarily the best for numerical evolution. Other fields of physics, in particular hydrodynamics, have developed very mature numerical methods that are specially designed to treat the well studied flux conservative, hyperbolic system of balance laws having the form

$$\partial_t \boldsymbol{u} + \partial_k F^k_- \boldsymbol{u} = S_- \boldsymbol{u} \qquad (18.11)$$

where the vector $\boldsymbol{u}$ displays the set of variables and both 'fluxes' F^k and 'sources' S are vector valued functions. In hydrodynamic systems, it often turns out that the characteristic matrix $\partial F / \partial \boldsymbol{u}$ projected into any spacelike direction can often be diagonalized, so that fields with definite propagation speeds can be identified (the eigenvectors and the eigenvalues of the projected characteristic matrix). One important point is that in (18.11) all spatial derivatives are contained in the flux terms, with the source terms in the equations containing no derivatives of the eigenfields. All of these features can be exploited in numerical finite difference schemes that treat each term in an appropriate way to preserve important physical characteristics of the solution.

Amazingly, the complete set of Einstein equations can also be put in this 'simple' form (the source terms still contain thousands of terms however). Building on earlier work by Choquet-Bruhat and Ruggeri [30], Bona and Massó began to study this problem in the late 1980s, and by 1992 they had developed a hyperbolic system for the Einstein equations with a certain specific gauge choice [31] (see below). Here by hyperbolic, we mean simply that the projected characteristic matrix has a complete set of eigenfields with real eigenvalues. This work was generalized recently to apply to a large family of gauge choices [32, 33]. The Bona–Massó system of equations is available in the 3D 'Cactus' code [34,35], as is the standard ADM system, where both are tested and compared on a number of spacetimes.

The Bona–Massó system is now one among many hyperbolic systems, as other independent hyperbolic formulations of Einstein's equations were developed [36–41] at about the same time as [42]. Among these other formulations only the one originally devised in [38] has been applied to spacetimes containing black holes [43], although still only in the spherically symmetry 1D case (a 3D version is under development [44].) Hence, of the many hyperbolic variants, only the Bona–Massó family and the formulations of York and co-workers have been tested in any detail in 3D numerical codes. Notably among the differences in the formulations, the Bona–Massó and Frittelli families contain terms equivalent to second time derivatives of the three metric γ_{ab}, while many other formulations go to a higher time derivative to achieve hyperbolicity. Another comment worth making is that for harmonic slicing, both the Bona–Massó and York families have characteristic speeds of either zero, or light speed. For maximal slicing, they both reduce to a coupled elliptic-hyperbolic system. The Bona–Massó system (at least) also allows for an additional family of explicit algebraic slicings, with the lapse proportional to an explicit function of the determinant of the 3-metric, and in those cases one can also identify gauge speeds which can be different from light speed (harmonic slicing is one example of this family where the gauge speed corresponds to light speed). Some of these slicings, such as '1 + log' [45], have been found to be very useful in 3D numerical evolutions. This information about the speed of gauge and physical propagation can be very helpful in understanding the system, and can also be useful in developing numerical methods. Only extensive numerical studies will tell if the various hyperbolic formulations live up to their promise.

Reula has recently reviewed, from the mathematical point of view, most of the recent hyperbolic formulations of the Einstein equations [46] (This article, in the online journal *Living Reviews in Relativity*, will be periodically updated). It is important to realize that the mathematical relativity field has been interested in hyperbolic formulations of the Einstein equations for many years and some systems that could have been suitable for numerical relativity were already published in the 1980s [30, 47]. However, these developments were generally not recognized by the numerical relativity community until recently.

18.3 Still newer formulations: towards a stable evolution system

Somewhere in between the standard ADM formulation and hyperbolic formulations are a class of formulations that have been getting significant attention lately, and which seem to be very promising and very stable.

As discussed above, the 3D evolution of Einstein's equations has proved very difficult, with instabilities developing on rather short timescales, even in cases of weakly gravitating, vacuum systems, such as low amplitude gravitational waves, as summarized in an important paper by Baumgarte and Shapiro [48]. In this work, it was shown how one can achieve highly improved stability by making a few key changes to the formulation of the ADM equations, most notably through a conformal decomposition and by rewriting certain terms in the 3D Ricci tensor to eliminate terms that spoil its elliptic nature. In fact, essentially these same tricks were already noticed a few years earlier by Shibata and Nakamura [49]. Hence, we refer to these formulations collectively as 'BSSN' after the four authors. These subtle changes to the standard ADM formalism have a very powerful stabilizing effect on the evolutions. Evolutions of weak waves that would develop instabilities and crash with the standard ADM formulation run much longer with the new system, and as shown in Alcubierre *et al* [50], the new system and variations allow for the first time the successful evolution of highly nonlinear gravitational waves to form a black hole in 3D while the standard ADM treatment would fail well in advance of black hole formation. Further work by the Palma group, showed the deep connection between the BSSN formulations and the Bona–Massó family of formulations [51], leading to the possibility of a fully hyperbolic, very stable formulation that shares advantages from many sides.

The Palma, Potsdam and Wash U groups also showed that these new formulations lead to much more stable black hole evolutions as well. While standard ADM formulations can evolve black holes very accurately for a short period of time, as described above, large peaks in metric functions caused by so-called 'grid-stretching' develop instabilities, which cause the codes to crash far too soon to study orbits of black holes. The new formulations can significantly extend the evolution times (by factors of two or much more) that can be achieved. In all cases, the evolutions are convergent, but seem to have larger error than the standard ADM or Bona–Massó system. These effects were recently analysed in a paper by Alcubierre *et al* [52]. We are now in the process of applying these new formulations to a series of interesting spacetimes, including pure gravitational waves, black holes, and neutron stars, some results of which are reported below.

In this and a companion paper [52] we focus on an alternative approach based on a conformal decomposition of the metric and the trace-free components of the extrinsic curvature. The conformal-tracefree (CT) approach was first devised by Nakamura in the 1980s in 3D calculations [22, 53], and then modified and applied to work on gravitational waves [49], and on neutron stars [9, 54]. This approach was not taken up by others in the community until a recent paper by Baumgarte and Shapiro [48], where a similar formulation was compared with the standard ADM approach and shown to be superior, in terms of both accuracy and stability, on tests involving weak gravitational waves, with geodesic and harmonic slicing. In a follow-up paper, Baumgarte, Hughes, and Shapiro [55] applied the same formulation to systems with given (analytically prescribed) matter sources, and found similar stability properties. More recently fully hydrodynamical simulations employing the CT approach have been reported in [56–58] in the context of collapse of rapidly-rotating (isolated) neutron stars and coalescence and merge of binary neutron stars.

In the companion paper [52] we perform an analytic investigation of the stability properties of the ADM and the CT evolution equations. Using a linearized plane wave analysis, we identify features of the equations that we believe are responsible for the difference in their stability properties.

18.3.0.1   Numerical techniques for the evolution equations

Most of what has been attempted in numerical relativity evolution schemes is built on explicit finite difference schemes. Implicit and iterative evolution schemes have been occasionally attempted, but the extra cost associated has made them less popular. We now describe the basic approach that has been tried for both the standard ADM formulation and more recent hyperbolic formulations of the equations.

ADM evolutions

The ADM system of evolution equations is often solved using some variation of the leapfrog method, similar to that described in [59]. The most extensively tested is the 'staggered leapfrog', detailed in axisymmetric cases in [59] and in 3D in [45], but other successful versions include full leapfrog implementations used in 3D by [60] and [34]. For the ADM system, the basic strategy is to use centred spatial differences everywhere, march forward according to some explicit time scheme, and hope for the best! Generally, this technique has worked surprisingly well until large gradients are encountered, at which time the methods often break down. The problem is that the equations in this ADM form are difficult to analyse, and hence *ad hoc* numerical schemes are often tried without detailed knowledge of how to treat specific terms in the equations, or how to treat instabilities when they arise. A recent development is that of the 'deloused' leapfrog, which amounts to filtering the solution [61].

Also recently, the iterative Crank–Nicholson (ICN) scheme has been found effective in suppressing some instabilities that occur [62–64]. ICN is an iterative, explicit version of the standard implicit Crank–Nicholson (CN) scheme [61,65]. The idea behind this method is to solve the implicit equations by an iterative procedure, where each iteration is an explicit operation depending only on previously computed data. Normally, this process is stopped after a certain number of iterations, or until some tolerance is achieved. For a linear equation (and, in particular, in one dimension), the iterative procedure can easily be much more computationally expensive than the matrix inversion required to solve the original implicit scheme. For a nonlinear system, however, solving the implicit scheme directly can prove to be extremely difficult.

The stability properties of the ICN scheme have been studied, with these important results.

- In order to obtain a stable scheme one must do *at least* three iterations, and not just the two one would normally expect (two iterations are enough to achieve second-order accuracy, but they are unstable!).
- The iterative scheme itself is only convergent if the standard Courant–Friedrichs–Lewy (CFL) stability condition is satisfied, otherwise the iterations diverge.

These two results taken together imply that there is no reason (at least from the point of view of stability) to ever do more that three ICN iterations. Three iterations are already second-order accurate, and provide us with a (conditionally) stable scheme. Increasing the number of iterations will not improve the stability properties of the scheme any further. In particular, we will never achieve the unconditional stability properties of the full implicit CN scheme, since if we violate the CFL condition the iterations will diverge.

Hyperbolic evolutions

The hyperbolic formulations are on a much firmer footing numerically than the ADM formulation, as the equations are in a much simpler form that has been studied for many years in computational fluid dynamics. However, the application of such methods to relativity is quite new, and hence the experience with such methods in this community is relatively limited. Furthermore, the treatment of the highly nonlinear source terms that arise in relativity is very much unexplored, and the source terms in Einstein's equations are much more complicated than those in hydrodynamics. Here we will just discuss the basic ideas in such schemes.

A standard technique for equations having flux conservative form is to split equation (18.11) into two separate processes. The transport part is given by the flux terms

$$\partial_t \boldsymbol{u} + \partial_k F^k_- \boldsymbol{u} = 0. \tag{18.12}$$

The source contribution is given by the following system of *ordinary* differential equations

$$\partial_t \boldsymbol{u} = S_- \boldsymbol{u}. \tag{18.13}$$

Numerically, this splitting is performed by a combination of both flux and source operators. Denoting by $E(\Delta t)$ the numerical evolution operator for system (18.11) in a single timestep, we implement the following combination sequence of subevolution steps:

$$E(\Delta t) = S(\Delta t/2)T(\Delta t)S(\Delta t/2) \tag{18.14}$$

where T, S are the numerical evolution operators for systems (18.12) and (18.13), respectively. This is known as 'Strang splitting' [66]. As long as both operators T and S are second-order accurate in Δt, the overall step of operator E is also second-order accurate in time.

This choice of splitting allows easy implementation of different numerical treatments of the principal part of the system without having to worry about the sources of the equations. Additionally, there are numerous computational advantages to this technique, as discussed in [67].

The sources can be updated using a variety of ODE integrators, and in 'Cactus' the usual technique involves second-order predictor-corrector methods. Higher order methods for source integration can be easily implemented, but this will not improve the overall order of accuracy. However, in special cases where the evolution is largely source driven [68], it may be important to use higher order source operators, and this method allows such generalizations. The details can be found in [34].

The implementation of numerical methods for the flux operator is much more involved, and there are many possibilities, ranging from standard choices to advanced shock capturing methods [33, 69, 70]. Among standard methods, the MacCormack method, which has proven to be very robust in the computational fluid dynamics field (see, e.g., [71] and references therein), and a directionally split Lax-Wendroff method have been implemented and tested extensively in 'Cactus'. These schemes are fully second order in space and time. Shock capturing methods have been shown to work extremely well in 1D problems in numerical relativity [32, 70], but their application in 3D is an active research area full of promise, but as yet, unfulfilled. The details of these methods, as they are applied to the Bona–Massó formulation of the equations, can be found in [34, 70].

18.3.0.2 The role of constraints

If the constraints are satisfied on the initial hypersurface, the evolution equations then guarantee that they remain satisfied on all subsequent hypersurfaces. Thus, once the initial value problem has been solved, one may advance the solution forward in time by using only the evolution equations. This is the same situation encountered in electrodynamics as discussed before. However, in a numerical solution, the constraints will be violated at some level due to numerical error. They hence provide useful indicators for the accuracy of the numerical spacetimes generated. Traditional alternatives to this approach, which is often referred to as 'free evolution', involve solving some or all of the constraint equations on each slice for certain metric and extrinsic curvature components, and then simply monitoring the 'left over' evolution equations. This issue is discussed further by Choptuik in [72], and in detail for the Schwarzschild spacetime in [73]. New approaches to this problem of constraint versus evolution equations are currently being pursued by Lee [74], among others [75]. This approach is to advance the system forward using the evolution equations, and then adjust the variables slightly so that the constraints are satisfied (to some tolerance), i.e. the solution is projected onto the constraint surface. Because there are many variables that go into the constraints, there is not a unique way to decide which ones to adjust and by how much. However, one can compute the 'minimum' perturbation to the system, which corresponds to projecting to the *closest* point on the constraint surface. Other approaches, similar in spirit to each other, have been suggested by Detweiler [76] and Brodbeck *et al* [77]. The Detweiler approach restricts the numerical evolution to the constraint surface by adding terms to the evolution equations (18.9) and (18.10) terms which are proportional to the constraints. Numerical tests of the scheme using gravitational wave spacetimes have recently been carried out, showing promising results [78].

18.3.0.3 Gauge conditions

Kinematic conditions for the lapse function α and shift vector β^i have to be specified for the evolution equations (18.9) and (18.10). With γ_{ab} and K_{ab} satisfying the constraint on the initial slice, the lapse and shift can be chosen *arbitrarily* on the initial slice and thereafter. These are referred to as gauge choices, analogous to the choice of the gauge function Λ in electrodynamics.

Einstein did not specify these quantities; they are up to the numerical relativist to choose at will!

Lapse

The choice of lapse corresponds to how one chooses 3D spacelike hypersurfaces in 4D spacetime. The 'lapse' of *proper* time along the normal vector of one slice to the next is given by $\alpha\, dt$, where dt is the *coordinate* time interval between slices. As $\alpha(x, y, z)$ can be chosen at will on a given slice, some regions of spacetime can be made to evolve farther into the future than others.

There are many motivations for particular choices of lapse. A primary concern is to ensure that it leads to a stable long-term evolution. It is easy to see that a naive choice of the lapse, for example, $\alpha = 1$, the so-called geodesic slicing, suffers from a strong tendency to produce coordinate singularities [79,80]. A related concern is that one would like to cover the region of interest in an evolution, say, where gravitational waves generated by some process could be detected, while avoiding troublesome regions, say, inside black holes where singularities lurk (the so-called 'singularity avoiding' time slicings). Another important motivation is that some choices of α allow one to write the evolution equations in forms that are especially suited to numerical evolution. Finally, computational considerations also play an important role in the choice of the lapse; one prefers a condition for α that does not involve great computational expense, while also providing smooth, stable evolution.

Some 'traditional' choices of the lapse used in the numerical construction of spacetimes are [81]: (1) Lagrangian slicing, in which the coordinates are following the flow of the matter in the simulation. This choice simplifies the matter evolution equations, but it is not always applicable, for example, in a vacuum spacetime or when the fluid flow pattern becomes complicated. (2) Maximal slicing, [79, 80] in which the trace of the extrinsic curvature is required to be zero always, i.e, $K(t = 0) = 0 = \partial_t K$. The evolution equations of the extrinsic curvature then lead to an elliptic equation for the lapse

$$D^a D_a \alpha - \alpha(R + K^2) = 0. \qquad (18.15)$$

The maximal slicing has the nice property of causing the lapse to 'collapse' to a small value at regions of strong gravity, hence avoiding the region where a curvature singularity is forming. It is one of the so-called 'singularity avoiding slicing conditions'. Maximal slicing is easily the most studied slicing condition in numerical relativity. (3) Constant mean curvature, where we let $K = $ constant differ from zero, a choice often used in constructing cosmological solutions. (4) Algebraic slicing, where the lapse is given as an algebraic function of the determinant of the 3-metric. Algebraic slicing can also be singularity avoiding [82]. As there is no need to solve an elliptic equation as in the case of maximal slicing, algebraic slicing is computationally efficient. Some algebraic slicings (e.g., the harmonic slicing in which α is set proportional to the square root of the determinant of the 3-metric g_{ab}) also make the mathematical structure of the evolution equations simpler. However, the local nature of the choice of the lapse could lead to noise in the lapse [45] and the formation of 'shock'-like features in numerical evolutions [83, 84]. The former problem can be dealt with by turning the algebraic slicing equation to an evolution equation with a diffusion term [45], but the latter problem does not seem to have a simple solution.

In addition to these most widely used 'traditional' choices of the lapse, there are also some newly developed slicing conditions whose use in numerical relativity though promising remain largely unexplored [85]: (5) K-driver. This is a generalization of the maximal slicing in which the extrinsic curvature, instead of being set to zero, is required to satisfy the condition

$$\partial_t K = -cK, \tag{18.16}$$

where c is some positive constant. This was first brought up by Eppley [86] and recently investigated in [87]. In this way the trace of the extrinsic curvature, when numerical inaccuracy causes it to drift away from zero, is 'driven' back to zero exponentially. When combined with the evolution equations, (18.16) again leads to an elliptic equation for the lapse. This choice of the lapse is shown in [87] to lead to a much more stable numerical evolution in cases where one wants to avoid large values of the extrinsic curvature. The optimal choice of the constant c as well as a number of variations on this 'driver' scheme are presently being studied. (6) γ-driver. This is another use of the 'driver' idea. In this case, the time rate of change of the determinant of the three metric $\det(g_{ab})$ is driven to zero [87]. In the absence of a shift vector or if the shift has zero divergence, this reduces to the K-driver. This choice of the lapse, which has the unique property of being able to respond to the choice of the shift, demands extensive investigations and evaluations.

Shift

The shift vector describes the 'shifting' of the coordinates from the normal vector as one moves from one slice to the next. If the shift vanishes, the coordinate point (x, y, z) will move normal to a given 3D time slice to the next slice in the future. (Refer to York [10] or Cook [88], for details and diagrams.) The choice of shift is perhaps less well developed than the choice of lapse in numerical relativity, and many choices need to be explored, particularly in 3D. The main purpose of the shift is to ensure that the coordinate description of the spacetime remains well behaved throughout the evolution. With an inappropriate or poorly chosen shift, coordinate lines may move toward each other, or become very stretched or sheared, leading to pathological behaviour of the metric functions that may be difficult to handle numerically. It may even cause the code to crash if, for example, two coordinate lines 'touch' each other creating a 'coordinate singularity' (i.e. the metric becomes singular as the distance ds between two coordinate lines goes to zero). Two important considerations for appropriate shift conditions are the ability to prevent large shearing or drifting of coordinates during an evolution, and the ability to control the coordinate location of a physical object, for example, the horizon of a black hole. These considerations are discussed below. The development of appropriate shift conditions for full 3D evolution, for systems without symmetries, is an important research area that needs much attention. Geometrical shift conditions that can be formulated without reference to specific coordinate systems or symmetries seem to be desirable. The basic idea is to develop a condition that minimizes the stretching, shearing, and drifting of coordinates in a general way. A few examples have been devised which partially meet these goals, such as 'minimal distortion', 'minimal strain' and variations [10], but much more investigations are needed. New gauge conditions, based on these earlier proposals, have recently been proposed but not yet tested in numerical simulations [28].

It is important to emphasize that the lapse and shift *only* change the way in which the slices are chosen through a spacetime and where coordinates are laid down on every slice, and do *not*, in principle, affect any physical results whatsoever. They *will* affect the value of the metric quantities, but not the physics derived from them. In this respect the freedom of choice in the lapse and shift is analogous to the freedom of gauge in electromagnetic systems.

On the other hand, it is also important to emphasize that proper choices of lapse and shift are crucial for the numerical construction of a spacetime in the Einstein theory of general relativity, in particular in a general 3D setting. In a general 3D simulation without symmetry assumption, there is no preferred choice of the form of the metric (e.g., a diagonal 3-metric, or $g_{\theta\theta} = r^2$ as in spherical symmetry), hence forcing us to deal with the gauge degree of freedom in relativity in full. This, when coupled with the inevitable lower resolution in 3D simulations, often leads to development of coordinate singularities, when evolved without a sophisticated choice of lapse and shift. Indeed the success of the 'driver' idea suggested [87] that in order to obtain a stable evolution over a long timescale, it is important to ensure that the coordinate conditions used are not only suitable for the geometry of the spacetime being evolved, but also that *the conditions themselves are stable*. That is, when the condition is perturbed, for example, by numerical inaccuracy, there is no long-term secular drifting. We regard the construction of an algorithm for choosing a suitable lapse and shift for a general 3D numerical simulation to be one of the most important issues facing numerical relativity at present.

18.3.1   General relativistic hydrodynamics

In order to make numerical relativity a tool for computational general relativistic astrophysics, it is important to combine numerical relativity with traditional tools of computational astrophysics, and, in particular, relativistic hydrodynamics. While a large amount of 3D studies in numerical relativity have been devoted to the *vacuum* Einstein equations, the spacetime dynamics with a non-vanishing source term remains a large uncharted territory. As astrophysics of compact objects that needs general relativity for its understanding is attracting increasing attention, general relativistic hydrodynamics will become an increasingly important subject as astrophysicists begin to study more relativistic systems, as relativists become more involved in studies of astrophysical sources. This promises to be one of the most exciting and important areas of research in relativistic astrophysics in the coming years.

Previously, most work in relativistic hydrodynamics has been done on fixed metric backgrounds. In this approximation the fluid is allowed to move in a relativistic manner in strong gravitational fields, say around a black hole, but its effect on the spacetime is not considered. Over the last years very sophisticated methods for general relativistic hydrodynamics have been developed by the Valencia group led by José M Ibáñez [89–92]. These methods are based on a hyperbolic formulation of the hydrodynamic equations, and are shown to be superior to traditional artificial viscosity methods for highly relativistic flows and strong shocks.

However, just fixed background approximation is inadequate in describing a large class of problems which are of most interest to gravitational-wave astronomy, namely those with substantial matter motion generating gravitational radiation, like the coalescences of neutron star binaries. We are constructing a multipurpose 3D code for the NASA Neutron Star Grand Challenge Project [93] that contains the full Einstein equations coupled to general relativistic hydrodynamics. The hydrodynamic part consists of both an artificial viscosity module, [94] and a module based on modern shock capturing schemes [95], containing three hydroevolution methods [95]: a flux split method, Roe's approximate Riemann solver [96] and Marquina's approximate Riemann solver [92, 97]. All are based on finite-difference schemes employing approximate Riemann solvers to account explicitly for the characteristic information of the equations. These schemes are particularly suitable for astrophysics simulations that involve matter in (ultra)relativistic speeds and strong shock waves.

In the flux split method, the flux is decomposed into the part contributing to the eigenfields with positive eigenvalues (fields moving to the right) and the part with negative eigenvalues (fields moving to the left). These fluxes are then discretized with one sided derivatives (which side depends on the sign of the eigenvalue). The flux split method presupposes that the equation of state of the fluid has the form $P = P(\rho, \epsilon) = \rho f(\epsilon)$, which includes, for example, the adiabatic equation of state. The second scheme, Roe's approximate Riemann solver [96] is by now a 'traditional' method for the integration of nonlinear hyperbolic systems of conservation laws [90, 91, 98]. This method makes no assumption on the equation of state, and, is more flexible than the flux split methods. The third method, the Marquina's method, is a promising new scheme [97]. It is based on a flux formula which is an extension of Shu and Osher's entropy-satisfying numerical flux [99] to systems of hyperbolic conservation laws. In this scheme there are no artificial intermediate states constructed at each cell interface. This implies that there are no Riemann solutions involved (either exact or approximate); moreover, the scheme has been proved to alleviate several numerical pathologies associated to the introduction of an *averaged* state (as Roe's method does) in the local diagonalization procedure (see [92, 97]). For a detailed comparison of the three schemes and their coupling to dynamical evolution of spacetimes, see [95].

The availability of the hyperbolic hydrotreatment and its coupling to the spacetime evolution code is particularly noteworthy. With the development of a hyperbolic formulation of the Einstein equations described above, the *entire* system can be treated as a single system of hyperbolic equations, rather than artificially separating the spacetime part from the fluid part.

18.3.2   Boundary conditions

Appropriate conditions for the outer boundary have yet to be derived for 3D numerical relativity. In 1D and 2D relativity codes, the outer boundary is generally placed far enough away that the spacetime is nearly flat there, and static or flat (i.e. copying data from the next-to-last zone to the outer edge) boundary conditions can usually be specified for the evolved functions. However, due to the constraints placed on us by limited computer memory, this is not currently possible in 3D. Adaptive mesh refinement will be of great use in this regard, but will not substitute for proper physical treatment. Most results to date have been computed with the evolved functions kept static at the outer boundary, even if the boundaries are too close for comfort in 3D!

There are several other approaches under development that promise to improve this situation greatly that we will not have room to explore in detail here, but should be mentioned. Generally, one has in mind using Cauchy evolution in the strong field, interior region where, say, black holes are colliding. The outer part of this region will be matched to some exterior treatment designed to handle what is primarily expected to be outgoing radiation.

Two major approaches have been developed by the NSF Black Hole Grand Challenge Alliance, a large USA collaboration working to solve the black hole coalescence problem, and other groups. First, by using perturbation theory, it is possible to identify quantities in the numerically evolved metric functions that obey the Regge–Wheeler and Zerilli wave equations that describe gravitational waves propagating on a black hole background. These can be used to provide boundary conditions on the metric and extrinsic curvature functions in an actual evolution, as described in a recent paper [100]. This is an excellent step forward in outer boundary treatments that should work to minimize reflections of the outgoing wave signals from the outer boundary. In tests with weak waves, a full 3D Cauchy evolution code has been successfully matched to the perturbative treatment at the boundary, permitting waves to escape from the interior region with very little reflection. Alternatively, 'Cauchy-characteristic matching' attempts to match spacelike slices in the Cauchy region to null slices at some finite radius, and the null slices can be carried out to null infinity. 3D characteristic evolution codes have progressed dramatically in recent years, and although the full 3D matching remains to be completed, tests of the scheme in specialized settings show promise [11]. One can also use the hyperbolic formulations of the Einstein equations to find eigenfields, for which outgoing conditions can in principle be applied [32] in 1D. In 3D this technique is still under development, but it shows promise for future work. Less sophisticated approaches that seem nonetheless rather successful are discussed in [63]. Finally, there is another hyperbolic approach which uses conformal rescaling to move the boundary to infinity [12–15]. These methods have different strengths and weaknesses, but all promise to improve boundary treatments significantly, helping to enable longer evolutions than are presently possible.

18.3.3 Special difficulties with black holes

The techniques described so far are generic in their application in numerical relativity. However, in this section we describe a few problems that are characteristic of black holes, and special algorithms under development to handle them. Black hole spacetimes all have in common one problem: singularities lurk within them, which must be handled numerically. Developing suitable techniques for doing so is one of the major research priorities of the community at present. If one attempts to evolve directly into the singularity, infinite curvature will be encountered, causing any numerical code to break down.

Traditionally, the singularity region is avoided by the use of 'singularity avoiding' time slices, that wrap up around the singularity. Consider the evolution shown in figure 18.1. A star is collapsing, a singularity is forming, and time slices are shown which avoid the interior while still covering a large fraction of the spacetime where waves will be seen by a distant observer. However, these slicing conditions by themselves do not solve the problem; they merely serve to delay the onset of instabilities. As shown in figure 18.1, in the vicinity of the singularity these slicings inevitably contain a region of abrupt change near the horizon, and a region in which the constant time slices dip back deep into the past in some sense. This behaviour typically manifests itself in the form of sharply peaked profiles in the spatial metric functions [80], 'grid stretching' [101] or large coordinate shift [73] on the BH throat, etc. Numerical simulations will eventually crash due to these pathological properties of the slicing.

18.3.3.1 Apparent horizon boundary conditions (AHBC)

Cosmic censorship suggests that in physical situations, singularities are hidden inside BH horizons. Because the region of spacetime inside the horizon is causally disconnected from the region of interest outside the horizon, one is tempted numerically to cut away the interior region containing the singularity, and evolve only the singularity-free region outside, as originally suggested by Unruh [102].

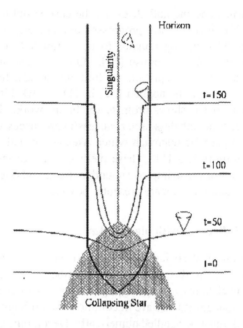

Figure 18.1. A spacetime diagram showing the formation of a BH, and time slices traditionally used to foliate the spacetime in traditional numerical relativity with singularity avoiding time slices. As the evolution proceeds, pathologically warped hypersurfaces develop, leading to unresolvable gradients that cause numerical codes to crash.

This has the consequence that there will be a region inside the horizon that simply has no numerical data. To an outside observer no information will be lost since the regions cut away are unobservable. Because the time slices will not need such sharp bends to the past, this procedure will drastically reduce the dynamic range, making it easier to maintain accuracy and stability. Since the singularity is removed from the numerical spacetime, there is in principle no physical reason why BH codes cannot be made to run indefinitely without crashing.

We spoke innocently about the BH horizon, but did not distinguish between the *apparent* and *event* horizon. These are very different concepts! While the event horizon, which is roughly a null surface that never reaches infinity and never hits the singularity, may hide singularities from the outside world in many situations, there is no guarantee that the apparent horizon, which is the (outermost) surface that has instantaneously zero expansion everywhere, even exists on a given slice! (By 'zero expansion' we mean that the surface area of outgoing bundles of photons normal to the surface is constant. Hence, the surface is 'trapped'.) Methods for finding event horizons in numerical spacetimes are now known, and will be discussed below. However, event horizons can only be found after examining the *history* of an evolution that has been already been carried out to sufficiently late times [103, 104]. Hence they are useless in providing boundaries as one integrates *forward* in time. On the other hand the apparent horizon, if it exists, can be found on any given slice by searching for closed two-surfaces with zero expansion. Although one should worry that in a generic BH collision, one may evolve into situations where no apparent horizon actually exists, let us cross that bridge if we come to it! Methods for finding apparent horizons will also be discussed below, but for now we assume that such a method exists.

Given these considerations, there are two basic ideas behind the implementation of the apparent horizon boundary condition (AHBC), also known as black hole excision:

(a) It is important to use a finite-differencing scheme which respects the causal structure of the spacetime. Since the horizon is a one-way membrane, quantities on the horizon can be affected only by quantities outside but not inside the horizon: all quantities on the horizon can in principle be updated solely in terms of known quantities residing on or outside the horizon. There are various technical details and variations on this idea, which is called 'causal differencing' [105] or 'causal reconnection' [106], but here we focus primarily on the basic ideas and results obtained to date.

(b) A shift is used to control the motion of the horizon, and the behaviour of the grid points outside the BH, as they tend to fall into the horizon if uncontrolled.

An additional advantage to using causal differencing is that it allows one to follow the information flow to create grid points with proper data on them, as needed inside the horizon, even if they did not exist previously. (Remember above that we have cut away a region inside the horizon, so in fact we have no data there.) One example is to let a BH move across the computational grid. If a BH is moving physically, it may also be desirable to have it move through coordinate space. Otherwise, all physical movement will be represented by the 'motion' of the grid points. For a single BH moving in a straight line, this may be possible (though complicated), but for spiralling coalescence this will lead to hopelessly contorted grids. The immediate consequence of this is that as a BH moves across the grid, regions in the wake of the hole, now in its exterior, must have previously been inside it where no data exist! However, with AHBC and causal differencing this need not be a problem.

Does the AHBC idea work? Preliminary indications are very promising. In spherical symmetry (1D), numerous studies show that one can locate horizons, cut away the interior, and evolve for essentially unlimited times ($t \propto 10^{3-4}M$, where M is the black hole mass). The growth of metric functions can be completely controlled, errors are reduced to a very low level, and the results can be obtained with a large variety of shift and slicing conditions, and with matter falling in the BH to allow for true dynamics even in spherical symmetry [105, 107–109].

In 3D, the basic ideas are similar but the implementation is much more difficult. The first successful test of these ideas to a Schwarzschild BH in 3D used horizon excision and a shift provided from similar simulations carried out with a

1D code [45]. The errors were found to be greatly reduced when compared even to the 1D evolution with singularity avoiding slicings. Another 3D implementation of the basic technique was provided by Brügmann [60].

This was a proof of principle, but more general treatments are following. Daues extended this work to a full range of shift conditions [6], including the full 3D minimal distortion shift [10]. He also applied these techniques to dynamic BHs, and collapse of a star to form a BH, at which point the horizon is detected, the region interior to the horizon excised, and the evolution continued with AHBC. The focus of this work has been on developing general gauge conditions for single BHs without movement through a grid. Under these conditions, BHs have been accurately evolved well beyond $t = 100M$. The NSF Black Hole Grand Challenge Alliance has been focused on the development of 3D AHBC techniques for moving Schwarzschild BHs [7]. In this work, analytic gauge conditions are provided, which are chosen to make the evolution static, although the numerical evolution is allowed to proceed freely. This moving hole is the first successful 3D test of populating grid points with data as they emerge in the BH wake. The recent, as yet unpublished, work of Huq *et al* has successfully evolved full 3D grazing collisions through about $t = 10M_{\mathrm{ADM}}$, including the topology change from two excision regions to a single one, while Alcubierre *et al* have recently evolved Schwarzschild black holes in 3D, with rather general gauge conditions, for well over $t = 4000M_{\mathrm{ADM}}$ with less than a few per cent error!

These new results are significant achievements, and show that the basic techniques outlined above are not only sound, but are also practically realizable in a 3D numerical code. However, there is still a significant amount of work to be done! The techniques have yet to be applied carefully to distorted BHs, with tests of the waveforms emitted. There are still clearly many steps to be taken before the techniques will be successfully applied to the general BH merger problem.

18.4   Tools for analysing the numerical spacetimes

We now turn to the description of several important tools that have been developed to analyse the results of a numerical evolution, carried out by some numerical evolution scheme. The evolution will generally provide metric functions on a grid, but as described above these functions are highly dependent on both the coordinate system and gauge in which the system is evolved. Determining *physical* information, such as whether a black hole exists in the data, or what gravitational waveforms have been emitted, are the subjects of this section.

18.4.1   Horizon finders

As described earlier, black holes are defined by the existence of an event horizon (EH), the surface of no return from which nothing, not even light, can escape. The event horizon is the boundary that separates those null geodesics that reach infinity from those that do not. The global character of such a definition implies that the position of an EH can only be found if the whole history of the spacetime is known. For numerical simulations of black hole spacetimes in particular, this implies that in order to locate an EH one needs to evolve sufficiently far into the future, up to a time where the spacetime has basically settled down to a stationary solution. Recently, methods have been developed to locate and analyse black hole horizons in numerically generated spacetimes, with a number of interesting results obtained [103, 104, 110–113].

In contrast, an apparent horizon (AH) is defined locally in time as the outermost marginally trapped surface [114], i.e. a surface on which the expansion of out-going null geodesics is everywhere zero. An AH can therefore be defined on a given spatial hypersurface. A well-known result [114] guarantees that if an AH is found, an EH must exist somewhere outside of it and hence a black hole has formed.

18.4.2 Locating the apparent horizons

The expansion Θ of a congruence of null rays moving in the outward normal direction to a closed surface can be shown to be [20]

$$\Theta = \nabla_i s^i + K_{ij} s^i s^j - \operatorname{tr} K, \tag{18.17}$$

where ∇_i is the covariant derivative associated with the 3-metric γ_{ij}, s^i is the normal vector to the surface, K_{ij} is the extrinsic curvature of the time slice, and $\operatorname{tr} K$ is its trace. An AH is then the outermost surface such that

$$\Theta = 0. \tag{18.18}$$

This equation is not affected by the presence of matter, since it is purely geometric in nature. We can use the same horizon finders without modification for vacuum as well as non-vacuum spacetimes. The key is to find a closed surface with normal vector s^i satisfying this equation.

18.4.2.1 Minimization algorithms

As apparent horizons are defined by the vanishing of the expansion on a surface, a fairly obvious algorithm to find such a surface involves minimizing a suitable norm of the expansion below some tolerance while adjusting test surfaces. Minimization algorithms for finding apparent horizons were among the first methods developed [115, 116]. More recently, a 3D minimization algorithm was developed and implemented by the Potsdam/NCSA/Wash U group, applied to a variety of black hole initial data and 3D numerically evolved black hole spacetimes [117–121]. Essentially the same algorithm was also implemented independently by Baumgarte *et al* [122].

The basic idea behind a minimization algorithm is to assume the surface can be represented by a function $F(x^i) = 0$, expand it in terms of some set of basis functions, and then minimize the integral of the square of the expansion Θ^2 over the surface. For example, one can parameterize a surface as

$$F(r, \theta, \phi) = r - h(\theta, \phi). \qquad (18.19)$$

The surface under consideration will be taken to correspond to the zero level of F. The function $h(\theta, \phi)$ is then expanded in terms of spherical harmonics:

$$h(\theta, \phi) = \sum_{l=0}^{l_{max}} \sum_{m=-l}^{l} a_{lm} Y_{lm}(\theta, \phi). \qquad (18.20)$$

Similar techniques were developed by [123].

At an AH the expansion integral over the surface should vanish, and we will have a global minimum. Of course, since numerically we will never find a surface for which the integral vanishes exactly, one must set a given tolerance level below which a horizon is assumed to have been found.

Minimization algorithms for finding AHs have a few drawbacks: First, the algorithm can easily settle down on a local minimum for which the expansion is not zero, so a good initial guess is often required. Moreover, when more than one marginally trapped surface is present, as is often the case, it is very difficult to predict which of these surfaces will be found by the algorithm. The algorithm can often settle on an inner horizon instead of the true AH. Again, a good initial guess can help point the finder towards the correct horizon. Finally, minimization algorithms tend to be very slow when compared with 'flow' algorithms of the type described in the next section. Typically, if N is the total number of terms in the spectral decomposition, a minimization algorithm requires of the order of a few times N^2 evaluations of the surface integrals (where in our experience 'a few' can sometimes be as high as ten).

This algorithm has been implemented in the 'Cactus' code for 3D numerical relativity [34]. For more details of the application of this algorithm, see [117–119, 122].

18.4.2.2   3D fast flow algorithm

A second method that has been implemented in the 'Cactus' code is the 'fast flow' method proposed by Gundlach [124]. Starting from an initial guess for the a_{lm}, it approaches the apparent horizon through the iteration

$$a_{lm}^{(n+1)} = a_{lm}^{(n)} - \frac{A}{1 + Bl(l+1)}(\rho\Theta)_{lm}^{(n)} \qquad (18.21)$$

where (n) labels the iteration step, ρ is some positive definite function ('a weight'), and $(\rho\Theta)_{lm}$ are the harmonic components of the function $(\rho\Theta)$. Various choices for the weight ρ and the coefficients A and B parametrize a family of such methods. The fast flow algorithm in Cactus uses

$$\rho = 2r^2 |\nabla F| [(g^{ij} - s^i s^j)(\bar{g}_{ij} - \nabla_i r \nabla_j r)]^{-1}, \qquad (18.22)$$

where $\bar{g}_{ij}$ is the flat background metric associated with the coordinates (r, θ, ϕ), and

$$A = \frac{\alpha}{l_{max}(l_{max} + 1)} + \beta, \quad B = \frac{\beta}{\alpha} \tag{18.23}$$

with $\alpha = c$ and $\beta = c/2$. Here c is a variable step size, with a typical value of $c \sim 1$. l_{max} is the maximum value of l one chooses to use in describing the surface. The iteration procedure is a finite-difference approximation to a parabolic flow, and the adaptive step size is chosen to keep the finite-difference approximation roughly close to the flow limit to prevent overshooting of the true apparent horizon. The adaptive step size is determined by a standard method used in ODE integrators: we take one full step and two half steps and compare the resulting a_{lm}. If the two results differ too much one from another, the step size is reduced.

Other methods for finding apparent horizons, based directly on computing the Jacobian of the finite differenced horizon equation, have been developed [125, 126] and successfully used in 3D. For details, see these references.

18.4.3 Locating the event horizons

The AH is defined locally in time and hence is much easier to locate than the event horizon (EH) in numerical relativity. The EH is a global object in time; it is traced out by the path of outgoing light rays that *never* propagate to future null infinity, and *never* hit the singularity. (It is the boundary of the causal past of future null infinity $\dot{\mathcal{J}}^-(\mathcal{I}^+)$.) In principle, one needs to know the entire time evolution of a spacetime in order to know the precise location of the EH. However, in spite of the global properties of the EH, hope is not lost for finding it very accurately, even in a numerical simulation of finite duration. Here we discuss a method to find the EH, given a numerically constructed black hole spacetime that eventually settles down to an approximately stationary state at late times. In principle, as the EH is a null surface, it can be found by tracing the path of null rays through time. Outward going light rays emitted just outside the EH will diverge away from it, escaping to infinity, and those emitted just inside the EH will fall away from it, towards the singularity. In a numerical integration it is difficult to follow accurately the evolution of a horizon generator forward in time, as small numerical errors cause the ray to drift and diverge rapidly from the true EH. It is a physically unstable process. However, we can actually use this property to our advantage by considering the time-reversed problem. In a global sense in time, any outward going photon that begins near the EH will be *attracted* to the horizon if integrated *backward* in time [103, 117]. In integrating backwards in time, it turns out that it suffices to start the photons within a fairly broad region where the EH is expected to reside. Such a horizon-containing region, as we call it, is often easy to determine after the spacetime has settled down to approximate stationarity. The crucial point is that when integrated backward in time along null geodesics, this horizon-containing region shrinks rapidly in 'thickness', leading to a very accurate determination of the location of the EH at earlier times. Note that it is the earlier time when the black hole is under highly dynamical evolution that we are really interested in.

Although one can integrate individual null geodesics backwards in time, we find that there are various advantages to integrating the entire null surface backwards in time. A null surface, if defined by $f(t, x^i) = 0$ satisfies the condition

$$g^{\mu\nu} \partial_\mu f \partial_\nu f = 0. \tag{18.24}$$

Hence the evolution of the surface can be obtained by a simple integration,

$$\partial_t f = \frac{-g^{ti} \partial_i f + \sqrt{(g^{ti} \partial_i f)^2 - g^{tt} g^{ij} \partial_i f \partial_j f}}{g^{tt}}. \tag{18.25}$$

The inner and outer boundary of the horizon containing region when integrated backwards in time, will rapidly converge to practically a single surface to within the resolution of the numerically constructed spacetime, i.e. a small fraction of a grid point. An accurate location of the event horizon is hence obtained. We henceforth shall represent the horizon surface as the function $f_H(t, x^i)$. Aside from the simplicity of this method, there are a number of technical advantages as discussed in [103]. One particularly noteworthy point is that this method is capable of giving the caustic structure of the event horizon if there is any; for details see [103].

The function $f_H(t, x^i)$ provides the complete coordinate location of the EH through the spacetime (or a very good approximation of it, as shown in [104]). This function by itself directly gives us the topology and location of the EH. When combined with the induced metric function on the surface, which is recorded throughout the evolution, it gives the intrinsic geometry of the EH. When further combined with the spacetime metric, all properties of the EH including its embedding can be obtained. Moreover, as the normal of $f_H(t, x^i) = 0$ gives the null generators of the horizon, it is an easy further step to determine the null generators, and hence the complete dynamics of the horizon in this formulation.

As described in a series of papers, the event horizon, once found with such a method, can be analysed to provide important information about the dynamics of black holes in a numerically generated spacetime [103, 104, 110–113].

18.4.4 Wave extraction

The gravitational radiation emitted is one of the most important quantities of interest in many astrophysical processes. The radiation is generated in regions of strong and dynamic gravitational fields, and then propagated to regions far away where it will someday be detected. We take the approach of computing the generation and evolution of the fields in a fully nonlinear way, while analysing the radiation with a perturbation formulation in the regions where it can be so treated.

The theory of black hole perturbations is well developed. One identifies certain perturbed metric quantities that evolve according to wave equations on the black hole background. These perturbed metric functions are also dependent of the gauge in which they are computed. We use a *gauge-invariant* prescription for isolating wave modes on black hole background, developed first by Moncrief [127]. The basic idea is that although the perturbed metric functions transform under coordinate transformations (gauge transformations), one can identify certain linear combinations of these functions that are invariant to first order of the perturbation. These gauge-invariant functions are the quantities that carry true physics, which does not depend on coordinate systems. They obey the wave equations describing waves propagating on the fixed black hole background. There are two independent wave modes, even- and odd-parity, corresponding to the two degrees of freedom, or polarization modes, of the waves.

A *waveform extraction* procedure has been developed that allows one to process the metric and to identify the wave modes. The gravitational wavefunction (often called the 'Zerilli function' for even-parity or the 'Regge–Wheeler function' for odd-parity) can be computed by writing the metric as the sum of a background black hole part and a perturbation:

$$g_{\alpha\beta} = \overset{o}{g}_{\alpha\beta} + h_{\alpha\beta}(Y_{\ell,m}), \qquad (18.26)$$

where the perturbation $h_{\alpha\beta}$ is expanded in spherical harmonics and their tensor generalizations and the background part $\overset{o}{g}_{\alpha\beta}$ is spherically symmetric. To compute the elements of $h_{\alpha\beta}$ in a numerical simulation, one integrates the numerically evolved metric components $g_{\alpha\beta}$ against appropriate spherical harmonics over a coordinate 2-sphere surrounding the black hole, making use of the orthogonality properties of the tensor harmonics. This process is performed for each ℓ, m mode for which waveforms are desired. The resulting functions $h_{\alpha\beta}(Y_{\ell,m})$ can then be combined in a gauge-invariant way, following the prescription given by Moncrief [127]. For each ℓ, m mode, this gauge-invariant gravitational waveform can be extracted when the wave passes through 'detectors' at some fixed radius in the computational grid. This procedure has been described in detail in [128–130], and more generally in [121, 131, 132]. It works amazingly well, allowing extraction of waves that carry very small energies (of order $10^{-7}M$ or less, with M being the mass of the source) away from the source. The procedure should apply to any isolated source of waves, such as colliding black holes, neutron stars, etc. The spherical perturbation theory (with a few minor modifications) has also been applied to distorted rotating black holes with satisfactory results [128–130].

18.5   Computational science, numerical relativity, and the 'Cactus' code

18.5.1   The computational challenges of numerical relativity

Before we describe our computational methods in the following subsections, we summarize the computational challenges of numerical relativity discussed above. It is in response to these challenges that we have devised the computational methods.

- Computational challenges due to the complexity of the physics involved. The Einstein equations are probably the most complex partial differential equations in all physics, forming a system of dozens of coupled, nonlinear equations, with thousands of terms, of mixed hyperbolic, elliptic, and even undefined types in a general coordinate system. The evolution has elliptic constraints that should be satisfied at all times. In simulations without symmetry, as would be the case for realistic astrophysical processes, codes can involve hundreds of 3D arrays, and ten of thousands of operations per grid point per update. Moreover, for simulations of astrophysical processes, we will ultimately need to integrate numerical relativity with traditional tools of computational astrophysics, including hydrodynamics, nuclear astrophysics, radiation transport and magnetohydrodynamics, which govern the evolution of the source terms (i.e. the right-hand side) of the Einstein equations. This complexity requires us to push the frontiers of massively parallel computation.
- Challenge in collaborative technology. The integration of numerical relativity into computational astrophysics is a multidisciplinary development, partly due to the complexity of the Einstein equations, and partly due to the physical systems of interest. Solving the Einstein equations on massively parallel computers involves gravitational physics, computational science, numerical algorithm and applied mathematics. Furthermore, for the numerical simulations of realistic astrophysical systems; many physics disciplines, including relativity, astrophysics, nuclear physics, and hydrodynamics are involved. It is therefore essential to have the numerical code software engineered to allow codevelopment by different research groups *and* groups with different expertise.
- The multiscale problem. Astrophysics of strongly gravitating systems inherently involves many length and timescales. The microphysics of the shortest scale (the nuclear force), controls macroscopic dynamics on the stellar scale, such as the formation and collapse of neutron stars (NSs). On the other hand, the stellar scale is at least ten times *less* than the wavelength of the gravitational waves emitted, and many orders of magnitude less than the astronomical scales of their accretion disk and jets; these larger scales provide the directly observed signals. Numerical studies of these systems, aiming at direct comparison with observations, fundamentally require the

capability of handling a wide range of dynamical time and length scales.

• Challenge in interactive computational science. In spite of the incredible advances in computational science, most simulations are still done in a very old-fashioned way. Jobs are submitted in batch mode, data are output, results are studied, and the process starts over again. This is a very time consuming and cumbersome process. In order to really take advantage of large-scale simulation as a tool for computational scientists, it is necessary to develop new techniques that allow one to conveniently make use of their computational resources, wherever they may be, to interactively monitor the simulations with advanced visualization tools, perhaps in conjunction with their colleagues in different parts of the world, and to interactively adjust the simulation based on the observed results.

All of these issues lead to important research questions in computational science. Here we give an overview of some of our effort in these directions, focusing on performance and coding issues on parallel machines, and on the development of a community code that incorporates all the mathematical and computational techniques described above (and many more), in a collaborative infrastructure for numerical relativity.

18.6 Cactus computational toolkit

The computational and collaborative needs of numerical relativity are clearly immense. To develop a basic 3D code with all the different modules, including parallel layers, adaptive mesh refinement, elliptic solvers, initial value solvers, gauge conditions, black hole excision modules, analysis tools, wave extraction, hydrodynamics modules, visualization tools, etc, require dozens of person years of effort from many different disciplines (in fact, such a feat has still not been done by the entire community!). Different groups often needlessly repeat each other's effort, further slowing the progress of the field. The NSF Black Hole Grand Challenge was a first attempt to address this problem, and an outgrowth of that effort led to the development of the 'Cactus' Computational Toolkit (CCTK), developed by the Potsdam group, in collaboration first with NCSA and Washington University, and now with a growing number of international collaborators in various disciplines. Originally designed to solve Einstein's equations, the CCTK has grown into a general purpose parallel environment for solving complex PDEs [133–135] that is being picked up by various communities in computational science. Here we focus on its application to Einstein's equations.

Cactus is designed to minimize barriers to the community development and use of the code, including the complexity associated with both the code itself and the networked supercomputer environments in which simulations and data analysis are performed. This complexity is particularly noticeable in large multidisciplinary simulations such as ours, because of the range of disciplines that must contribute to code development (relativity, hydrodynamics, astrophysics, numerics and computer science) and because of the geographical distribution of the people and computer resources involved in simulation and data analysis.

The name Cactus comes from the design of a central core (or *flesh*) which connects to application modules (or *thorns*) through an extensible interface. Thorns can implement custom developed scientific or engineering applications, such as the Einstein solvers, or other applications such as computational fluid dynamics. Other thorns from a standard computational toolkit provide a range of capabilities, such as parallel I/O, data distribution, or checkpointing.

Cactus runs on many architectures. Applications, developed on standard workstations or laptops, can be seamlessly run on clusters or supercomputers. Parallelism and portability are achieved by hiding the driver layer and features such as the I/O system and calling interface under a simple abstraction API. The Cactus API supports C/C++ and F77/F90 programming languages for the thorns. Thus, thorn programmers can work in the language they find most convenient, and are not required to master the latest and greatest computing paradigms. This makes it easier for scientists to turn existing codes into thorns which can then make use of the complete Cactus infrastructure, and in turn be used by other thorns within Cactus.

Cactus provides easy access to many cutting edge software technologies being developed in the academic research community, such as the Globus Metacomputing Toolkit, HDF5 parallel file I/O, the PETSc scientific computing library, adaptive mesh refinement, web interfaces, and advanced visualization tools.

So, how does a user use the code? A detailed user guide is available with the code (see http://www.cactuscode.org), but in a nutshell, one specifies which physics modules, and which computational/parallelism modules, are desired in a configuration file, and makes the code on the desired architecture, which can be any one of a number of machines from SGI/Cray Origin or T3E, Dec Alpha, Linux workstations or clusters, NT clusters, and others. The make system automatically detects the architecture and configures the code appropriately. Control of run parameters is then provided through an input file. Additional modules can be selected from a large community-developed library, or new modules may be written and used in conjunction with the pre-developed modules.

Our experiences with Cactus up to now suggest that these techniques are effective. It allows a code of many tens of thousands of lines, but with a compact flesh that is possible to maintain despite the large number of people contributing to it. The common code base has enhanced the collaborative process, having important and beneficial effects on the flow of ideas between remote groups. This flexible, open code architecture allows, for example, a relativity expert to contribute to the code without knowing the details of, say, the computational layers (e.g., message passing or AMR libraries) or other components (e.g., hydrodynamics). We encourage users from throughout the relativity and astrophysics communities to make use of this freely downloadable code infrastructure and physics modules, either for their own use, or as a collaborative tool to work with other groups in the community.

18.6.1 Adaptive mesh refinement

3D simulations of Einstein's equations are very demanding computationally. In this section we outline the computational needs, and techniques designed to reduce them. We will need to resolve waves with wavelengths of order $5M$ or less, where M is the mass of the BH or the neutron star. Although for Schwarzschild black holes, the fundamental $\ell = 2$ quasinormal mode wavelength is $16.8M$, higher modes, such as $\ell = 4$ and above, have wavelengths of $8M$ and below. The BH itself has a radius of $2M$. More importantly, for very rapidly rotating Kerr BHs, which are expected to be formed in realistic astrophysical BH coalescence, the modes are shifted down to significantly shorter wavelengths [3,4]. As we need at least 20 grid zones to resolve a single wavelength, we can conservatively estimate a required grid resolution of about $\Delta x = \Delta y = \Delta z \approx 0.2M$. For simulations of timescales of order $t \propto 10^2 - 10^3 M$, which will be required to follow coalescence, the outer boundary will probably be placed at a distance of roughly $R \propto 100M$ from the coalescence, requiring a Cartesian simulation domain of about $200M$ across. This leads to about 10^3 grid zones in each dimension, or about 10^9 grid zones in total. As 3D codes to solve the full Einstein equations have typically 100 variables to be stored at each location, and simulations are performed in double precision arithmetic, this leads to a memory requirement of order 1000 Gb or more! (In fairness to some groups that use spectral methods instead of finite differences (e.g., the Meudon group), we should point out highly accurate 3D simulations can now be achieved on problems that are well suited to such techniques, using much less memory! [136].)

The largest supercomputers available to scientific research communities today generally have only a fraction of this capacity, and machines with such capacity will not be routinely available for some years. Furthermore, if one needs to double the resolution in each direction for a more refined simulation, the memory requirements increase by an order of magnitude. Although such estimates will vary, depending on the ultimate effectiveness of inner or outer boundary treatments, gauge conditions, etc, they indicate that barring some unforeseen simplification, some form of adaptive mesh refinement (AMR) that places resolution only where it is required is not only desirable, but essential. The basic idea of AMR is to use some set of criteria to evaluate the quality of the solution on the present time step. If there are regions that require more resolution, then data are interpolated onto a finer grid in those regions; if less resolution is required, grid points are destroyed. Then the evolution proceeds to the next time step on this hierarchy of grids, where the process is repeated. These rough ideas have been refined and applied in many applications now in computational science.

There are several efforts ongoing in AMR for relativity. Choptuik was the early pioneer in this area, developing a 1D AMR system to handle the resolution requirements needed to follow scalar field collapse to a BH [137]. As an initially regular distribution of scalar field collapses, it will require more and more resolution as its density builds up. The grid density required to resolve the initial distribution may not even see the final BH. Further, as pulses of radiation propagate back out from the origin, they too may have to be resolved in regions where there was previously a coarse grid. Choptuik's AMR system, built on early work of Berger and Oliger [138], was able to track dynamically features that develop, enabling him to discover and accurately measure BH critical phenomena that have now become so widely studied [139].

Based on this success and others, and on the general considerations discussed above, full 3D AMR systems are under development to handle the much greater needs of solving the full set of 3D Einstein equations. A large collaboration, begun by the NSF Black Hole Grand Challenge Alliance, has been developing a system for distributing computing on large parallel machines, called Distributed Adapted Grid Hierarchies, or DAGH. DAGH was developed to provide MPI-based parallelism for the kinds of computations needed for many PDE solvers, and it also provides a framework for parallel AMR. It is one of the major computational science accomplishments to come out of the Alliance. Developed by Manish Parashar and Jim Browne, in collaboration with many subgroups within and without the Alliance, it is now being applied to many problems in science and engineering. One can find information about DAGH online at http://www.cs.utexas.edu/users/dagh/.

At least two other 3D software environments for AMR have been developed for relativity: one is called HLL, or Hierarchical Linked Lists, developed by Wild and Schutz [140]; another, called BAM, was the first AMR application in 3D relativity developed by Brügmann [60]. The HLL system has recently been applied to the test problem of the Zerilli equation (discussed above) describing perturbations of black holes [141]. This nearly 30 year old linear equation is still providing a powerful model for studying BH collisions, and it is also being used as a model problem for 3D AMR. In this work, the 1D Zerilli equation is recast as a 3D equation in Cartesian coordinates, and evolved within the AMR system provided by HLL. Even though the 3D Zerilli equation is a single linear equation, it is quite demanding in terms of resolution requirements, and without AMR it is extremely difficult to resolve both the initial pulse of radiation, the blue shifting of waves as they approach the horizon, and the scattering of radiation, including the normal modes, far from the hole.

18.7   Recent applications and progress

18.7.1   Evolving pure gravitational waves

With the new formulations of the Einstein equations discussed above, we are now able to study the nonlinear dynamics of pure gravitational waves with much more stability than ever before. This allows us to use numerical relativity to probe general relativity in highly nonlinear regime. Can one form a black hole in full 3D from pure gravitational waves? Does one see critical phenomena in full 3D? These inherently nonlinear phenomena have been investigated in 1D and 2D studies, but little is known about generic 3D behaviour.

In our investigations, we take as initial data a pure Brill-type gravitational wave [142], later studied by Eppley [86, 116] and others [143]. The metric takes the form

$$ds^2 = \Psi^4[e^{2q}(d\rho^2 + dz^2) + \rho^2\,d\phi^2] = \Psi^4 d\hat{s}^2, \tag{18.27}$$

where q is a free function subject to certain boundary conditions. Following [120, 132, 144], we choose q of the form

$$q = a\rho^2 e^{-r^2}\left[1 + c\frac{\rho^2}{(1+\rho^2)}\cos^2(n\phi)\right], \tag{18.28}$$

where a, c are constants, $r^2 = \rho^2 + z^2$ and n is an integer. For $c = 0$, these data sets reduce to the Holz [143] axisymmetric form, recently studied in full 3D Cartesian coordinates [145]. Taking this form for q, we impose the condition of time-symmetry, and solve the Hamiltonian constraint numerically in Cartesian coordinates. An initial data set is thus characterized only by the parameters (a, c, n). For the case $(a, 0, 0)$, we found in [145] that no AH exists in initial data for $a < 11.8$, and we also studied the appearance of an AH for other values of c and n.

We have surveyed a large range of this parameter space, but here we discuss two cases of interest: (a) a subcritical (but highly nonlinear) case where after a violent collapse of the self-gravitating waves, there is a subsequent rebound and after a few oscillations the waves all disperse; and (b) a supercritical case where the waves collapse in on themselves and immediately form a black hole.

The subcritical case studied in [50] has parameters $(a = 4, c = 0, n = 0)$ in the notation above. It is a rather strong axisymmetric Brill wave (BW). The evolution of this data set shows that part of the wave propagates outward while part implodes, re-expanding after passing through the origin. However, due to the nonlinear self-gravity, not all of it immediately disperses out to infinity; again part re-collapses and bounces again. After a few collapses and bounces the wave completely disperses out to infinity. This behaviour is shown in figure 18.2(a), where the evolution of the central value of the lapse is given for simulations with three different grid sizes: $\Delta x = \Delta y = \Delta z = 0.16$ (low resolution), 0.08 (medium resolution) and 0.04 (high resolution), using 32^3, 64^3 and 128^3 grid points respectively. At late times, the lapse returns to 1 (the log returns to 0). Figure 18.2(b) shows the evolution of the log of the central value of the Riemann invariant J for the same resolutions. At late times J settles on a constant value that converges rapidly to zero as we refine the grid. With these results, and direct verification that the metric functions become stationary at late times, we conclude that spacetime returns to flat (in non-trivial spatial coordinates; the metric is decidedly non-flat in appearance!).

The same simulation carried out with the standard ADM systems crashes far earlier than in the present case with the BSSN systems, which essentially run

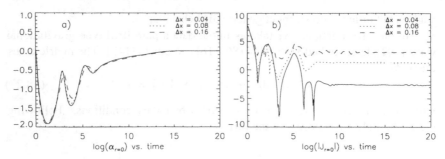

Figure 18.2. (*a*) Evolution of the log of the lapse α at $r = 0$ for the axisymmetric data $(4, 0, 0)$. The dashed/dotted/full curves represent simulations at low/medium/high resolution. (*b*) Evolution of the Riemann invariant J at $r = 0$. The wave disperses after dynamic evolution, leaving flat space behind.

forever. With this experience, we next try the case of an even stronger amplitude wave, which in this case will actually collapse on itself and form a black hole. In figure 18.3, we study the development of the data set ($a = 6$, $c = 0.2$, $n = 1$), a full 3D data set, and watch it collapse to form a black hole (the first such 3D simulation). The figure also compares this black hole formation to results obtained with an axisymmetric data set. The system clearly collapses on itself and rapidly forms a black hole. The waveform extraction shows that the newly formed hole then rings at its quasinormal mode frequency. High quality images and movies of these simulations can be found at http://jean-luc.aei-potsdam.mpg.de.

These results are exciting examples of how numerical relativity can act as a laboratory to probe the nonlinear aspects of Einstein's equations. Pure gravitational waves are clearly a rich and exciting research area that allows one to study Einstein's equations as a nonlinear theory of physics. With these new capabilities of accurate 3D evolution that can follow the implosion of waves to a black hole, there is much more physics to study, including the structure of horizons, full 3D studies of critical phenomena, and much more. Further, this study of pure vacuum waves has helped us to understand the importance of developing and testing new formulations of Einstein's equations for numerical purposes. Without the new formulations, these results simply could not be obtained. Further, we have run literally hundreds of simulations like these in order to determine which variation on the 'BSSN' families of formulations perform best. With this new knowledge, we turn to the problem of 3D black holes.

18.7.2 Black holes

Having tested these new formulations of Einstein's equations on the problem of pure gravitational waves, we now apply what we have learned to the considerably more complex problem of black hole evolutions. We first applied these new formulations to black hole spacetimes that have been very carefully tested in

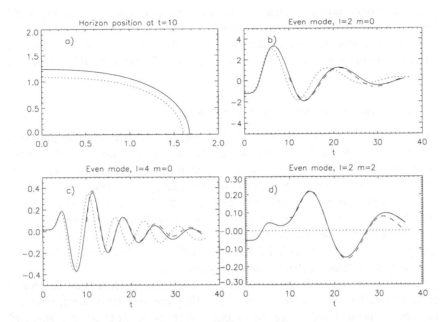

Figure 18.3. (*a*) The full (dotted) curve is the AH for the full 3D data set (6, 0.2, 1) ((6, 0, 0)) at $t = 9$ in the x–z plane. (*b*) The $\{l = 2, m = 0\}$ waveform for the 3D (6, 0.2, 1) case at $r = 4$ (full curve) is compared to the axisymmetric (6, 0, 0) case (dotted curve). The chain curve shows the fit of the 3D case to the two lowest lying QNMs for a BH of mass 0.99. (*c*) Same comparison for the $\{l = 4, m = 0\}$ waveform. (*d*) Same comparison for the non-axisymmetric $\{l = 2, m = 2\}$ waveform.

axisymmetry and with 3D nonlinear numerical codes, but with standard $3 + 1$ formulations, as well as with perturbative methods. In summary, these new formulations are able to extend the evolutions by a considerable amount, since instabilities that develop during grid stretching caused by singularity avoiding slicings are highly suppressed. 3D black hole evolutions that crashed previously by $t = 20$–$30M$ now routinely extend several times longer. However, by such late times the pathological peaks in metric functions associated with singularity avoiding slicings cannot be resolved, and then of course evolutions become very inaccurate. This is actually a major improvement! Previously, with traditional formulations, instabilities would develop as large gradients were created near the black hole, causing the codes to crash prematurely. As discussed above and shown in many papers [121, 131, 132, 146], the evolutions can actually be very accurate— allowing the extraction of very delicate waveforms from a large spectrum of modes—and remain so until the code crashes. With the new formulations, we find that we can break far through the former crash barrier, but as features become under-resolved the results naturally become less accurate. For a fuller discussion of these results, and mathematical analysis giving insight into the improved behaviour of the new formulations, see [52].

As an example of what we are now able to do with these new formulations, we now turn to an advanced application of a fully 3D 'grazing collision' of two black holes.

18.7.2.1   True 3D grazing black hole collision

In the previous sections we have shown that using the standard ADM formulations of Einstein's equations in 3D Cartesian coordinates, it is possible to perform very accurate evolutions of gravitational-wave and black hole spacetimes. However, with these formulations, there are instabilities that appear when large gradients develop, leading to premature crashing of the code. On the other hand, the new 'BSSN' family of formulations are much more robust. Having allowed us to break through the barriers seen in evolving pure nonlinear gravitational waves, we now apply these formulations to full 3D grazing collisions of black holes of the type originally considered by Brügmann a few years ago.

The initial data sets we use here for binary BH systems were developed originally by Brandt and Brügmann [25]. They are very convenient, since no isometry is needed and hence the elliptic solver can be applied on a standard Cartesian grid without the need to apply boundary conditions on strangely shaped (e.g., non-planar!) surfaces.

A few of these data sets were first evolved by Brügmann [147] using the standard ADM formulation. This was a first pioneering attempt to go beyond the highly symmetric black hole collisions that had been studied previously, combining for the first time unequal mass, spinning black holes with linear and orbital angular momentum. Brügmann was able to use nested grids to provide reasonable resolution near the holes, while putting the boundary reasonably far away. The result was that for selected data sets he was able to carry out the evolution far enough to observe a merger of the two apparent horizons. However, the difficulties of the ADM formulation, discussed above, coupled with poor resolution achievable at that time limited these evolutions to about $t = 7M$, and it was not possible to extract detailed physics, such as horizon masses, waveforms, energies, spins, etc.

Now we apply all we have learned in the last few years, together with the advanced computational infrastructure developed in Cactus and the accessibility of much larger computers, and revisit this same problem. This work is still in progress, and calculations are presently underway to refine the waveforms and the energy accounting, but we can report the following preliminary results.

For initial data of the type described in [147], we follow Brügmann and choose individual mass parameters $M_1 = 1.5$, and $M_2 = 1.0$, and linear and spin momenta on each hole such that the overall mass and angular momentum of the initial slice are measured from asymptotic properties to be

$$M_{\text{ADM}} = 3.1 \qquad (18.29)$$

$$J = 6.7 \quad \text{so that } a/m = J/M_{\text{ADM}}^2 = 0.70. \tag{18.30}$$

On a 256 processor Origin 2000 machine at NCSA we are able to run simulations of 387^3, which take roughly 100 Gb of memory. Still, with sufficient resolution to carry out long-term evolutions, the boundaries are still rather close, at roughly $x = 12M$. We use the 'BSSN' formulations to carry out the evolutions, coupled with vanishing shift and either maximal or algebraic slicings (of the '1 + log' family [45]) and with a three-step Crank–Nicholson method. Further details are in preparation for publication. Under these conditions, we find that we are able to evolve the black hole merger far beyond the time at which the horizons merge, beyond $t = 30M$, at which time the simulations become fairly inaccurate. (We must point out that we have to date only studied the apparent horizons. The event horizons can also be located by techniques developed in [103]. At present it is not known whether a single event horizon is present on the initial slice in this data set.) Depending on computational parameters, the simulations can be carried out far beyond this time without crashing, in stark contrast to earlier attempts which were doomed to crash far earlier.

Of course, the 'time to crash' is not a measure of success of a code! What we are really interested in is whether we are able to extract meaningful physics from such simulations. We are in the process of analysing such simulations in great detail, and the results are very encouraging. First, for the example discussed above, we begin with qualitative measurements of the physics we extract. In figure 18.4 we show a sequence of visualizations of simulations near the time just before, during, and after the merger of the two holes. The coordinate locations of the apparent horizons (AH) are shown as coloured surfaces. The colourmap represents the local Gaussian curvature of the surface, computed from the induced 2-metric on the horizon. As the holes approach each other and merge, a global AH develops. Meanwhile, a burst of gravitational waves, indicated by the coloured wisps emanating from the BH system develops and propagates away. The Newman–Penrose quantity Ψ_4, computed fully nonlinearly, is used to indicate the gravitational waves. As this system has no symmetries, and includes rotation, all ℓ-m-modes and both even- and odd-parity polarizations of the waves are present, leading to a much more complex structure in the wave patterns than one is used to seeing in such simulations. However, this is now moving much closer to what one expects to see in nature, and it, too, will be rather complicated! A full multipolar analysis of the waves is in progress, and it is clear that quasinormal mode ringing of the final BH is present, as expected.

These results are preliminary, but indicate that for the first time we are indeed now able to simulate the late merger stages of two black holes colliding, with rather general spin, mass and momenta, and that we can now begin to study the fine details of the physics. A quantitative analysis of the horizon evolution, mass of the final black hole, the energy emitted, the total angular momentum, etc, are underway, and preliminary results indicate that much detailed physics can be accurately extracted from these simulations. Without more advanced

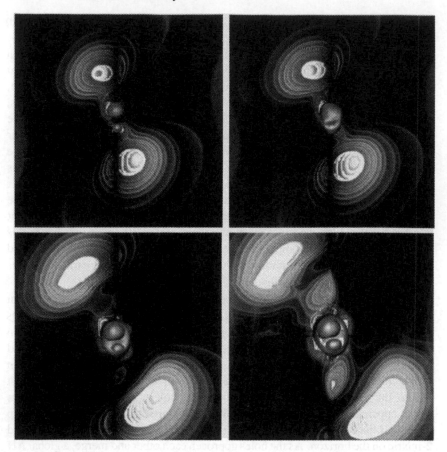

Figure 18.4. We show a sequence of visualizations of the merger of two black holes with unequal mass and spin. The apparent horizons are shown as the surfaces at the centre of the image, and the colours represent the Gaussian curvature. The waves, shown emanating from the merger, are visualizations of the real part of the Newman–Penrose quantity Ψ_4. The top left-hand panel shows the system just before the merger, while the bottom right-hand panel shows the system much later.

techniques, such as black hole excision, these simulations will be limited to the final merger phase of black hole coalescence. Hence, it is important that the community continue to focus on this long term solution. However, while that is under development, we can take advantage of our capabilities and explore this phase of the inspiral now. Our goal is several fold: (a) to explore new black hole physics of the 'final plunge' phase of the binary BH merger; (b) to try to determine some useful information relevant for gravitational-wave astronomy; and (c) to provide a strong foundation of knowledge for this process that will be useful when more advanced techniques, such a black hole excision, are fully developed. When these techniques are used to extend the ability of the community to handle the earlier orbital phase, it will be important to have an understanding of details of this most violent phase in advance, both as a testbed to ensure that results are correct, and because the understanding we gain may be useful in devising the appropriate techniques for longer-term evolution.

18.8 Summary

In this article we have attempted to review the essential mathematical and computational elements needed for a full-scale numerical relativity code that can treat a variety of problems in relativistic astrophysics and gravitation. Various formulations of the Einstein equations for evolving spacelike time slices, techniques for providing initial data, the basic ideas of gauge conditions, several important analysis tools for discovering the physics contained in a simulation, and the numerical algorithms for each of these items have been reviewed. Unfortunately, we have only been able to cover the basics of such a program, and in addition many promising alternative approaches have necessarily been left out.

As one can see, the solution to a single problem in numerical relativity requires a huge range of computational and mathematical techniques. It is truly a large-scale effort, involving experts in computer and computational science, mathematical relativity, astrophysics, and so on. For these reasons, aided by collaborations such as the NSF Black Hole Grand Challenge Alliance and the NCSA/Potsdam/Wash U collaboration, there has been a great focusing of effort over the last years.

A natural byproduct of this focusing has been the development of codes that are used and extended by large groups. A code must have a large arsenal of modules at its disposal: different initial data sets, gauge conditions, horizon finders, slicing conditions, waveform extraction, elliptic equation solvers, AMR systems, boundary modules, different evolution modules, etc. Furthermore, these codes must run efficiently on the most advanced supercomputers available. Clearly, the development of such a sophisticated code is beyond any single person or group. In fact, it is beyond the capability of a single community! Different research communities, from computer science, physics and astrophysics, must work together to develop such a code.

As an example of such a project, the 'Cactus' code has been developed by a large international collaboration [133, 134]. This code is an outgrowth of the last decade of 3D numerical relativity development primarily at NCSA/Potsdam/Wash U, and builds heavily on the experience gained in developing previous generation codes [34, 45, 148]. Cactus has a very modular structure, allowing different physics, analysis, and computational science modules to be plugged in. In fact, versions of essentially all the modules listed above are already developed for the code. For example, several formulations of Einstein's equations, including the ADM formalism and the Bona–Massó hyperbolic formulation, can be chosen as input parameters, as can different gauge conditions, horizon finders, hydrodynamics evolvers, etc. Cactus was also designed as a community code. After first developing and testing it within our rather large community of collaborators, it is available with full documentation. By having an entire research community using and contributing to such a code, we hope to accelerate the maturation of numerical relativity. Information about the code is available online, and can be accessed at http://www.cactuscode.org

Acknowledgments

It is a pleasure to acknowledge many friends and colleagues who have contributed to the work described in this article, some of which was derived from papers we have written together. I especially thank Wai-Mo Suen, who helped write previous reviews on which this article is partly based. The Cactus code was originally started by Joan Massó and Paul Walker at AEI. Without the contributions from people at many institutions, the work described here would not have been possible. This work has been supported by AEI, NCSA, NSF Grant No PHY-96-00507, NASA HPCC/ESS Grand Challenge Applications Grant No NCCS5-153 and NSF MRAC Allocation Grant No MCA93S025.

18.9   Further reading

Here are some references that we think will help fill in the details of many issues we can only gloss over. These references are clearly not complete; it was just easier for us to heavily bias this list towards work coming out of our own group, or closely associated groups! Our apologies to many others who have written fine papers on these subjects, but we have tried to give references that are current and relevant to the most important topics in numerical relativity, if incomplete and biased.

18.9.1   Overviews/formalisms of numerical relativity

For basic $3 + 1$ formalism, see [10] and the PhD thesis of Cook [88]. This provides the basics in a very clear, readable way. A somewhat more recent York article describes many 'miscellaneous' topics, such as more modern initial data and apparent horizon conditions, etc [20]. However, there are no details on more recent reformulations of Einstein's equations for numerical relativity, which are becoming very important. This is a breaking research area, with new papers every month, but some that stand out for hyperbolicity are [32–34, 46, 52, 63, 70, 149, 150, and references therein]. Even these are being overtaken by some recent developments! Every month there is new excitement in some variations on all the above that seem to have very good stability properties: [48,51] are among the more recent ones.

For the most recent overviews of numerical relativity especially related to black holes, see the review articles by the author and Wai-Mo Suen: [1, 27, 151–154].

18.9.2   Numerical techniques

An old but still very useful primer on numerical techniques for numerical relativity can be found in a little article by Smarr in [155]. More modern treatments for solving PDEs are available in *Numerical Recipes* [66]. For hyperbolic systems, the one we learned everything from is [69].

18.9.3   Gauge conditions

For gauge conditions, we recommend the classics: York [10] and Smarr and York [79, 80] for standard maximal slicing and variational principle shift conditions (minimal distortion, etc). For more modern views on how to actually implement such conditions more effectively, including the 'driver' ideas, see [87] and the very recent paper [51]. For work on so-called algebraic slicing conditions, see [33, 45], and for problems that can develop with such conditions, see [83, 84]. For the most recent ideas on shift conditions, see [156]. There is still a lot of work to do here, especially on shift conditions: please publish some ideas yourselves! This is a crucial area of needed research in numerical relativity that has not received much attention, especially in 3D.

18.9.4   Black hole initial data

There are by now many black hole initial data sets. There are early references by Misner [157,158], and Brill and Lindquist [115] and then Bowen and York [159], but more recent ones cover the same older material sufficiently. Take a look at [21,26] and references therein for the classic work. [160] looks at some physics of initial data sets. For the very large family of distorted black hole plus Brill wave data sets, check out [161], or including rotation [128]. These were extended to 3D and discussed briefly in [132]. More recently, important new ways of determining initial data were developed by Brandt and Brügmann [25]; see also [18, 162]. There are many others!

18.9.5   Black hole evolution

18.9.5.1   Spherical and distorted (axisymmetric) black holes

There have been extensive studies of numerical evolution of distorted black holes. For that we would check out [73] for 1D (spherical, undistorted), [59, 163, 164] for 2D, and finally [128, 129] for 2D rotating black holes.

18.9.5.2   Colliding black holes

For colliding Misner black holes, the original work of Smarr and others is remarkable: see [165–167]. For a more advanced attempt see [168, 169]. For boosted black holes, see [170], and for unequal mass black hole collisions see [171]. For an alternative type of black hole collision, with particles forming black holes, see [110] and references therein.

18.9.5.3   3D black holes

In 3D, so far little has been published. 3D Schwarzschild was evolved numerically by [45, 60]. Distorted 3D black holes were studied numerically, and compared to perturbation theory for the first time, in [131, 132].

Colliding black holes in 3D were studied in [172], and the first full 3D collisions were performed in [2, 147].

For a completely alternative approach to that considered here, characteristic evolution techniques have been used very successfully to evolve black holes in [8].

18.9.6   Black hole excision

For black hole excision, see [105] for an early 1D success (but the idea was floating around long before). Then [173] followed up with a more advanced treatment. Other successful 1D work includes [43, 109, 174]. [6, 7, 45, 60] give the first successful, and increasingly complex, 3D attempts.

18.9.7   Perturbation theory and waveform extraction

For work on the perturbative/numerical synergy approach, there are many papers. The original that really started it all, for Misner data, was [175], followed by [176], and with boosted holes, [170]. An extraordinary paper that carries the Misner problem to second order is [177]. [178] gives a review. For the move towards the Teukolsky formalism for rotating black holes, see [179–181], among others.

For the first attempts to use perturbation theory for 3D distorted black holes, showing the incredible accuracy one can achieve in waveforms, see [132].

18.9.8   Event and apparent horizons

For a nice description of the apparent and event horizons, see [182]. Techniques to find apparent horizons abound; the most recent variations can be found in [124, 125, 145, 183, 184].

Two different event horizon finding methods are described in [103, 110]; the latter method is used now by all groups we are aware of. More details of the method can be found in [104].

Physics that can be extracted by studying numerically generated horizons has been detailed in [110, 111, 113, 164, 185, 186].

18.9.9 Pure gravitational waves

We have focused almost exclusively on black holes in these lectures, but we also touched on recent and very exciting work in pure gravitational wave evolutions. For the original classic (a 'must read paper'), see [142]. Many followers of the 'Brill wave' school are [86, 116, 143, 187], and most recently, for the first true, long-term 3D evolutions, see [50].

18.9.10 Numerical codes

For Cactus, see [34, 133–135, 188]. Cactus, a full 3D numerical relativity code, is available as a community code this year. Please feel free to test it out and contribute to it!

References

[1] Seidel E and Suen W-M 1999 *J. Comput. Appl. Math.*
[2] Seidel E 1999 *Proc. Yukawa Conf. (Kyoto)*
[3] Flanagan É É and Hughes S A 1998 *Phys. Rev.* D **57** 4566 (gr-qc/9710129)
[4] Flanagan É É and Hughes S A 1998 *Phys. Rev.* D **57** 4535 (gr-qc/9701039)
[5] For a description of the project, see
    http://www.aei-potsdam.mpg.de/research/astro/eu_network/index.html
[6] Daues G E 1996 *PhD Thesis* Washington University, St Louis, Missouri
[7] Cook G B *et al* 1998 *Phys. Rev. Lett.* **80** 2512
[8] Gomez R *et al* 1998 *Phys. Rev. Lett.* **80** 3915 (gr-qc/9801069)
[9] Nakamura T and Oohara K 1999 *Preprint* gr-qc/9812054
[10] York J 1979 *Sources of Gravitational Radiation* ed L Smarr (Cambridge: Cambridge University Press)
[11] Bishop N *et al* 1998 *On the Black Hole Trail* ed B Iyer and B Bhawal (Dordrecht: Kluwer) (gr-qc/9801070)
[12] Friedrich H 1981 *Proc. R. Soc.* A **375** 169
[13] Friedrich H 1981 *Proc. R. Soc.* A **378** 401
[14] Friedrich H 1996 *Class. Quantum Grav.* **13** 1451
[15] Hübner P 1996 *Phys. Rev.* D **53** 701
[16] Bernstein D and Holst M Private communication
[17] Laguna P Private communication
[18] Thornburg J 1998 *Phys. Rev.* D **59** 104007
[19] Miller M Private communication
[20] York J 1989 *Frontiers in Numerical Relativity* ed C Evans, L Finn and D Hobill (Cambridge: Cambridge University Press) pp 89–109
[21] Cook G B *et al* 1993 *Phys. Rev.* D **47** 1471
[22] Nakamura T and Oohara K 1989 *Frontiers in Numerical Relativity* ed C Evans, L Finn and D Hobill (Cambridge: Cambridge University Press) pp 254–80

[23] Ashby S F, Manteuffel T A and Saylor P E 1990 *SIAM J. Num. Anal.* **27** 1542
[24] Cook G B 1989 *Frontiers in Numerical Relativity* ed C Evans, L Finn and D Hobill (Cambridge: Cambridge University Press) and references therein
[25] Brandt S and Brügmann B 1997 *Phys. Rev. Lett.* **78** 3606
[26] Cook G B 1991 *Phys. Rev.* D **44** 2983
[27] Seidel E 1996 *Relativity and Scientific Computing* ed F Hehl (Berlin: Springer)
[28] Brady P R, Creighton J D E and Thorne K S 1998 *Phys. Rev.* D **58** 061501
[29] Arnowitt R, Deser S and Misner C W 1962 *Gravitation: An Introduction to Current Research* ed L Witten (New York: Wiley) pp 227–65
[30] Choquet-Bruhat Y and Ruggeri T 1983 *Commun. Math. Phys.* **89** 269
[31] Bona C and Massó J 1992 *Phys. Rev. Lett.* **68** 1097
[32] Bona C, Massó J, Seidel E and Stela J 1995 *Phys. Rev. Lett.* **75** 600
[33] Bona C, Massó J, Seidel E and Stela J 1997 *Phys. Rev.* D **56** 3405
[34] Bona C, Massó J, Seidel E and Walker P 1998 *Preprint* gr-qc/9804052
[35] Alcubierre M *et al* in preparation
[36] Frittelli S and Reula O 1994 *Commun. Math. Phys.* **166** 221
[37] Choquet-Bruhat Y and York J 1995 *C. R. Acad. Sci., Paris* **321** 1089
[38] Abrahams A, Anderson A, Choquet-Bruhat Y and York J 1995 *Phys. Rev. Lett.* **75** 3377
[39] Frittelli S and Reula O 1996 *Phys. Rev. Lett.* **76** 4667
[40] van Putten M H and Eardley D 1996 *Phys. Rev.* D **53** 3056
[41] Abrahams A, Anderson A, Choquet-Bruhat Y and York J 1997 *Class. Quantum Grav.* A **9**
[42] Bona C, Massó J and Stela J 1995 *Phys. Rev.* D **51** 1639
[43] Scheel M *et al* 1997 *Phys. Rev.* D **56** 6320
[44] Cook G and Scheel M Private communication
[45] Anninos P *et al* 1995 *Phys. Rev.* D **52** 2059
[46] Reula O 1998 *Living Reviews in Relativity* **1**
[47] Friedrich H 1985 *Commun. Math. Phys.* **100** 525
[48] Baumgarte T W and Shapiro S L 1999 *Phys. Rev.* D **59** 024007
[49] Shibata M and Nakamura T 1995 *Phys. Rev.* D **52** 5428
[50] Alcubierre M *et al* 2000 *Phys. Rev.* D **D61** 041501 (gr-qc/9904013)
[51] Arbona A, Bona C, Massó J and Stela J 1999 *Phys. Rev.* D **60** 104014 (gr-qc/9902053)
[52] Alcubierre M *et al* 1999 *Preprint* gr-qc/9908079
[53] Nakamura T, Oohara K and Kojima Y 1987 *Prog. Theor. Phys. Suppl.* **90** 1
[54] Shibata M 1999 *Prog. Theor. Phys.* **101** 1199 (gr-qc/9905058)
[55] Baumgarte T W, Hughes S A and Shapiro S L 1999 *Phys. Rev.* D **59** 024007
[56] Shibata M 1999 *Phys. Rev.* D **60** 104052 (gr-qc/9908027)
[57] Shibata M and Uryu K 2000 *Phys. Rev.* D **61** 064001 (gr-qc/9911058)
[58] Shibata M, Baumgarte T W and Shapiro S L 2000 *Phys. Rev.* D **61** 044012
[59] Bernstein D *et al* 1994 *Phys. Rev.* D **50** 5000
[60] Brügmann B 1996 *Phys. Rev.* D **54** 7361
[61] New K C B, Watt K, Misner C W and Centrella J M 1998 *Phys. Rev.* D **58** 064022 (gr-qc/9801110)
[62] Huq M Private communication
[63] Alcubierre M *et al* 2000 *Phys. Rev.* D **62** 044034
[64] Teukolsky S 2000 *Phys. Rev.* D **61** 087501

[65] Gustafsson B, Kreiss H-O and Oliger J 1995 *Time Dependent Problems and Difference Methods* (New York: Wiley)

[66] Press W H, Flannery B P, Teukolsky S A and Vetterling W T 1986 *Numerical Recipes* (Cambridge: Cambridge University Press)

[67] Clune T, Massó J, Miller M and Walker P *Technical Report* National Center for Supercomputing Applications, unpublished, in preparation

[68] Massó J 1992 *PhD Thesis* University of the Balearic Islands

[69] Leveque R J 1992 *Numerical Methods for Conservation Laws* (Basel: Birkhauser)

[70] Bona C 1996 *Relativity and Scientific Computing* ed F Hehl (Berlin: Springer)

[71] Yee H C 1989 *Computational Fluid Dynamics* (NASA Ames Research Center, CA: von Karman Institute for Fluid Dynamics)

[72] Choptuik M 1991 *Phys. Rev.* D **44** 3124

[73] Bernstein D, Hobill D and Smarr L 1989 *Frontiers in Numerical Relativity* ed C Evans, L Finn and D Hobill (Cambridge: Cambridge University Press) pp 57–73

[74] Lee S 1993 *PhD Thesis* University of Illinois Urbana-Champaign

[75] Ashby S *et al* 1995 *Elsevier Sci.*

[76] Detweiler S 1987 *Phys. Rev.* D **35** 1095

[77] Brodbeck O, Frittelli S, Hübner P and Reula O A 1998 gr-qc/9809023

[78] C W L *et al* 1998 *Preprint* Physics Department, Chinese University of Hong Kong

[79] Smarr L and York J 1978 *Phys. Rev.* D **17** 1945

[80] Smarr L and York J 1978 *Phys. Rev.* D **17** 2529

[81] Piran T 1983 *Gravitational Radiation* ed N Deruelle and T Piran (Amsterdam: North-Holland)

[82] Bona C and Massó J 1988 *Phys. Rev.* D **38** 2419

[83] Alcubierre M 1997 *Phys. Rev.* D **55** 5981

[84] Alcubierre M and Massó J 1998 *Phys. Rev.* D **57** 4511

[85] Tobias M 1997 *PhD Thesis* Washington University, Saint Louis, MO

[86] Eppley K 1979 *Sources of Gravitational Radiation* ed L Smarr (Cambridge: Cambridge University Press) p 275

[87] Balakrishna J *et al* 1996 *Class. Quantum Grav.* **13** L135

[88] Cook G 1990 *PhD Thesis* University of North Carolina at Chapel Hill, Chapel Hill, North Carolina

[89] Martí J M, Ibáñez J M and Miralles J A 1991 *Phys. Rev.* D **43** 3794

[90] Font J, Ibáñez J, Martí J and Marquina A 1994 *Astron. Astrophys.* **282** 304

[91] Banyuls F *et al* 1997 *Ap. J.* **476** 221

[92] Donat R, Font J A, Ibáñez J M and Marquina A 1998 *J. Comput. Phys.* **146** 58

[93] For the NASA Neutron Star Grand Challenge Project, see, e.g., http://wugrav.wustl.edu/Relativ/nsgc.html.

[94] Wang E and Swesty F D 1998 *Proc. 18th Texas Symp. on Relativistic Astrophysics* ed J F A Olinto and D Schramm (Singapore: World Scientific)

[95] Font J A, Miller M, Suen W M and Tobias M 2000 *Phys. Rev.* D **61** 044011 (gr-qc/9811015)

[96] Roe P L 1981 *J. Comput. Phys.* **43** 357

[97] Donat R and Marquina A 1996 *J. Comput. Phys.* **125** 42

[98] Eulderink F and Mellema G 1994 *Astron. Astrophys.* **284** 652

[99] Shu C W and Osher S J 1989 *J. Comput. Phys.* **83** 32

[100] Abrahams A M *et al* 1998 *Phys. Rev. Lett.* **80** 1812 (gr-qc/9709082)

[101] Shapiro S L and Teukolsky S A 1986 *Dynamical Spacetimes and Numerical Relativity* ed J M Centrella (Cambridge: Cambridge University Press) pp 74–100
[102] Thornburg J 1987 *Class. Quantum Grav.* **14** 1119 (Unruh is cited here by Thornburg as originating AHBC)
[103] Anninos P *et al* 1995 *Phys. Rev. Lett.* **74** 630
[104] Libson J *et al* 1996 *Phys. Rev.* D **53** 4335
[105] Seidel E and Suen W-M 1992 *Phys. Rev. Lett.* **69** 1845
[106] Alcubierre M and Schutz B 1994 *J. Comput. Phys.* **112** 44
[107] Anninos P *et al* 1995 *Phys. Rev.* D **51** 5562
[108] Scheel M A, Shapiro S L and Teukolsky S A 1995 *Phys. Rev.* D **51** 4208
[109] Marsa R and Choptuik M 1996 *Phys. Rev.* D **54** 4929
[110] Hughes S *et al* 1994 *Phys. Rev.* D **49** 4004
[111] Matzner R *et al* 1995 *Science* **270** 941
[112] Massó J, Seidel E, Suen W-M and Walker P 1999 *Phys. Rev.* D **59** 064022 (gr-qc/9804059)
[113] Shapiro S, Teukolsky S and Winicour J 1995 *Phys. Rev.* D **52** 6982
[114] Hawking S W and Ellis G F R 1973 *The Large Scale Structure of Spacetime* (Cambridge: Cambridge University Press)
[115] Brill D and Lindquist R 1963 *Phys. Rev.* **131** 471
[116] Eppley K 1977 *Phys. Rev.* D **16** 1609
[117] Anninos P *et al* 1998 *Phys. Rev.* D **58** 024003
[118] Libson J, Massó J, Seidel E and Suen W-M 1996 *The Seventh Marcel Grossmann Meeting: On Recent Developments in Theoretical and Experimental General Relativity, Gravitation, and Relativistic Field Theories* ed R T Jantzen, G M Keiser and R Ruffini (Singapore: World Scientific) p 631
[119] Libson J 1994 *Numerical Relativity Conference* ed P Laguna (Urbana, IL: Grand Challenge Meeting, NCSA, Champaign)
[120] Camarda K 1998 *PhD Thesis* University of Illinois at Urbana-Champaign, Urbana, Illinois
[121] Camarda K and Seidel E 1999 *Phys. Rev.* D **59** 064026 (gr-qc/9805099)
[122] Baumgarte T W *et al* 1996 *Phys. Rev.* D **54** 4849
[123] Nakamura T, Kojima Y and Oohara K 1984 *Phys. Lett.* A **106** 235
[124] Gundlach C 1998 *Phys. Rev.* D **57** 863 (gr-qc/9707050)
[125] Thornburg J 1996 *Phys. Rev.* D **54** 4899
[126] Huq M Private communication, unpublished
[127] Moncrief V 1974 *Ann. Phys.* **88** 323
[128] Brandt S and Seidel E 1996 *Phys. Rev.* D **54** 1403
[129] Brandt S and Seidel E 1995 *Phys. Rev.* D **52** 856
[130] Brandt S and Seidel E 1995 *Phys. Rev.* D **52** 870
[131] Allen G, Camarda K and Seidel E 1998 gr-qc/9806014
[132] Allen G, Camarda K and Seidel E 1998 gr-qc/9806036
[133] Allen G, Goodale T and Seidel E 1999 *7th Symposium on the Frontiers of Massively Parallel Computation-Frontiers 99* (New York: IEEE)
[134] Allen G *et al* 1999 *IEEE Comput.* **32**
[135] http://www.cactuscode.org
[136] Bonazzola S, Gourgoulhon E and Marck J-A 1998 *Preprint* astro-ph/9803086
[137] Choptuik M 1989 *Frontiers in Numerical Relativity* ed C Evans, L Finn and D Hobill (Cambridge: Cambridge University Press)

[138] Berger M and Oliger J 1984 *J. Comput. Phys.* **53** 484
[139] Choptuik M 1993 *Phys. Rev. Lett.* **70** 9
[140] Wild L and Schutz B in preparation
[141] Papadopoulos P, Seidel E and Wild L 1998 *Phys. Rev.* D **58** 084002 (gr-qc/9802069)
[142] Brill D S 1959 *Ann. Phys.* **7** 466
[143] Holz D, Miller W, Wakano M and Wheeler J 1993 *Directions in General Relativity: Proc. 1993 Int. Symp., Maryland; Papers in Honor of Dieter Brill* ed B Hu and T Jacobson (Cambridge: Cambridge University Press)
[144] Brandt S, Camarda K and Seidel E 2000 *Proc. 8th M Grossmann Meeting: On Recent Developments in Theoretical and Experimental Relativity* ed T Piran (Singapore: World Scientific)
[145] Alcubierre M *et al* 2000 *Class. Quantum Grav.* **17** 2159
[146] Camarda K and Seidel E 1998 *Phys. Rev.* D **57** R3204 (gr-qc/9709075)
[147] Brügmann B 1999 *Int. J. Mod. Phys.* D **8** 85
[148] Anninos P *et al* 1997 *Phys. Rev.* D **56** 842
[149] Anderson A and York J 1999 *Preprint* gr-qc/9901021
[150] Alcubierre M, Brügmann B, Miller M and Suen W-M 1999 *Phys. Rev.* D **60** 064017 (gr-qc/9903030)
[151] Seidel E 1997 *Gravitation and Cosmology* ed S Dhurandhar and T Padmanabhan (Dordrecht: Kluwer) pp 125–44
[152] Seidel E 1996 *Helv. Phys. Acta* **69** 454
[153] Seidel E 1998 *Proc. GR15* ed N Dadich
[154] Seidel E 1998 *Bad Honnef Meeting on Black Holes* ed F Hehl (Berlin: Springer)
[155] Smarr L 1979 *Sources of Gravitational Radiation* ed L Smarr (Cambridge: Cambridge University Press) p 139
[156] Thorne K 1999 *Phys. Rev.* D (gr-qc/9808024)
[157] Misner C W 1963 *Ann. Phys.* **24** 102
[158] Misner C 1960 *Phys. Rev.* **118** 1110
[159] Bowen J and York J W 1980 *Phys. Rev.* D **21** 2047
[160] Cook G B 1994 *Phys. Rev.* D **50** 5025
[161] Bernstein D, Hobill D, Seidel E and Smarr L 1994 *Phys. Rev.* D **50** 3760
[162] Matzner R A, Huq M F and Shoemaker D 1999 *Phys. Rev.* D **59** 024015
[163] Anninos P *et al* 1997 *Computational Astrophysics: Gas Dynamics and Particle Methods* ed W Benz *et al* (New York: Springer) in press
[164] Anninos P *et al* 1993 *IEEE Comput. Graphics Appl.* **13** 12
[165] Smarr L, Čadež A, DeWitt B and Eppley K 1976 *Phys. Rev.* D **14** 2443
[166] Smarr L 1977 *Ann. N. Y. Acad. Sci.* **302** 569
[167] Smarr L 1979 *Sources of Gravitational Radiation* ed L Smarr (Cambridge: Cambridge University Press) p 245
[168] Anninos P *et al Technical Report* No 24, National Center for Supercomputing Applications, unpublished
[169] Anninos P *et al* 1995 *Phys. Rev.* D **52** 2044
[170] Baker J *et al* 1997 *Phys. Rev.* D **55** 829
[171] Anninos P and Brandt S 1998 *Phys. Rev. Lett.* **81** 508
[172] Anninos P, Massó J, Seidel E and Suen W-M 1996 *Physics World* **9** 43
[173] Anninos P *et al* 1996 *The Seventh Marcel Grossmann Meeting: On Recent Developments in Theoretical and Experimental General Relativity, Gravitation, and Relativistic Field Theories* ed R T Jantzen, G M Keiser and R Ruffini

(Singapore: World Scientific) p 637

[174] Scheel M A *et al* 1998 *Phys. Rev.* D **58** 044020

[175] Price R H and Pullin J 1994 *Phys. Rev. Lett.* **72** 3297

[176] Anninos P *et al* 1995 *Phys. Rev.* D **52** 4462

[177] Gleiser R J, Nicasio C O, Price R H and Pullin J 1996 *Phys. Rev. Lett.* **77** 4483

[178] Pullin J 1998 *Proc. GR15* ed N Dadhich and J Narlikar (Puna: Inter Univ. Centre for Astr. and Astrop.) p 87

[179] Campanelli M and Lousto C O 1998 *Phys. Rev.* D **58** 024015

[180] Campanelli M, Krivan W and Lousto C O 1998 *Phys. Rev.* D **58** 024016

[181] Campanelli M *et al* 1998 *Phys. Rev.* D **58** 084019

[182] Hawking S W 1973 *Black Holes* ed C DeWitt and B S DeWitt (New York: Gordon and Breach) pp 1–55

[183] Anninos P *et al* 1998 *Phys. Rev.* D **58** 24003

[184] Shibata M 1997 *Phys. Rev.* D **55** 2002

[185] Anninos P *et al* 1994 *Phys. Rev.* D **50** 3801

[186] Anninos P *et al* 1995 *Austral. J. Phys.* **48** 1027

[187] Miyama S M 1981 *Prog. Theor. Phys.* **65** 894

[188] Seidel E 1999 *Forschung und wissenschaftliches Rechnen* ed T P and P Wittenburg (München: Max-Planck-Gesselschaft)

Index

Milton Keynes UK
Ingram Content Group UK Ltd.
UKHW021833071024
449327UK00021B/1488

9 780367 397609